Chlorophyll Organization and
Energy Transfer in Photosynthesis

The Ciba Foundation for the promotion of international cooperation in medical and chemical research is a scientific and educational charity established by CIBA Limited—now CIBA-GEIGY Limited—of Basle. The Foundation operates independently in London under English trust law.

Ciba Foundation Symposia are published in collaboration with Excerpta Medica in Amsterdam

Excerpta Medica, P.O.Box 211, Amsterdam

Chlorophyll Organization and Energy Transfer in Photosynthesis

Ciba Foundation Symposium 61 (new series)

1979

Excerpta Medica

Amsterdam · Oxford · New York

ISBN Excerpta Medica 90 219 4067 1
ISBN Elsevier/North-Holland 0 444 90044 6

Published in January 1979 by Excerpta Medica, P.O.Box 211, Amsterdam and Elsevier/North-Holland, Inc., 52 Vanderbilt Avenue, New York, N.Y. 10017.

Suggested series entry for library catalogues: Ciba Foundation Symposia.
Suggested publisher's entry for library catalogues: Excerpta Medica

Ciba Foundation Symposium 61 (new series)

388 pages, 104 figures, 22 tables

Library of Congress Cataloging in Publication Data

Symposium on Chlorophyll Organization and Energy
        Transfer in Photosynthesis, Ciba Foundation 1978.
        Chlorophyll organization and energy transfer in photosynthesis.

        (Ciba Foundation symposium; 61 (new ser.))
        Proceedings of the symposium held Feb. 7–9, 1978 in London.
        Bibliography: p.
        Includes indexes.
        1. Photosynthesis–Congresses. 2. Chlorophyll–Congresses. 3. Energy transfer–Congresses. I. Title. II. Series: Ciba Foundation. Symposium; new ser., 61.
        QK882.S946 1978   581.1'3342   78–12368

ISBN 0–444–90044–6

Printed in The Netherlands by Casparie, Heerhugowaard.

Contents

v

Participants

Symposium on Chlorophyll Organization and Energy Transfer in Photosynthesis *held at the Ciba Foundation, London, 7–9th February, 1978*

Chairman: SIR GEORGE PORTER   The Royal Institution, 21 Albemarle Street, London W1X 4BS, UK

J. AMESZ   Department of Biophysics, Huygens Laboratory, State University of Leiden, P.O. Box 9504, Leiden, The Netherlands

JAN M. ANDERSON   Division of Plant Industry, Commonwealth Scientific and Industrial Research Organization, P.O. Box 1600, Canberra City, A.C.T. 2601, Australia

J. BARBER   Department of Botany, Imperial College of Science and Technology, Prince Consort Road, London SW7 2BB, UK

G. S. BEDDARD   Davy Faraday Research Laboratory, The Royal Institution, 21 Albemarle Street, London W1X 4BS, UK

J. BRETON   Department of Biology, Centre d'Etudes Nucléaires de Saclay, B.P. No. 2, Gif-sur-Yvette 91190, France

W. L. BUTLER   Department of Biology, B-022, University of California, San Diego, La Jolla, California 92093, USA

R. K. CLAYTON   Section of Genetics, Development & Physiology, Plant Science Building, Cornell University, Division of Biological Sciences, Ithaca, New York 14853, USA

R. COGDELL   Department of Botany, University of Glasgow, Glasgow G12 8QQ, UK

L. N. M. DUYSENS   Department of Biophysics, Huygens Laboratory, State University of Leiden, P.O. Box 9504, Leiden, The Netherlands

P. JOLIOT   Institut de Biologie Physico-Chimique, Fondation Edmond de Roth-schild, 13 rue Pierre et Marie Curie, 75005 Paris, France

W. JUNGE   Max-Volmer-Institut für Physikalische Chemie und Molekular-biologie, Technische Universität Berlin, Strasse des 17. Juni 135, D1000 Berlin 12, West Germany

J. J. KATZ   Chemistry Division, Argonne National Laboratory, 9700 South Cass Avenue, Argonne, Illinois 60439, USA

R. S. KNOX   Department of Physics and Astronomy, The University of Rochester, Rochester, New York 14627, USA

M. LUTZ   Department of Biology, Centre d'Etudes Nucléaires de Saclay, B.P. No. 2, Gif-sur-Yvette 91190, France

R. MALKIN   Department of Cell Physiology, University of California, Berkeley, California 94750, USA

G. PAILLOTIN   Department of Biology, Centre d'Etudes Nucléaires de Saclay, B.P. No. 2, Gif-sur-Yvette 91190, France

G. W. ROBINSON   Department of Chemistry, Texas Tech University, PO Box 4260, Lubbock, Texas 79409, USA

G. F. W. SEARLE   Department of Botany, Imperial College of Science and Technology, Prince Consort Road, London SW7 2BB, UK

G. R. SEELY   Charles F. Kettering Research Laboratory, 150 East South College Street, Yellow Springs, Ohio 45387, USA

L. A. STAEHELIN   Department of Molecular, Cellular and Developmental Biology, University of Colorado at Boulder, Boulder, Colorado 80309, USA

J. P. THORNBER   Department of Biology, University of California, Los Angeles, California 90024, USA

C. J. TREDWELL   Davy Faraday Research Laboratory, The Royal Institution, 21 Albemarle Street, London W1X 4BS, UK

J. S. C. WESSELS   Philips Research Laboratories, N.V. Philips' Gloeilampen-fabrieken, Eindhoven, The Netherlands

Editors: GORDON WOLSTENHOLME (*Organizer*) and DAVID W. FITZSIMONS

Chairman's opening remarks

SIR GEORGE PORTER

The Royal Institution, London

At this meeting we are going to look at the primary processes of the most important biological application of photochemistry. The meeting was conceived so as to give chemists, physicists and biologists the opportunity of talking to each other about this important developing area of science —chlorophyll organization and energy transfer in photosynthesis—because it is a subject with a language which is obscure to many chemists and physicists. Although most people here are working full time in the field of photosynthesis, many have come into the field from other disciplines. I hope that at the end of the symposium we shall leave with a clearer understanding of the photosynthetic apparatus and how it does its wonderful chemistry. In so far as there are gaps in the picture, as there must be at this stage, we want to know what they are and to what extent there is agreement about the areas which seem to have been well mapped out.

Our theme is both timely and important. It is timely because it is developing so rapidly that it seems possible that a fairly complete understanding of the structure and function of the photosynthetic unit may be obtained in just a few more years. It is timely also because many people from outside the privileged cabal of photosynthetic research have become interested in it since the new techniques that they use (for example picosecond laser-pulse spectroscopy) are so admirably matched to studies of the primary processes. These processes, with rates that fall so tantalizingly within their time range will probably never be resolved and fully understood without direct studies on that time scale. That explains the increasing attention which is being given to mechanistic and kinetic aspects by direct picosecond studies.

On the organizational side of the process this symposium is equally timely because rapid advances have been made in the past few years in the isolation and characterization of chlorophyll–protein complexes. One has been crystallized and its structure has been determined. The hope in our minds is that complex particles of the green leaf will be subjected to similar precise characterization. Even without this, the grosser structure of the photosynthetic unit as a whole, and the part that these complexes play in it, will figure greatly in our discussions, as will the smaller scale organization of the chlorophyll molecules and their oligomers.

I also said that the subject is important. It is; not only because it is one of the basic processes in nature and in which nature can teach the photochemist and photophysicist many skills as yet impossible *in vitro*, but also because over the past few years the importance of photosynthesis to the survival of man and his modern technology has become increasingly apparent. To live and even to survive in the modern world man needs principally two things: food and fuel. Both of these depend entirely on the process of photosynthesis. It is our hope that by understanding better the organization and mechanism of the natural photosynthetic process we may be better able to improve on it in agriculture and perhaps even to adapt it specifically to some of man's energy needs, by replacing the fossil fuels when the wells run dry at the end of the century. We shall not discuss this here but what we shall discuss is certainly not irrelevant to these practical purposes.

The subjects in the papers are interrelated in such a way that we shall inevitably oscillate between green plants and bacteria, between theory and experiment, and between the organization and the kinetics of energy transfer in the photosynthetic unit. We shall start with structure, and then discuss how this structure operates in the first steps of energy transfer in photosynthesis. We shall go no further than this; we shall stop our considerations after about the first nanosecond, when the chemistry begins. In the organization part, some of the questions to which I should like to know answers are the following: what is the state of the chlorophyll molecule (i.e. when is it a monomer and when a dimer) in the reaction centres of photosystems I and II and in the light-harvesting unit? Secondly, how do we account for the many different apparent states which are observed in the absorption spectrum? Can we account for them in terms of the dimer and of solvation differences? Thirdly, how is chlorophyll incorporated into the membrane, into the lipid, and into the protein complexes? Is it partly exposed to the lipid? What size are the basic protein complexes? How many chlorophylls are there in each unit? How are these individual units arranged with respect to each other and in the membrane? How do the chlorophyll molecules in these units manage to overcome the concentration quenching which occurs *in vitro*? Fourthly, how do the chlorophyll–protein complexes, arranged in the way we shall have discussed, transfer energy between themselves (i.e. the units as opposed to the molecules within the complex) and at what rate? Finally how are the whole systems dispersed in the membrane and how does our picture of this account for the electron-microscope photographs of the membrane and the particles which we see in it?

If we can formulate answers to those questions before we pass on to discussion of the kinetics of energy transfer within these structures, we shall already have made great progress.

In the kinetics section there are more questions to be answered. First, what are the experimentally determined laws of fluorescence decay of chlorophyll in the chloroplast? Is the rate-determining process one of energy transfer or of trapping? In the latter case an exponential decay is expected; in the former the decay would be non-exponential and—if Förster-type kinetics are followed—an $\exp(-kt^{\frac{1}{2}})$ dependence of fluorescence on time might be expected. Second, now that the fluorescence of the different light-harvesting pigments can be time-resolved, can the results be reconciled with what is known of the structural arrangement in particles such as phycobilisomes, for example? Do they now allow us to distinguish between 'lake' and 'puddle' models of the photosynthetic unit? And, again, we have to ask how the kinetics *in vivo* can be reconciled with those *in vitro* where fluorescence lifetimes, at comparable concentrations, are so much shorter.

All these things happen in the first nanosecond of photosynthesis and if we can understand them in the three days available to us, we shall have done very well indeed.

Structure and function of photoreaction-centre chlorophyll

J. J. KATZ, L. L. SHIPMAN and J. R. NORRIS

Chemistry Division, Argonne National Laboratory, Argonne, Illinois

Abstract Evidence from electron paramagnetic resonance (e.p.r.) studies suggests that the unpaired spin in oxidized $P700^{+\cdot}$ or $P865^{+\cdot}$ is shared by two special chlorophyll *a* (Chl *a*) or bacteriochlorophyll *a* (Bchl *a*) molecules respectively. Three classes of models have been proposed for special pair reaction centre chlorophyll: asymmetric, in which one Chl *a* (or Bchl *a*) acts as electron donor to a second acting as acceptor; models with translational symmetry only; and models with C_2 symmetry. Models with C_2 symmetry have been synthesized *in vitro* with two chlorophyll macrocycles tied together by a covalent link. The singlet and triplet states of the *in vitro* models have been characterized by e.p.r., nuclear magnetic resonance, and optical studies involving absorption, emission, and lasing behaviour. The fact that lasing occurs only from the folded configuration of the linked dimers suggests the availability of a highly effective non-radiative decay path from the S_1 state of the excited open dimer. A radical-pair mechanism that accounts for the unusual spin polarization of the special pair triplet is proposed for the primary photochemistry in the reaction centre.

THE PHOTOSYNTHETIC UNIT

Green plants and certain bacteria can carry out the process of photosynthesis in which the energy of sunlight is converted into chemical energy. The input of chemical energy makes it possible for photosynthetic organisms to do chemical reactions that otherwise would not proceed spontaneously. Crucial to the ability of photosynthetic organisms to use light energy for chemical purposes are the chlorophylls, a small group of closely related compounds (Fig. 1) that are deeply implicated in all aspects of the primary act of light conversion. Chlorophylls are the primary photoacceptors; they are the principal energy-transfer agents; they form the energy trap and they are the primary electron donor in photosynthesis. Almost 50 years ago, Emerson & Arnold

1

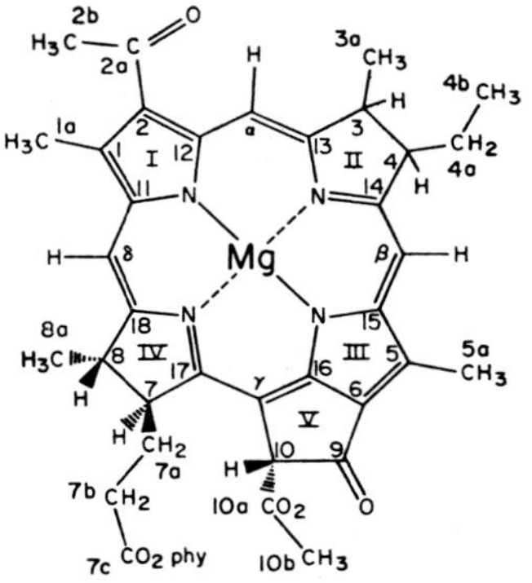

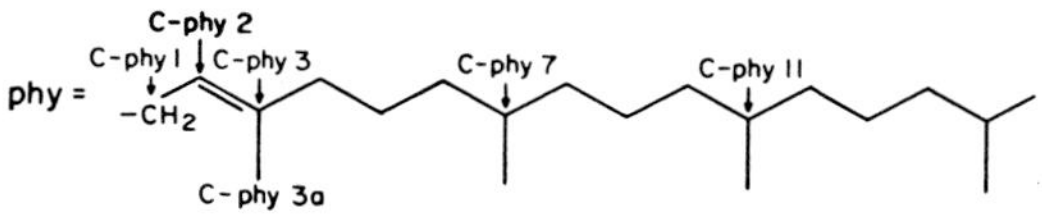

FIG. 1. Structures and numbering system of (1) chlorophyll *a* and (2) bacteriochlorophyll *a*.

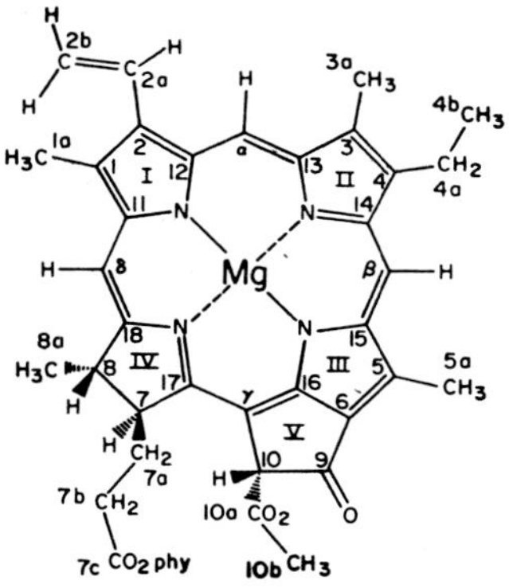

(1931, 1932) proposed that chlorophyll function in photosynthesis is a cooperative phenomenon. Many chlorophyll molecules are involved in the conversion of a single photon into an electron (a reducing agent) and a 'positive hole' (an oxidizing agent). [Nearly all chlorophyll molecules in the photosynthetic apparatus have a light-gathering or antenna function; these chlorophyll molecules act as the primary photoacceptors of electromagnetic radiation. The (electronic) excitation energy of a particular chlorophyll in the antenna array caused by absorption of a photon is then transferred to a few chlorophyll molecules in a photoreaction centre where energy is trapped and conversion occurs. The antenna and photoreaction-centre chlorophyll, together with auxiliary pigments and electron-transport chains, comprise a photosynthetic unit. *In vivo* antenna and photoreaction-centre chlorophylls have different physical properties, and differ from each other and from an *in vitro* solution of chlorophyll in a polar solvent in such important respects as visible absorption maxima (electronic transition) and fluorescence. Nevertheless, chlorophyll of the same molecular structure may be used to construct both the antenna and the photoreaction centre. A central problem in photosynthetic research, then, has been to provide a structural (or environmental) basis for the various species of chlorophyll that occur in the photosynthetic unit that rationalizes the anomalous properties of chlorophyll *in vivo*. The magnitude of the anomaly can be judged from the fact that solutions of chlorophyll *a* (Chl *a*) and bacterio-chlorophyll *a* (Bchl *a*) in polar solvents absorb light in the red region of the spectrum at about 665 nm and about 770 nm, respectively, and the solutions are intensely fluorescent, whereas *in vivo* chlorophylls have their red absorption maxima substantially shifted to the red and are only feebly fluorescent.

Here we shall be concerned only with the photoreaction centres I (PS I) in green plants and with bacterial photoreaction centres containing bacterio-chlorophyll *a*. It is convenient to discuss PS I and bacterial reaction centres together. Although there are fundamental differences between green plant and bacterial photosynthesis, the essential features of PS I and bacterial photo-reaction centres appear to be similar if not identical.

Progress in the elucidation of the structure and function of reaction centres has been greatly accelerated by the successful procedures developed by Clayton (1963), Clayton & Wang (1971), Loach & Sekura (1967), and Feher (1971) for the isolation of reaction centres from photosynthetic bacteria. These preparations of bacterial reaction centres are functional entities of relatively simple composition. They are free of antenna Bchl *a* and have the optical and redox properties of photoreaction centre Bchl *a* present in intact photosynthetic bacteria. Isolated bacterial reaction centres contain several Bchl *a* molecules as well as some bacteriopheophytin *a* (Bpheo *a*) (the Mg-free derivative of Bchl *a*). For reasons discussed below, it appears that not all the Bchl *a* in the isolated reaction centre is involved in the production of electrons in the primary light-conversion event. With respect to the primary electron-production event, however, there appears to be a great deal of similarity in both structure and function between PS I in green plants and the bacterial reaction centre. As many aspects of reaction-centre behaviour can as yet be studied only in reaction-centre preparations, experiments with bacterial reaction centres make an im-portant contribution to studies on green plants. Progress in the preparation of reaction centres from green plants has so far been slower and most of what we shall say about green plant PS I centres is based on observations in intact photosynthetic organisms.

There are obviously many different vantage points from which photosynthetic reaction centres can be viewed, and many different levels at which interpretation can be attempted. Our objective is to provide an interpretation *on the molecular level* of the structure and function of photoreaction centres in terms of the molecular structure and physical properties of their constituent chlorophyll.

E.p.r. and optical properties of photoreaction-centre chlorophyll

It will facilitate our subsequent discussion to summarize some of the salient e.p.r. and optical properties of the reaction centres of green plants and bacteria. Commoner *et al.* (1956) made the important discovery that free radicals (paramagnetic entities with an unpaired electron) are produced in the light-energy conversion step of photosynthesis. Because of the great sensitivity of e.p.r. spectroscopy, the photo-e.p.r. signal is readily detected and serves as the most informative experimental probe of photoreaction-centre activity now available. The e.p.r. signal is composite. Its most prominent component (generally called Signal I) is rapidly reversible and has a *g*-value of 2.0025, indicative of an unpaired electron delocalized over a large π-system. The line-shape of e.p.r. Signal I is Gaussian and has a peak-to-peak line-width of about 7.0 G. The corresponding e.p.r. signal in photosynthetic bacteria is also reversible, Gaussian, and has a peak-to-peak line-width of about 9.5 G (Androes *et al.* 1962). In fully deuteriated algae, Signal I is narrowed to about 3 G (Kohl *et al.* 1965) and in fully deuteriated bacteria to 3–4 G (Kohl *et al.* 1965; McElroy *et al.* 1969). In all cases, the photo-e.p.r. signal has no observable hyperfine structure. The availability of fully deuteriated photosynthetic organisms, as well as of organisms highly enriched in ^{13}C, ^{15}N and ^{25}Mg, has considerably enhanced the applicability of e.p.r. to the study of photosynthesis.

The origin of Signal I in green plants and the photo-e.p.r. signal in photosynthetic bacteria has been established by correlation of the kinetics of formation and decay of the photo-e.p.r. signal with optical transients that can also be associated with the light-conversion event, and by comparison of the *in vivo* e.p.r. signals with those of chlorophyll free radicals produced in the laboratory in defined systems. Kok (1956, 1957) observed that a decrease in the intensity of light absorption (photobleaching) occurs at 702–705 nm during active photosynthesis, and that the photobleaching is reversed in the dark, and Duysens (1952; Duysens *et al.* 1956) observed reversible photobleaching in photosynthetic bacteria at about 870 nm. From these optical transients it was deduced that the photoreaction centres in green plants and in photosynthetic bacteria absorb light at about 700 nm and about 865 nm, respectively, and the photoreaction-centre chlorophylls were assigned the symbols P700 and P865. The photobleaching was interpreted as a photooxidation. The paramagnetic (free radical) species produced in the photoreaction centres are then P700+· and P865+·, formed by ejection of an electron during the light-conversion event. The conclusion that the photobleaching is an oxidation is reinforced by the observation that the optical changes and the concomitant e.p.r. signal produced by the chemical oxidant potassium ferricyanide are similar to those produced *in vivo* by light. Evidence that the chlorophyll free radicals produced in the conversion step are cationic free radicals is derived from the important *in vitro* studies of Fuhrhop & Mauzerall (1969) on porphyrins and by Borg *et al.*

(1970) on chlorophyll. (Although the conclusion that the *in vitro* chlorophyll free radicals are cationic free radicals is conclusive, no similar evidence establishes that the free radicals P700$^{+\cdot}$ and P865$^{+\cdot}$ are charged species. To be sure, P700$^{+\cdot}$ and P865$^{+\cdot}$ are doublet states, but they could be neutral species. Nevertheless we shall follow the usual convention and use the symbols P700$^{+\cdot}$ and P865$^{+\cdot}$ to indicate the paramagnetic photooxidized photoreaction centres.)

The assignment of Signal I to P700$^{+\cdot}$ is supported by studies that show the kinetics of the e.p.r. signal and of the photobleaching are similar (Warden & Bolton 1972, 1973). In photosynthetic bacteria, the kinetics of photobleaching of P865$^{+\cdot}$ at both 4 K and room temperature are similar to the kinetics of the e.p.r. signal (McElroy *et al.* 1974). The identity of the kinetics of the photochemistry and the e.p.r. signal is evident on even the fastest time scale on which relevant observations can be made.

Careful quantitative comparisons of the quantum yield for free-radical formation have shown that the ratio of light-induced spins in e.p.r. Signal I to bleached P700 in green plants is within experimental error 1 : 1 (Warden & Bolton 1972, 1973). Similar quantitative experiments on reaction-centre preparations from photosynthetic bacteria also show that the ratio of photobleached P865 centres to the number of spins is essentially 1 : 1 (Bolton *et al.* 1969; Loach & Sekura 1967; Wraight & Clayton 1973). The experimental evidence is thus convincing that the quantum yield for free-radical formation in both green plant chloroplasts and in isolated bacterial reaction centres is close to unity, that is, for each photon trapped in the reaction centre one electron is ejected leaving the reaction centre with one unpaired spin.

The characteristics of the e.p.r. signals from the photooxidized reaction centres are consistent with the oxidation of a large aromatic molecule. The Gaussian line-shape and the free-electron *g*-value suggest that many interactions occur between the unpaired spin and carbon and hydrogen nuclei for both the *in vivo* and *in vitro* chlorophyll free radicals. This supposition is buttressed by a comparison between free radicals produced in organisms of ordinary isotopic composition and those in photosynthetic organisms of unnatural isotopic composition containing ^{2}H in place of ^{1}H. In fully deuteriated algae or bacteria, the e.p.r. line-width is reduced by about 60%, reflecting the considerably weaker electron–nuclear hyperfine interaction of ^{2}H. The ^{2}H effect in simple aromatic molecules reduces the line-width by a maximum of about 4 G. The ^{2}H effect in deuteriated organisms of only about 2.4 G can be accounted for by additional interactions in these systems with the nitrogen atoms present in the chlorophyll macrocycle. That the ^{2}H effect on the e.p.r. line-width is the same for both *in vivo* and *in vitro* chlorophyll systems is itself good proof that a chlorophyll species is the origin of the *in vivo* signal.

TABLE 1

Comparison of e.p.r. properties of P700$^{+\cdot}$ and P865$^{+\cdot}$ in selected photosynthetic organisms

System	$\Delta H_{pp}{}^a$	R^b
[^{1}H]Chl $a^{+\cdot}$	9.3 $\pm$ 0.3	2.4
[^{2}H]Chl $a^{+\cdot}$	3.8 $\pm$ 0.2	
[^{1}H]Bchl $a^{+\cdot}$	12.8 $\pm$ 0.5	2.4
[^{2}H]Bchl $a^{+\cdot}$	5.4 $\pm$ 0.2	
[^{1}H]*Syneccochocus lividus*	7.1 $\pm$ 0.2	2.4
[^{2}H]*Syneccochocus lividus*	2.95 $\pm$ 0.5	
[^{1}H]*Rhodospirillum rubrum*	9.5 $\pm$ 0.5	2.3
[^{2}H]*Rhodospirillum rubrum*	4.2 $\pm$ 0.3	

^{a}All lines are Gaussian and have $g = 2.0025 \pm 0.0002$.
^{b}Ratio of the line-width of the ^{1}H-system to the ^{2}H-system: [^{1}H]ΔH_{pp}/[^{2}H]ΔH_{pp}.

Thus, the features of the e.p.r. Signal I in green plants and the corresponding signal from photosynthetic bacteria and bacterial reaction centre are consistent with the formation of Chl $a \cdot L_1^{+\cdot}$ or Bchl $a \cdot L_1^{+\cdot}$ (L_1 is a ligand nucleophile). The exception is the line-width. The *in vivo* signals are about 40% narrower than the e.p.r. signals from monomeric Chl $a \cdot L_1^{+\cdot}$ or Bchl $a \cdot L_1^{+\cdot}$. (For convenience in comparison, the relevant e.p.r. data on *in vitro* and *in vivo* signals are collected in Table 1.) The discrepancy in line-width makes it impossible to equate P700$^{+\cdot}$ or P865$^{+\cdot}$ with monomeric chlorophyll free radicals. In a similar fashion, the optical properties of P700 and P865 are not satisfactorily accounted for in terms of monomeric chlorophylls. To account for the discrepancies, it is necessary to invoke the participation of more than one chlorophyll molecule in the photooxidation of P700 or P865, and this leads to new views about the structure and function of the photoreaction centre.

THE CHLOROPHYLL SPECIAL PAIR

The e.p.r. data discussed above make it plausible that chlorophyll is the primary electron donor in the photoreaction centre (Katz & Norris 1973). Discrepancies between the line-width of the e.p.r. signal and the visible absorption of P700 and P865 and those of monomeric Chl a and Bchl a make it impossible to identify the *in vivo* primary electron donor with monomeric chlorophyll. Attribution of the discrepancy to the consequences of biological environment of an unspecified nature is no longer satisfactory. The unusual photo-e.p.r. signal that can be elicited from *in vitro* P740, however, points a ywa to the resolution of the dilemma.

E.p.r. and the $\sqrt{2}$ effect

The chlorophyll–water adduct absorbing maximally at 740 nm (P740) has an extraordinarily narrow e.p.r. signal with a line-width of about 1 G, far narrower than the signals from either P700$^{+\cdot}$ or Chl $a \cdot L_1^{+\cdot}$. The unusual line-width of the P740$^{+\cdot}$ species can be rationalized by delocalization of the unpaired spin over the entire assembly of chlorophyll molecules in the aggregate. The delocalization can be viewed as a rapid process of spin migration between equivalent sites. Given a sufficiently high rate of spin migration, the e.p.r. signal from the effectively delocalized electron collapses to a narrow line. When a 'free' electron is delocalized over an aggregate of N molecules, it can readily be proved (Norris *et al.* 1971) that ΔH_N, the line-width when the unpaired spin is delocalized over N equivalent chlorophyll molecules, is given by equation (1),

$$\Delta H_N = \Delta H_M / N^{1/2} \tag{1}$$

where ΔH_M is the line-width of the monomeric chlorophyll free radical. A value of $N = 2$ accounts with considerable precision for the 40% narrowing of the P700$^{+\cdot}$ and P865$^{+\cdot}$ signals relative to Chl $a \cdot L_1^{+\cdot}$ and Bchl $a \cdot L_1^{+\cdot}$ (Table 2). The $\sqrt{2}$ narrowing in the *in vivo* P700$^{+\cdot}$ and P865$^{+\cdot}$ signals is analogous to that observed in organic dimeric cationic free radicals where the unpaired spin is shared by molecules (for a review, see Bard *et al.* 1976). The $\sqrt{2}$ narrowing in line-width holds reasonably well for all photoreaction centres containing Chl a or Bchl a. The relationship applies from ambient temperatures down to

TABLE 2

The $\sqrt{2}$ e.p.r. line-width effect in plants and photosynthetic bacteria

Organism	ΔH_{pp} (G)[a]	Calculated special-pair ΔH_{pp} (G)[b]	R[c]
[^{1}H]*S. lividus*	7.1 ± 0.2	6.6 ± 0.3	1.08 ± 0.06
[^{2}H]*S. lividus*	2.95 ± 0.1	2.7 ± 0.1	1.10 ± 0.05
[^{1}H]*C. vulgaris*	7.0 ± 0.2	6.6 ± 0.3	1.06 ± 0.05
[^{2}H]*C. vulgaris*	2.7 ± 0.1	2.7 ± 0.1	1.00 ± 0.05
[^{1}H]*S. obliquus*	7.1 ± 0.2	6.6 ± 0.8	1.08 ± 0.06
[^{2}H]*S. obliquus*	2.7 ± 0.1	2.7 ± 0.1	1.00 ± 0.05
[^{1}H]HP700	7.0 ± 0.2	6.66 ± 0.3	1.06 ± 0.05
[^{1}H]*R. rubrum*	9.5 ± 0.5	9.1 ± 0.4	1.05 ± 0.07
[^{2}H]*R. rubrum*	4.2 ± 0.3	3.8 ± 0.1	1.10 ± 0.09

[a]From Norris *et al.* (1971).
[b]Calculated from equation (1) with $N = 2$.
[c]$R = \Delta H_{\text{in vitro}} / \Delta H_{\text{in vivo}}$.

1.8 K and is followed equally well by the intact living organism and by isolated reaction-centre preparations. It is equally valid for green plants and for photosynthetic bacteria containing Bchl *a*. The almost universal occurrence of the $\sqrt{2}$ effect in photosynthetic reaction centres strongly implies that the primary donor in the photoreaction centre is a special pair of chlorophyll molecules. We hesitate to call the pair of chlorophyll molecules acting as the primary donor a 'dimer'. The term dimer has been preempted to describe a true chlorophyll dimer formed by a keto $C=O \cdots Mg$ coordination interaction between two chlorophyll molecules. The electronic transition spectra, redox properties, and the geometry of the true dimer differ in major respects from those of the two chlorophylls that act as donor in the reaction centre. In addition, there is evidence to suggest that the geometry of the two chlorophylls results from the intervention of a bifunctional nucleophile. To avoid confusion, we therefore refer to the primary donor in the reaction centre as a chlorophyll special pair, Chl_{sp} or $Bchl_{sp}$.

Endor and the one-half effect

Electron-nuclear double resonance (Endor) spectroscopy, a high-resolution variant of e.p.r. (Feher 1956), has made a valuable contribution to establishing the special-pair nature of P865 (Norris *et al.* 1973, 1974, 1975; Feher *et al.* 1973, 1975). In an aggregate of size N over which an unpaired electron is shared equally (effectively delocalized), the electron–proton hyperfine coupling constants are related to those in the monomer by the equation (2),

$$a_{Ni} = a_{Mi}/N \tag{2}$$

where a_{Mi} is the electron–nuclear hyperfine coupling constant for the ith nucleus in the monomer, and a_{Ni} is the coupling constant for that site in an aggregate of size N. For a Chl_{sp} where $N = 2$, the hyperfine coupling constant will be halved relative to those in the monomer, i.e. equation (3).

$$a_{2i} = a_{1i}/2 \tag{3}$$

One such equation applies to each different nuclear site in the molecule which makes up the aggregate. Consequently, a comparison of proton–electron hyperfine coupling constants in $Chl_{sp}^{+\cdot}$ or $Bchl_{sp}^{+\cdot}$ and $Chl\ a \cdot L_1^{+\cdot}$ or $Bchl$ $a \cdot L_1^{+\cdot}$ is a much more rigorous and demanding test of the special-pair hypothesis than is line-shape analysis. Assignment of the coupling constants accounting for $> 80\%$ of the line-width of $Chl\ a^{+\cdot}$ has been done by endor spectroscopy on a suite of isotopically substituted derivatives of Chl *a* (Scheer *et al.* 1977). Table 3 lists the aggregation numbers deduced from endor experiments

TABLE 3

Endor evidence for special-pair chlorophyll[a]

Protons	Hyperfine coupling constants (G)		Aggregation number	
	Bchl+·	Chl a+·	R. rubrum	S. lividus
(α, β, δ, 10)	0.49	0.24	1.7	
1a	1.88		2.4	
(1a, 3a, 4a)		1.13		1.9
		1.31		2.2
5a	3.46	2.63	2.1	2.0
7, 8	4.95	4.17	2.0	2.2
		Average	2.0	2.1

[a]Data taken from Norris et al. (1974).

on P700+· and P865+· in vivo. In both cases the in vivo coupling constants are (approximately) halved relative to the monomer free radical, thus providing convincing support for the Chl$_{sp}$ model.   For photosynthetic bacteria the assignment of the in vivo endor spectra is straightforward and is compatible with the Chl$_{sp}$ model. For green plants, the endor spectra are more complicated and the interpretation is not as direct.   Nevertheless, here also the simplest interpretation of the endor data requires a pair of chlorophyll molecules.

Evidence from electron spin-echo spectroscopy

Electron spin-echo spectroscopy is a pulsed form of e.p.r. spectroscopy in which resonance is detected by 'spin-echo' from the free radical excited by a suitable sequence of high-intensity radio-frequency pulses (Mims 1972).   This relatively new technique can be applied to good advantage to the special-pair problem.   Electron spin-echo spectroscopy has been used to study nitrogen hyperfine interactions that cannot be observed by endor.   Fig. 2 shows two sets of spin-echo envelopes in which P700+· is compared with Chl a·L$_1$+· and P865+· with Bchl a·L$_1$+·.   The pulse spin-echo envelopes have superimposed on them a modulation pattern caused by interactions of the nitrogen atom with the unpaired electron.   It can be deduced immediately from these patterns that the simple monomer chlorophyll cation cannot be responsible for the in vivo signals.   There are such large differences in the spin-echo envelope modulations between the in vivo and in vitro nitrogen environments experienced by the unpaired spins in P700+· and Chl a·L$_1$+· as to eliminate the possibility that

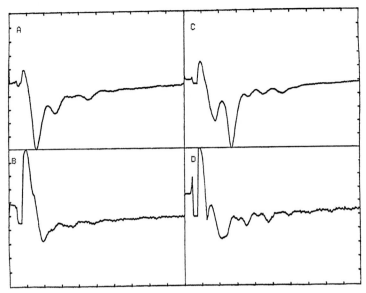

FIG. 2. Comparison of *in vivo* and *in vitro* spin-echo envelope modulations: spin-echo intensity (ordinate in arbitrary units) is plotted against time (abscissa, one division corresponds to 40 ns). A, *in vitro* [^{2}H]Bchl $a \cdot L_1^{+\cdot}$; B, *in vivo* P865$^{+\cdot}$ in [^{2}H]$R. rubrum$; C, *in vitro* [^{2}H]Chl $a \cdot L_1^{+\cdot}$; D, *in vivo* P700$^{+\cdot}$ in [^{2}H]$C. vulgaris$.

Chl $a \cdot L_1^{+\cdot}$ is the origin of the *in vivo* signal (J. R. Norris & M. K. Bowman, unpublished work).

A possible alternative to the Chl$_{sp}$ explanation for the endor data is a 'special environment' effect. For example, the differences in the endor of P700$^{+\cdot}$ and Chl $a \cdot L_1^{+\cdot}$ could conceivably arise from distortions in the geometry of a monomer cation produced by some aspect of the *in vivo* environment, or rotation of methyl groups in the chlorophyll could be hindered. Either of these two possible *in vivo* perturbations could be adequate to explain the differences between the *in vivo* and *in vitro* endor data. The nitrogen atoms, on the other hand, are embedded in the conjugated system and thus are not nearly so sensitive to geometry. Consequently the nitrogen hyperfine interactions with the unpaired spin provide, in many respects, an even better test for the validity of the Chl$_{sp}$ model than does endor. The best evidence yet against a monomeric chlorophyll primary donor in green plant reaction centres comes from the electron spin-echo experiments.

Evidence from the bacterial triplet state

The discovery by Dutton *et al.* (1972, 1973) that Bchl *a* triplet states could be detected by e.p.r. or optically detected magnetic resonance in intact photosynthetic bacteria or bacterial reaction centres when the normal course of forward photosynthesis is blocked has provided a new approach to the study of the structure and function of the bacterial reaction centre. The extensive literature on chlorophyll triplets and their significance for photosynthesis has recently been thoroughly reviewed (Levanon & Norris 1978). Here we shall only say that comparison of the properties of *in vitro* monomeric ^{3}Bchl $a \cdot L_1$ with ^{3}P865 in the reaction centre again rules out monomeric ^{3}Bchl *a* as the origin of the e.p.r. triplet signal. The zero-field splitting parameters and the unusual spin polarization in the triplet spectra from the *in vivo* reaction centre exclude monomer ^{3}Bchl *a* but, since the triplet-state parameters are sensitive to geometry, the triplet results do not distinguish unambiguously between a triplet state confined to only two Bchl *a* molecules and a triplet state involving several Bchl *a* molecules. The triplet results, however, do furnish valuable information about some of the important details of special pair function; these are briefly described later (see pp. 28–31).

Evidence from optical properties for Chl$_{sp}$

It has been a perplexing question why P700 and P865 are red-shifted relative to the red absorption maxima of Chl $a \cdot L_1$ and Bchl $a \cdot L_1$. Earlier investigators were impressed by the red shifts that can be readily observed in solid chlorophyll films, concentrated solutions in non-polar solvents, and in colloidal dispersions. Consequently, chlorophyll aggregation has been advanced as an explanation alternative to a 'biological environment' or a simple 'protein–chlorophyll interaction' for the *in vivo* red shifts. A red shift can be expected on theoretical grounds whenever chlorophyll molecules are forced into close proximity with their transition moments aligned. The red shift in chlorophyll aggregates arises in part because of the electronic perturbations induced by chlorophyll acting as a donor in coordination or hydrogen-bonding interactions, and in part by transition dipole–transition dipole interactions between closely positioned chlorophyll molecules. The optical consequences of chlorophyll aggregation are perhaps most vividly illustrated by the Chl *a*–water aggregate that absorbs maximally at 740 nm. The large optical shift relative to Chl $a \cdot L_1$ ($\lambda_{max} = 660$ nm) in this aggregate can readily be rationalized (Shipman *et al.* 1976; Shipman & Katz 1977) by a combination of environmental and transition-dipole interactions using intermolecular distances derived from the Strouse X-ray crystal structure of ethyl chlorophyllide $a.2H_2O$ and the formalism developed by Shipman *et al.* (1976). It can be shown that a stack of two (or three) Chl *a* molecules arranged with one water molecule between them and having the same geometrical arrangement as in the Strouse (1974) linear stack is expected to have an absorption maximum near 700 nm. Thus, to a first approximation, a pair of Chl *a* molecules arranged as in the ethyl chlorophyllide $2H_2O$ crystal structure would reasonably be expected to have the optical properties of P700. As we shall show (pp. 13–16) there are several ways of arranging two Chl *a* molecules so that they have an absorption maximum at about 700 nm. For a $Bchl_{sp}$ the situation is more complex. On the basis of the experimental evidence now available, orienting two Bchl *a* molecules in the same geometry as in ethyl chlorophyllide $a.2H_2O$ does not produce the required optical red shift to 865 nm. However, purified reaction-centre preparations from photosynthetic bacteria contain at least four Bchl *a* and two Bpheo *a* molecules, which makes it plausible that it is the further interaction of the additional Bchl *a* and Bpheo *a* with a $Bchl_{sp}$ that is responsible for the red shift to 865 nm. In any event, monomeric Chl $a \cdot L_1$ and Bchl $a \cdot L_1$ have optical properties inconsistent with P700 and P865, but on both experimental and theoretical grounds Chl_{sp} and $Bchl_{sp}$ have optical properties entirely consistent with those required for a valid model for P700 or P865.

We have summarized a sizable body of optical and magnetic resonance data that casts serious doubt on a role for monomeric Chl *a* or Bchl *a* as the primary electron donor in P700 or P865, but does support the view that a special pair of chlorophyll molecules functions as the primary donor in light-energy conversion. We can now consider specific models for Chl_{sp} and $Bchl_{sp}$.

MODELS FOR Chl_{sp} AND $Bchl_{sp}$

Various models for Chl_{sp} and $Bchl_{sp}$ have been advanced, which have much in common but which also have important differences. All the models use two chlorophyll molecules with the macrocycle planes arranged in a parallel orientation, but they differ in such details as symmetry, the relative orientations of the two chlorophyll molecules, the distance between the macrocycle planes, and the functional groups and nucleophiles used to cross-link the two chlorophylls to form the special pair.

There are several requirements that must be met by a valid special pair model. The lowest energy $S_0 \rightarrow S_1$ electronic transition (called Q_y in the literature) must be red-shifted relative to the Q_y transition in the antenna to assure effective trapping of the singlet excitation energy in the reaction centre. The necessary shift in the Q_y transition band by the special pair requires a parallel alignment and the shortest practicable distance between the two Q_y transition moments. An appropriate Chl_{sp} model should also provide the necessary redox properties: that is, it must be more easily and rapidly oxidizable than either monomeric or antenna chlorophyll. Such will be the case when the highest occupied molecular orbitals (HOMOs) of the two chlorophylls in the special pair mix to form two 'supermolecular' HOMOs (see p. 18), from the higher of which it will be easier to remove an electron than from the HOMOs of the monomer chlorophylls. And finally, the arrangement of the two chlorophylls in the special pair model must provide for overlap of the π-systems of the macrocycles and for equality of corresponding sites in the two chlorophyll molecules to make possible the equal sharing of the unpaired spin in $Chl_{sp}^{+\cdot}$.

Model of Shipman et al. (1976)

The most satisfactory model so far proposed, or at least the one that has the best basis in experiment, is that of Shipman *et al.* (1976) (see Fig. 3). A similar but less-detailed model has been proposed by Boxer & Closs (1976). In the former model, the two Chl *a* molecules are held together by two molecules of

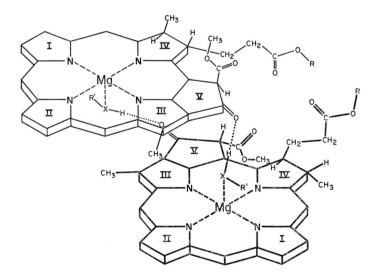

FIG. 3. Model of Chl_{sp} proposed by Shipman *et al.* (1976): note the hydrogen-bonding to the keto C=O functions. Many hydrogen-bonding nucleophiles can act as cross-linking agents.

a bifunctional ligand. The bifunctional ligand, in addition to an electron lone pair available for coordination to Mg, must have hydrogen-bonding properties. Ligands of the general type $R'XH$, where $R' = H$ or alkyl group and $X = O$, NH or S are suitable. Typical nucleophiles of this general class are water, HOH, or ethanol, $CH_3 \cdot CH_2 \cdot OH$. The electron lone pair on the oxygen atom of these ligands is coordinated to the Mg atom of one of the Chl *a* molecules and hydrogen-bonded to the keto carbonyl group of the other Chl *a* molecule in the pair. The arrangement in Fig. 3 sets the macrocycles at a π–π stacking distance of 0.36 nm, a distance that just brings the π-systems into contact and provides optimum π-overlap. The extent of π-overlap in this special-pair configuration assures spin delocalization in the special-pair cationic free radical.

The optical properties expected for this model are also consistent with the requirement for a 700 nm absorption maximum. In the orientation of Fig. 3, the Q_y transitions are parallel and, from exciton theory, the red-shifted Q_y exciton transition will have all the oscillator strength. If each of the monomer Chl *a* molecules is considered to be environmentally shifted to 686 nm by the hydrogen-bond interaction at its keto carbonyl group and if π–π stacking is taken into account, then the model of Shipman *et al.* (1976) is calculated to have its Q_y transition at 700 nm. An environmental shift to 686 nm is required in a Chl *a* molecule strongly hydrogen-bonded at its keto carbonyl function to account for the experimentally observed red shift in the 740 nm-absorbing Chl *a*–water adduct. Thus this model appears to have the necessary red shift in its Q_y transition.

One of the interesting features of this model is that a considerable variety of nucleophiles can be used to form it. The geometrical arrangement in Fig. 3 is such that the cross-linking nucleophile is not restricted to a small ligand such as water. Nucleophilic groups present in the generic class $R'XH$ are characteristically also present in protein side-chains. We can contemplate the use of the OH groups of serine or threonine, the NH_2 group of lysine, the SH group of cysteine etc. to form the Shipman *et al.* (1976) special pair. The proposed structure is open enough to allow the entry of large nucleophilic groups without difficulty. The possibility that special pairs could be formed by the intervention or participation of protein is of considerable interest. The composition of the bacterial reaction-centre preparations (polypeptides and chlorophyll) and the current activity in the isolation of chlorophyll–protein complexes from photosynthetic membranes or organelles make such a possibility more immediate. A chlorophyll–protein interaction in the sense described here to form a Chl_{sp} does not involve formation of covalent bonds. Extraction of the chlorophyll with organic solvents from a chlorophyll–protein complex would be accompanied by a change in protein conformation, thereby providing a rationale for the difficulty encountered up to now in reconstituting a disassembled photo-synthetic membrane. The kind of interaction suggested here would provide a reasonable explanation for a red shift in a chlorophyll–protein complex in terms of chlorophyll–chlorophyll interactions mediated but not directly caused by interaction with the protein. Protein participation in special-pair formation also raises the possibility that there could be many reaction centres *in vivo* formed with different nucleophiles or combinations of nucleophiles which would have essentially the same optical properties and the same ability to share an unpaired spin, but which might have significantly different redox properties. A Chl_{sp} linked by two water molecules might have excited-state properties different from those of one formed from, say, one water molecule and one seryl OH group, or from a special pair formed from a lysyl NH_2 group and a cysteinyl SH group. Other ways in which the nucleophilic groups in protein side-chains could be used to form Chl_{sp} should also be considered. Given two nucleophilic groups (selected from OH, NH_2, SH etc.) in a poly-peptide it is conceivable that coordination of a nucleophile to either of the fifth coordination sites of the Mg atoms of two chlorophyll molecules could position the two chlorophylls with a geometry appropriate to a special pair. In this mode of organization by protein, coordination would be from the rear, and no nucleophiles would be present between the chlorophylls. The question here would be whether a peptide structure could be made sufficiently rigid to maintain the geometry of the special pair formed in this way. Hydrophobic interactions involving the phytyl group of the chlorophylls and the hydrophobic regions of the protein or polypeptide could also be considered as a possible mode of interaction in the formation of organized chlorophyll species. All these possibilities suggest new experimental initiatives to the general question of chlorophyll–protein interactions and the possible role of protein in the formation of Chl_{sp} (and antenna chlorophyll too for that matter).

Symmetry considerations in Chl_{sp} models

The models of Shipman *et al.* (1976) and Boxer & Closs (1976) have C_2 symmetry which makes the two chlorophyll molecules identical. Fong (1974*a*) originally raised the issue of symmetry in the special pair and proposed the first model with C_2 symmetry. The symmetry requirement for Fong's special-pair model was introduced to satisfy the presumed requirements of a singlet–triplet (up-conversion) scheme for photosynthesis (Fong 1974*a, b*). Whatever the merits of the up-conversion scheme (see, for example, Menzel 1976; Warden 1976; Govindjee & Warden 1977), Fong's Chl_{sp} model has some serious problems. Cross-linking is exclusively by hydrogen-bonding to the ring V methoxycarbonyl groups of the chlorophylls, and the keto carbonyl functions are not used in any way. From molecular models the spacing between the macrocycle planes in the Fong structure is 0.57 nm, a substantially greater distance than the van der Waals' contact distance of 0.34–0.36 nm that is optimum for porphyrin and chlorophyll π-system overlap. Molecular overlap between the macrocycles in the Fong structure is expected to be small because orbital overlap falls off exponentially with distance. Further, when the methoxy-carbonyl C=O groups are used, the transition moments in the two macro-cycles are at a 60° angle, not parallel as is required for strong exciton coupling. Exciton considerations indicate that the optical red shifts in this structure are too small to be compatible with a 700 nm absorption requirement for the Chl_{sp} and, moreover, predict a blue shift. Models formed by hydrogen-bonding to the keto carbonyl function, however, avoid the intractable problems in the Fong structure (see previous section).

Symmetry in the special pair, as we see it, may be important not because of its possible contribution to a long-lived triplet state for $Chl_{sp}^{+\cdot}$ but because it makes for better geometry and promotes the near-equivalency of corre-sponding sites in the two chlorophyll molecules of the Chl_{sp}. A more complete discussion of the possible participation of the triplet state in photosynthesis can be found in a review by Katz *et al.* (1978*a*).

Lower symmetry models for Chl$_{sp}$

Although the C_2 models of Shipman *et al.* (1976) and Boxer & Closs (1976) have an undeniable attraction, other configurations for the Chl_{sp} structure cannot be excluded on the basis of the experimental evidence now available. The original model for the chlorophyll special pair proposed by the Argonne group (Ballschmiter & Katz 1968; Katz & Norris 1973) was based on the structure inferred at that time for the P740 Chl *a*–water adduct. In this structure two Chl *a* molecules are held together by a single water molecule coordinated to the Mg of one Chl *a* molecule and simultaneously hydrogen-bonded to the keto and methoxycarbonyl C=O groups of the second Chl *a* in the Chl_{sp}. The principal problems with this model are the distance between the two macrocycles (because of the presence of the hydrogen-bonded methoxycarbonyl group) and the unfavourable angles for hydrogen-bonding. The two Chl *a* molecules are not identical as one functions as a donor, the other as acceptor with respect to hydrogen bonding. There is, therefore, a question whether the two Chl *a* molecules are sufficiently equivalent to assure delocalization of an unpaired electron.

A modification of the asymmetric model that avoids its worrisome aspects

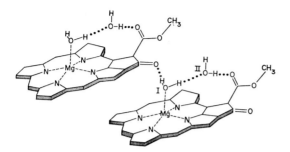

FIG. 4. A model for Chl$_{sp}$ with only translational symmetry. Only the keto carbonyl group of the top chlorophyll molecule participates in an hydrogen-bond, making the two chlorophylls non-equivalent. The orientation is that in a Strouse (1976) stack.

can be derived from the Strouse (Chow *et al.* 1975) ethyl chlorophyllide *a*.2H$_2$O crystal structure. The first two Chl *a* molecules in a Chl *a*–water stack that is the fundamental unit in the Strouse crystal structure provide an asymmetric model for Chl$_{sp}$ that appears to have an acceptable geometry and optical properties suitable for a Chl$_{sp}$ model (Fig. 4). A second, loosely-held, water molecule satisfies the hydrogen-bonding requirements of water without insertion of the methoxycarbonyl groups between the macrocycles. The two Chl *a* molecules are positioned as in the Strouse structure, which places them at their van der Waals' radii and brings their π-systems into contact at essentially the same distance as in the Shipman *et al.* (1976) model. Exciton calculations (Shipman & Katz 1977) suggest that this model will have its Q_y maximum at about 695 nm. Whether charge separation will be facilitated enough to compensate for the detrimental effects on spin sharing because of loss of full equivalence remains to be settled. Nor can the possibility be excluded that the *in vivo* environment may tend to make asymmetric Chl$_{sp}$ molecules more equivalent either by hydrogen-bond formation to the acceptor Chl *a* in the pair (see below) or by interaction between the keto carbonyl groups of the acceptor Chl *a* with the Mg atom of an antenna chlorophyll.

Another Chl$_{sp}$ model can be derived from a Strouse stack of three Chl *a* molecules (Fig. 5). A stack of three is expected from exciton calculations (Shipman & Katz 1977) to have its Q_y absorption maximum at 704 nm. The two Chl *a* molecules within the dashed line are fully equivalent and are related by translational symmetry. In this model, three Chl *a* molecules are responsible for the optical shift but an unpaired spin would be expected to be delocalized mainly over the two nearly equivalent Chl *a* molecules. The model with translational symmetry is formed from the asymmetric model in Fig. 4 by the ad-

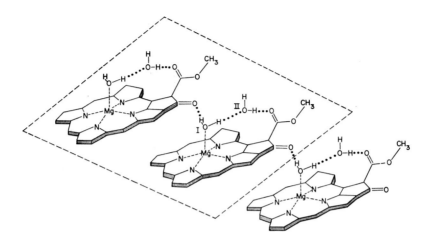

FIG. 5. Model with translational symmetry only: the two chlorophylls in the box are perturbed in much the same way and are nearly equivalent. The geometry is the same as in the asymmetric model of Fig. 4. In the models of Figs. 3–5 the macrocycles are at their van der Waals' radii—about 0.36 nm.

dition of an additional Chl a.H_2O whose hydrogen-bonding perturbation makes two of the three Chl a molecules equivalent.

Existing experimental evidence (see p. 21) supports the conclusion that the C_2 model (Fig. 3) is an adequate representation of *in vivo* Chl$_{sp}$ properties. It is a sufficient but not necessarily a unique solution to the problem of Chl$_{sp}$ structure *in vivo*. Which, if any, of the proposed models exists *in vivo*, and whether a Chl$_{sp}$ structure formed in the same way occurs in all photosynthetic organisms are questions that cannot be answered on *a priori* considerations. These and the many other questions that still remain can only be resolved by more information on both *in vivo* and *in vitro* reaction-centre systems.

Why a Chl$_{sp}$?

The Chl$_{sp}$ can be considered a supermolecule, with electronic properties more suitable to reaction-centre requirements than those of monomeric Chl a or Bchl a. We can now consider in more detail how the supermolecular properties of a Chl$_{sp}$ can be expected to affect its redox, e.p.r., endor, and optical properties.

When one electron is removed from chlorophyll by chemical oxidation or photooxidation, molecular-orbital theory predicts that two low-lying π-cation radical doublet states (D_1 and D_2) are produced and each of these states has two spin sublevels corresponding to $m_s = \pm 1/2$. In state D_1, which is the lowest-energy doublet state of the special-pair supermolecule, the distribution of the unpaired electron is x on Chl 1 and $(1 - x)$ on Chl 2, where $0 \leqslant x \leqslant 1$. If the $D_1 \rightarrow D_2$ transition energy is large relative to kT, then the e.p.r. and endor spectra (see above) measure the electrically-forbidden but magnetically-allowed transition between the m_s sublevels of D_1. If the $D_1 \rightarrow D_2$ transition energy is not large compared to kT, then transitions between the m_s sublevels of D_2 may also contribute to the observed e.p.r. and endor spectra. Following the analysis of Norris et al. (1971), we find that the e.p.r. signal from state D_1 has a peak-to-peak width (ΔH_{sp}) that is related to the peak-to-peak width (ΔH_M) of the monomeric $Chl \cdot L_1^{+ \cdot}$ radical e.p.r. signal by equation (4).

$$\Delta H_{sp}/\Delta H_M = [x^2 + (1 - x)^2]^{1/2} \qquad (4)$$

The endor spectrum resolves the hyperfine coupling between the nuclear spins of the protons adjacent to the macrocycle π-system and the spin of the unpaired electron in the macrocycle π-system. For both states D_1 and D_2 two sets of hyperfine coupling constants are to be expected: one set is reduced by x and the other is reduced by $(1 - x)$ with respect to the corresponding hyperfine coupling constants for monomeric $Chl^{+ \cdot}$. The observed $2^{1/2}$ narrowing of the e.p.r. signal width and the observed factor of $1/2$ reduction in the hyperfine coupling constants (see above) of in vivo $P700^{+ \cdot}$ or $P865^{+ \cdot}$ compared to monomeric chlorophyll in vivo implies that $x \approx 1/2$ in equation (4); that is, the unpaired electron is delocalized almost equally over both Chl_{sp} molecules in states D_1 and D_2.

Is the $D_1 \rightarrow D_2$ electron-transition electric dipole allowed? To answer this question, we must explicitly consider the molecular orbital structure of the special pair that has a highest-occupied canonical Hartree–Fock molecular orbital, H_1 and H_2 on chlorophylls 1 and 2, respectively. When two chlorophyll molecules are brought together to form the special pair, H_1 and H_2 add and subtract to form the two highest-occupied molecular orbitals (H_+ and H_-) in the special-pair supermolecule (equations 5 and 6), where $-1 \leqslant a \leqslant 1$ is a

$$H_+ \approx (1 - a^2)^{1/2} H_1 + a H_2 \qquad (5)$$

$$H_- \approx a H_1 - (1 - a^2)^{1/2} H_2 \qquad (6)$$

mixing parameter. If the special pair is oxidized by removal of one electron, the unpaired electron left behind has a distribution of a^2 on one chlorophyll and $(1 - a^2)$ on the other. Thus a^2 in equations (5) and (6) is equal to x in equation (4). Using the information from e.p.r. and endor that the unpaired electron is shared almost equally by both molecules, we have equations (7) and (8).

$$H_+ \approx 2^{-1/2}(H_1 + H_2) \tag{7}$$

$$H_- \approx 2^{-1/2}(H_1 - H_2) \tag{8}$$

Without loss of generality, let us assume that the orbital energy of H_+ is lower than that of H_-. State D_1 corresponds to a doubly-occupied H_+ and singly-occupied H_- and state D_2 corresponds to a singly-occupied H_+ and a doubly-occupied H_-. This is illustrated in Fig. 6 where (a) and (b) are the $m_s = +1/2$ and $m_s = -1/2$ spin sublevels of state D_1, respectively, and (c) and (d) are the $m_s = +1/2$ and $m_s = -1/2$ spin sublevels of state D_2, respectively. The transitions from (a) to (d) and from (b) to (c) are spin-forbidden and therefore the transition dipole is null. The transitions from (a) to (c) and from (b) to (d) are electric-dipole allowed and spin-allowed, and the transition dipole, μ (in Debyes), is given by equations (9) and (10), where $\mathbf{R}$ (Å) is the dipole-

$$\mu = 4.803 \langle \Psi_{D_1} | \mathbf{R} | \Psi_{D_2} \rangle \tag{9}$$

$$= 2.402(\langle H_1 | \mathbf{R} | H_1 \rangle - \langle H_2 | \mathbf{R} | H_2 \rangle) \tag{10}$$

moment operator; $<H_1|\mathbf{R}|H_1>$ is the average vector position of an electron in H_1, and this is approximately at the centre of the chlorophyll 1 macrocycle (Spangler *et al.* 1977); and similarly $<H_2|\mathbf{R}|H_2>$ is approximately the vector position of the centre of the chlorophyll 2 macrocycle. Thus, we conclude that the $D_1 \rightarrow D_2$ transition is electric-dipole-allowed and that the dipole strength, μ^2 (Debye^2), is 5.77 r^2, where r (Å) is the distance between the centres of the macrocycles 1 and 2. If this electronic transition predicted for $Chl_{sp}^{+\cdot}$ can be detected by experiment, then it should provide valuable information about the geometry of the Chl_{sp}, because the dipole strength of the predicted transition

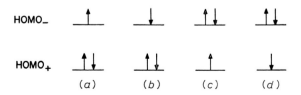

FIG. 6. Electron configuration in a $Chl_{sp}^{+\cdot}$ supermolecule. See text for details.

would be a sensitive measure of the distance between the centres of the two macrocycles in the special pair.

We should point out that the experimentally observable quantity is the oscillator strength, f (unitless), and this is related to the dipole strength, μ^2, by a constant times the transition energy. Since the transition energy for this kind of interaction falls off as intermolecular molecular orbital overlap (an exponential dependence on distance), it follows that the distance-dependence of the oscillator strength is approximately r^2 times an exponential.

SYNTHETIC CHLOROPHYLL SPECIAL PAIRS

The explicit nature of the models for the Chl_{sp} described above has stimulated efforts to produce them in the laboratory. The goal has been to make well-defined chlorophyll systems or species that have the optical and e.p.r. properties of P700 and P865. This goal has been fully reached for P700 and partially for P865. Two approaches have been successfully used.

Stensby & Rosenberg (1961) and Brody (1958; Brody & Broyde 1968) had observed that Chl a dissolved in ethanol developed a new absorption maximum near 700 nm at low temperatures. More recently, Fong & Koester (1976) noted increased absorption near 700 nm at cryogenic temperatures in Chl a solutions in aliphatic hydrocarbon solvents containing an indeterminate amount of water. As visible absorption and fluorescence spectroscopy were the only techniques used to observe the Chl a species formed at low temperatures, little if any structural information could be extracted from the spectral data. Shipman et al. (1976) and Cotton et al. (1978) have reexamined the red-shifted Chl a species formed at low temperature in the presence of hydrogen-bonding nucleophiles by visible absorption spectroscopy but have also made infrared (i.r.) studies of the red-shifted species that considerably facilitate structural assignments. Cotton prepared solutions of 0.1M-Chl a containing 0.15M-ethanol in toluene (Shipman et al. 1976; Cotton et al. 1978) whose room-temperature absorption maximum at 668 nm is almost completely shifted to 702 nm at 77 K. (Ethanol is a more advantageous nucleophile than water for these experiments because the much higher solubility of ethanol in non-polar solvents allows much more concentrated Chl a solutions to be used in the experiments.) I.r. spectroscopy shows little change in the area or peak position of ester carbonyl absorptions at low temperature. The keto carbonyl absorption peak is strongly decreased in area and shifts from 1697 cm^{-1} to 1686 cm^{-1} as the 702 nm absorption peak grows and a new peak at 1658 cm^{-1} appears in the low-temperature spectra. This peak is logically assigned to a hydrogen-bonded keto carbonyl group. (The i.r. data alone appear to eliminate the structure proposed by Fong & Koester (1976) from their low-temperature experiments.) The model of Shipman *et al.* (1976), however, which requires strongly hydrogen-bonded keto carbonyl groups is fully compatible with the i.r. data.

With tetranitromethane as an electron acceptor, the species which absorbs at about 700 nm and is formed in a Chl *a*–ethanol system at low temperatures undergoes photooxidation on illumination with 700 nm light. An e.p.r. signal with a line-width of 7.5 G can be detected in the photooxidized system. The e.p.r. signal in this *in vitro* system is practically indistinguishible in shape, line-width, and temperature-dependence from the e.p.r. signal resulting from the photooxidation of P700 in *Chlorella vulgaris*.

This system, then, replicates the properties of P700 to a considerable extent: it has the correct visible absorption maximum, it is oxidizable with red light, and it shares a narrowed e.p.r. line, indicative of spin sharing by two Chl *a* molecules. It is, however, a somewhat awkward system to work with. The photoactive species is immobilized in a frozen matrix at 77 K, and the structure of the photoactive species is by no means unambiguously established. The i.r. evidence does not preclude the presence of short linear stacks such as the one shown in Fig. 5. We consider it likely that these Chl *a* systems at low temperature are heterogeneous in composition and, rigorously speaking, the structure of the species absorbing at about 700 nm is ambiguous. Although this synthetic reaction-centre system is not without interest or utility, other routes to a synthetic reaction centre that is better defined in structure and more amenable to manipulation are needed.

One such approach is the synthesis of covalently-linked special pairs. The problem with inducing two molecules of Chl *a*–ethanol to assume the pose shown in Fig. 3 is that the entropy of association of the monomer units is highly unfavourable, as indicated by the requirements of high Chl *a* concentrations and low temperatures. Essentially the monomers must be forced from solution, and the structures that result depend not only on concentration, final temperature, nucleophile, and solvent, but on variables more difficult to control such as rate of cooling, presence or absence of adventitious nucleophiles, and so forth. A C_2-special-pair structure is not formed at room temperature because the forces leading to aggregation are not able to compete with diffusion. Diffusion of the two halves of a special pair away from each other can be permanently prevented by linking two chlorophyll macrocycles with a covalent bond (Fig. 7). A covalently linked dimer of pyrochlorophyll *a* (a Chl *a* derivative minus the methoxycarbonyl group at C-10) has been prepared by Boxer & Closs (1976), and Wasielewski *et al.* (1976, 1977) have succeeded in preparing covalently-linked dimers of Chl *a* and Bchl *a*.

In Fig. 7 two chlorophyll macrocycles are shown in their folded configuration.

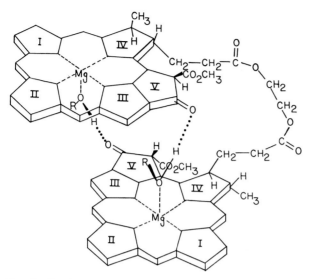

FIG. 7. Covalently linked chlorophyll molecules in the folded (Chl$_{sp}$) form: ...., hydrogen bonds. The link (in the upper right quadrant) prevents the two macrocycles from diffusing away from each other. Any of several hydrogen-bonding nucleophiles can be used to cause folding. Boxer & Closs (1976) have proposed a similar procedure for covalently linked pyrochlorophyll a dimer.

The phytyl group normally present at the propionic side-chain in the chlorophyll molecules has been removed, and the carbonyl groups of two such macrocycles have been re-esterified with the two hydroxy groups of ethanediol, thus fashioning a practically-indissoluble link between the two macrocycles. In solution in polar media, such as acetone or pyridine, the linked dimers exist in an open configuration with a molecule of the solvent ligated to each of the two Mg atoms in the open-linked dimer. However, in a non-nucleophilic solvent, such as benzene, introduction of water or alcohol in the appropriate amount causes the linked dimer to fold and to assume the structure shown in Fig. 7, which is identical, except for the link, with that of the Shipman *et al.* (1976) Chl$_{sp}$ model (Fig. 3). ^{1}H n.m.r. furnishes convincing evidence that the folded linked pyrochlorophyll a, Chl a, and Bchl a dimers have the structure shown in Fig. 7. The optical and e.s.r. properties of the open and folded linked dimers are summarized in Table 4.

The linked Chl a dimer mimics green plant P700 to a remarkable degree. The linked Chl a dimer in dry CCl$_4$ has an intense absorption maximum at 677 nm, similar to that of (Chl a)$_2$. Introduction of water causes a marked red-

TABLE 4

Comparison of optical and e.p.r. properties of synthetic photoreaction centres with those of P700 and P865

Species	λ_{max}/nm	$\Delta H_{pp}/G^a$
In vivo P700	about 700	7.0[b]
Bis(chlorophyllide *a*) ethylene glycol diester		
Open	662[e] (666)[k]	(9.3)[g]
Folded	697[d] (696)[k]	7.5[h]
Cryogenic (Chl *a*·ethanol) adduct[e]	702[f]	7.1[i]
In vivo P865	about 865	9.6
Bis(bacteriochlorophyllide *a*) ethylene glycol diester		
Open	760[c]	(12.8)[g]
Folded	803[d]	10.6[f]

[a]E.p.r. peak-to-peak line-width of the oxidized species.
[b]Oxidized with $FeCl_3$ or I_2 in methanol solution.
[c]In benzene solution containing >0.1M-ethanol.
[d]In benzene or CCl_4 containing water or <0.1M-ethanol.
[e]Formed in a 0.1M-Chl *a* solution in toluene containing 0.15M-ethanol.
[f]At 175 K.
[g]Line-width of monomer Chl $a^+\cdot$ or Bchl $a^+\cdot$.
[h]Photooxidized with I_2 in H_2O-saturated CCl_4.
[i]Photooxidized at 96 K with tetranitromethane (100 mmol/l) in a toluene solution containing 1mM-Chl *a*, 1.5mM-methanol.
[j]Photooxidized in wet benzene with tetranitromethane.
[k]Data of Boxer & Closs (1976) for open and folded linked pyrochlorophyll *a* dimer.

shift in the absorption maximum to 697 nm. (A small residual absorption at 667 nm is attributed to the diastereoisomeric form of Chl *a* always present in laboratory systems and whose geometry prevents folding.) Oxidation of the folded dimer in CCl_4 by iodine or by red light causes the 697 nm absorption to bleach, and the free radical so produced has a strong e.p.r. signal suitably narrowed by the factor $1/2^{\frac{1}{2}}$ to a linewidth of 7.5 G. The oxidation potential of P700 in spinach chloroplasts is generally estimated to be +0.43 V. Oxidation of the folded linked Chl *a* dimer indicates an oxidation potential close to 0.5 V and distinctly lower than that of monomeric Chl *a*.

The optical, e.p.r. and redox properties of the covalently linked Chl *a* dimer in its folded configuration are thus seen to resemble those of P700 with high fidelity. The synthetic 'reaction centre' is a remarkably-faithful imitation of an essential component of the light-conversion apparatus of the green plant.

The synthetic bacterial reaction centre is still under investigation. Synthesis of a covalently linked Bchl *a* dimer, a considerably more formidable task than the synthesis of the linked pyrochlorophyll *a* or even the linked true Chl *a* dimers, has been successfully completed (Wasielewski *et al.* 1977; full details of the other linked dimers will be published elsewhere by M. R. Wasielewski).

The folded Bchl *a* linked dimer absorbs maximally at 803 nm rather than at the 865 nm characteristic of the *in vivo* bacterial reaction centre. The folded covalently-linked Bchl *a* dimer is photooxidized in water-saturated toluene containing an equivalent amount of iodine by red-light; the absorption peak at 803 nm is completely bleached in a few seconds, and a pair of absorption peaks near 1150 nm appears. This compares with the peak at 1250 nm formed by the oxidation of P865 and which has been attributed to the formation of P865$^{+\cdot}$. The photooxidized folded linked Bchl *a* dimer gives a Gaussian e.p.r. signal with a line-width of 10.6 G. A comparison of this with the e.p.r. line-width of about 13 G for monomeric Bchl $a^{+\cdot}\cdot L_1$ makes it evident that the synthetic Bchl$_{sp}^{+\cdot}$ has the ability to share an unpaired spin.

From the foregoing it appears that two Bchl *a* molecules in the configuration of Fig. 3 are sufficient to ensure spin sharing in the doublet state but that the interactions between the only two Bchl *a* molecules is not sufficient to provide the red shift required for a P865 model system. This is one of the most interesting results so far to emerge from studies with the synthetic reaction centres. As it is well established that bacterial reaction-centre preparations contain at least four Bchl *a* and two Bpheo *a* molecules, it seems plausible that a red shift to 865 nm must involve interactions of the Bchl$_{sp}$ with the other Bchl *a* and Bpheo *a* molecules. ^{1}H n.m.r. spectroscopy shows clearly that the acetyl carbonyl groups are not involved in the formation of the folded configuration of the linked Bchl *a* dimer. Additional interactions between the special pair and other Bchl *a* molecules could occur by coordination interactions with the unused acetyl groups in the linked Bchl *a* dimer. The possibility that different numbers of Bchl *a* molecules may be involved in spin sharing and in optical red shifts should be seriously considered when thinking about the structure of P865.

Another way in which the synthetic reaction centres have contributed to the understanding of the excited states of P700 and P865 is from laser studies. The recent finding that with only a few exceptions all the chlorophylls and their derivatives can emit coherent laser light on excitation (Hindman *et al.* 1977; Mory *et al.* 1976) has provided a new tool for studying the photophysics of chlorophyll excited states (Hindman *et al.* 1978). The fluorescence and lasing properties of covalently linked dimers of Chl *a* and pyrochlorophyll *a* in their open and folded configurations have been examined with the object of acquiring information about their excited states that cannot be obtained only from normal absorption and fluorescence spectroscopy.

TABLE 5

Absorption, fluorescence and lasing characteristics of covalently linked chlorophyll (Chl) *a* and pyrochlorophyll (pyrochl) *a* dimers (Hindman *et al.* 1978)

Compound	Absorption maximum (nm)	Fluorescence maximum (nm)	Lasing maximum (nm)	Fluorescence yield, ϕ_f	Fluorescence lifetime, τ_f (ns)
Chl a^a	671	684, 736	680	0.35	7.3 ± 0.1
Linked Chl *a* (open)^a	670	683, 734	NL^c	0.21	5.8 ± 0.3
Linked pyrochl *a* (open)^a	669	685, 735	NL^c	0.13	5.7 ± 0.2
Chl a^b	665	678, 725	675	0.2	6.0 ± 0.3
Linked Chl *a* (folded)^b	695	730	735		3.2 ± 0.1
Linked pyrochl *a* (folded)^b	695	730	731	0.3	4.4 ± 0.3

^{a}In pyridine.
^{b}In toluene + 0.1M-ethanol.
^{c}No lasing observed to a maximum pump power of 50 MW/cm^2.

Table 5 lists the absorption, fluorescence and lasing properties of the co-valently linked dimers of Chl *a* and pyrochlorophyll *a* at room temperature. The effect of laser excitation is significantly different for the open and folded dimers. No laser emission was observed in solutions of polar solvents of the open linked dimers even at pump powers up to 50 MW/cm^2. This is remarkable as monomeric Chl $a \cdot L_1$ or Chl $a \cdot L_2$ or the folded dimer lase readily in polar solvents at pump powers of 2–4 MW/cm^2. The folded linked dimers lase readily at thresholds of about 3.5 MW/cm^2 in wet benzene, which is about a third of the power required to pump monomer Chl *a* in the same solvent. The fluorescence yield of the open dimers is, however, comparable. The differences in properties of the open and folded dimers are also reflected in the effect of increasing pump power on the fluorescence lifetimes (Table 5). The creation of a significant S_1 population leads to shortening of the fluorescence lifetime by stimulated emission as pointed out by Lessing (1976). This effect is readily observed in monomeric Chl *a* and in the folded but not the open con-figuration of the linked Chl *a* dimer. The failure to observe lasing and lifetime-shortening for solutions of the open linked dimers in polar solvents suggests that the open dimer has molecular properties that mitigate against the build-up of a significant S_1 population.

One piece of information of considerable value that has come out of this work is the location of the emission band of the folded dimers. Although fluorescence is widely used as a probe of photosynthesis, assignment of the numerous emission bands of intact photosynthetic organisms has proved difficult. Now, on the basis of the experiments with the folded Chl *a* dimer it is possible to assign tentatively emission at 730 nm to PS I. This is especially useful for green plants as no reaction-centre preparation comparable to the ones from photosynthetic bacteria is as yet available. We can expect additional information of this kind to result from further work.

Although the synthesis of Chl_{sp} in the form of covalently linked dimers is a distinct step forward, much remains to be done before a truly-functional reaction centre can be synthesized. The *in vivo* P700 and P865 reaction centres have associated with their special chlorophyll pairs electron-transport chains, antenna, etc. With the synthetic chlorophyll special pairs as a starting point we can now contemplate the linking of electron acceptors and donors to the Chl_{sp}. Experiments in this direction should make a useful contribution to an expanded understanding of plant and bacterial photosynthesis, and should also provide a more adequate basis for biomimetic conversion of solar energy (Katz *et al.* 1978*b*).

THE RADICAL-PAIR MECHANISM FOR THE PRIMARY ACT IN PHOTOSYNTHESIS

Introduction

In the initial act of photosynthesis the Chl_{sp}, acting as the primary donor, ejects an electron to some nearby electron acceptor. From all the current evidence the first electron acceptor is diamagnetic before the initial photoreduction step. Thus, in the initial steps the formation of radical pairs (Thurnauer *et al.* 1975) between the chlorophyll special pair and some primary electron acceptor is expected. Such radical-pair states can be paramagnetic and can be a source of new information on the initial act of photosynthesis. Optical studies indicate that these initial intermediate states have a range of lifetimes of the order of a few hundred ps up to about 10 ns, depending on the redox potential. It is these short-lived initial species that are involved in the initial radical pairs. The initial paramagnetic states are, however, too short-lived to be investigated directly by magnetic resonance techniques. Fortunately the existence and properties of radical-pair states can be deduced from magnetic resonance observations of the longer-lived states that evolve from the original radical pair. We now discuss the observable longer-lived triplet states of photosynthetic bacteria in terms of a precursor radical-pair state.

Singlet–triplet intersystem crossing

When a molecule goes from a singlet state (spin zero) to a triplet state (spin one) or *vice versa*, electron spin angular momentum must change. Changes in electron spin angular momentum can be thought of loosely as 'spin flips'. Such spin flips always require intrinsic, internal magnetic fields that are provided by the orbital motion of the electrons, and it is these spin-orbit interactions that facilitate spin flips. Clearly the magnetic fields associated with the orbital motions and properties of the electrons are closely associated with the molecular framework of the molecule. Thus intersystem crossing is highly anisotropic (the rate of crossing depends on orientation with respect to an external field) and its parameters reflect this anisotropy and the molecular framework.

Radical-pair intersystem crossing

The situation is different for radical pairs that undergo intersystem crossing. Whereas in the ordinary case intersystem crossing in a large external magnetic field (e.g. the typical e.p.r. magnetic field) reflects the molecular framework, that of radical pairs in a magnetic field does not reflect the molecular framework. Instead intersystem crossing in radical pairs is best described in the external laboratory coordinate system. In other words, the orbital motion of the electrons is no longer important. Each of the two electrons of the radical pair are like tiny gyroscopes tracking the external magnetic field, H_0, and almost completely independent of the molecular-frame orientation with respect to the imposed field H_0. Thus the interacting electrons in the radical pair are associated with the laboratory coordinate system, not the molecular framework. Consequently intersystem crossing in radical pairs invokes explanations other than spin-orbit coupling. We now proceed to consider intersystem crossing in a weakly-interacting radical pair.

Intersystem crossing in a weakly-interacting radical pair

A weakly-interacting pair of electrons form a radical pair. The two electrons are typically located in different molecules, e.g., Chl_{sp}^{+}---X^{-}. The spins of the two electrons combine to form either a singlet state in which the spin multiplicity is zero and in which all spin angular momentum is absent, or all the spin angular momenta are combined constructively to form a triplet state ($S = 1$). Most often in real situations some combination 'singlet-triplet' state that is neither pure singlet or pure triplet is produced. We use the term 'weakly-

A. PURE SINGLET

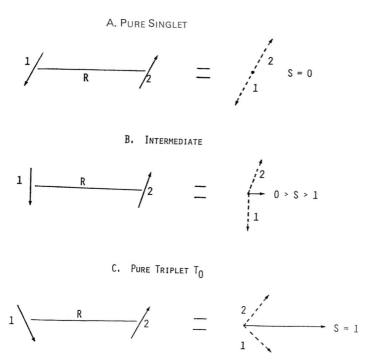

B. INTERMEDIATE

C. PURE TRIPLET T_0

FIG. 8. A vectorial representation of the radical-pair mechanism for $S_1 \rightarrow T_0$ intersystem crossing: A, initial radical pair in a pure singlet state; B, intermediate or mixed state with both singlet and triplet characters; C, pure triplet state.

interacting' to imply that the electrons are sufficiently far apart that the energy difference between the pure singlet-state level and the mean of the three pure triplet-state levels is small (i.e. the intermolecular exchange integral is small). In an externally applied magnetic field such a radical pair can have four energy levels, one for the singlet state (S_0) and three for the triplet state (T_+, T_0, T_-). The S_0 level and the T_0 level are close together in energy in the weak radical pair and intersystem crossing occurs between the T_0 and the S_0 levels.

We now describe explicitly the intersystem crossing $S_0 \leftarrow T_0$. The spins behave like vector quantities precessing about H_0. Suppose a radical pair is born from a singlet state (e.g., a photoexcited singlet special pair). Since there was no spin angular momentum in the precursor singlet state, the radical pair starts as a singlet state as indicated in Fig. 8A. However, after a short time one of the electrons of the radical pair (electron 1) may have precessed at a different rate from the other electron (electron 2) in such a way that the system develops net spin angular momentum (Figure 8B). The pair then has both a singlet nature and a triplet nature. Not until more time has passed can the radical pair state become a pure triplet as shown in Figure 8C. Given still more time the system will oscillate between singlet and triplet descriptions. What causes the unpaired electrons to precess at different rates? Any differences between the magnetic environment of electron 1 and electron 2 will cause the oscillation between singlet and triplet descriptions. For example, the differences in the magnetic moments of the nuclei of radical 1 and those of radical 2 are sufficient.

The important feature of this mechanism is that intersystem crossing occurs primarily to the T_0 level of the triplet and in no significant way does the molecular framework of a single molecule enter the description of intersystem crossing. Instead the crossing is influenced by the radical-pair distance R rather than the molecular framework of a single molecule. For all photo-excited organic triplets that have been studied the molecular framework has influenced intersystem crossing in such a way that it occurs to T_+ and T_- as well as to T_0. In fact, crossing to T_+ or T_- must be more important than to T_0 for at least some orientations that the triplet molecular framework makes with the external field. One and only one exception is known to this general rule, namely photosynthetic bacteria. The e.p.r. of the triplet state of photo-synthetic bacteria is interpretable only by a dominant intersystem crossing to the T_0 spin sublevel for *all* possible orientations of the triplet molecule to the external applied magnetic field. Ordinary single molecule intersystem crossing promoted by the spin-orbit mechanism is completely inadequate to explain the observations on bacteria. For bacterial triplets a radical pair mechanism seems to be the best answer.

In this view, the triplet state of Bchl *a* observed in photosynthetic bacteria or bacterial reaction centres as observed by e.p.r. has been preceded by a short-lived radical-pair intermediate. Note that the radical-pair state is not directly observed by e.p.r. Scheme (1) is suggested:

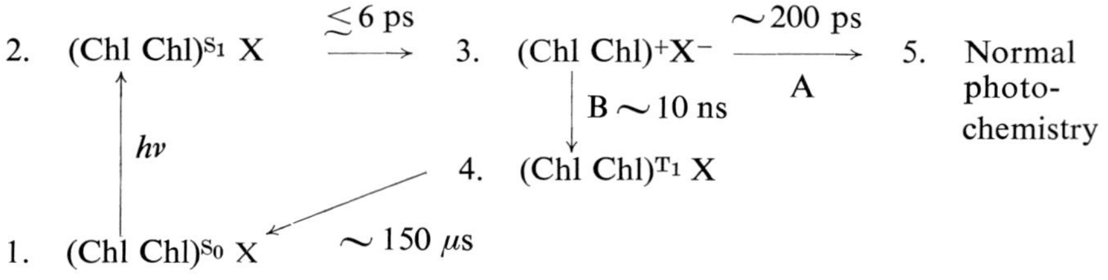

$$
\begin{array}{ccccc}
 & \overset{\leq 6\text{ ps}}{\longrightarrow} & & \overset{\sim 200\text{ ps}}{\longrightarrow} & \\
2.\ (\text{Chl Chl})^{S_1}\,X & & 3.\ (\text{Chl Chl})^+X^- & & 5.\ \text{Normal} \\
\end{array}
$$

2. (Chl Chl)^{S_1} X $\xrightarrow{\leq 6\text{ ps}}$ 3. (Chl Chl)$^+$X$^-$ $\xrightarrow[\text{A}]{\sim 200\text{ ps}}$ 5. Normal photo-chemistry

B ~ 10 ns

hv

4. (Chl Chl)^{T_1} X

1. (Chl Chl)^{S_0} X   ~ 150 μs

SCHEME 1. (A) Path in 'normal' photosynthesis; (B) path in 'blocked' photosynthesis (i.e., when the electron acceptor that follows X is saturated with electrons).

The triplet state 4 is experimentally observed only when normal electron transfer is blocked or the next electron acceptor in the electron-transport chain is removed or incapacitated and is primarily associated with $Bchl_{sp}$, because so long as triplet–triplet optical transitions occur, the 865 nm band of $Bchl_{sp}$ is bleached. The optical spectrum of state 3 is similar to the combination of $Bchl_{sp}^{+}\cdot$ and bacteriopheophytin anion, $Bpheo^{-}\cdot$. The overall implication of this interpretation is that the triplet state observed in photosynthetic bacteria is a side-reaction of normal photosynthesis, that a short-lived radical-pair state (3) precedes the first 'stable' charge-separation state (4), and that the radical pair is born from singlet $Bchl_{sp}$. These observations are based on magnetic resonance data and agree completely with the optical data as well.

The mechanism for Chl_{sp} function in light conversion that seems most compatible with all the experimental data is a radical-pair mechanism. Photochemistry proceeds from the first excited singlet state of Chl_{sp}. A radical-pair intermediate involving the two chlorophylls in the special pair, or as seems more probable in the case of photosynthetic bacteria, the radical pair involves $Bchl_{sp}^{+}\cdot$ and $Bpheo^{-}\cdot$. The triplet states of Chl_{sp} are not then on the main path of normal forward photosynthesis. We expect that synthetic Chl_{sp} (now being prepared) and the techniques of electron spin-echo spectroscopy will make a useful contribution to the clarification of the many problems that still remain to be solved in the mechanism of the primary step of light conversion.

ACKNOWLEDGEMENT

This work was done under the auspices of the Division of Basic Energy Sciences of the US Department of Energy.

References

ANDROES, G. M., SINGLETON, M. F. & CALVIN, M. (1962) EPR in chromatophores from *Rhodospirillum rubrum* and in quantasomes from spinach chloroplasts. *Proc. Natl. Acad. Sci. U.S.A. 48,* 1022–1031

BALLSCHMITER, K. & KATZ, J. J. (1968) Long wavelength forms of chlorophyll. *Nature (Lond.) 220,* 1231–1233

BARD, A. J., LEDWITH, A. & SHINE, H. J. (1976) Formation, properties, and reactions of cation radicals in solution. *Adv. Phys. Org. Chem. 13,* 156–277

BOLTON, J. R., CLAYTON, R. K. & REED, D. W. (1969) An identification of the radical giving rise to the light-induced electron spin resonance signal in photosynthetic bacteria. *Photochem. Photobiol. 9,* 209–218

BORG, D. C., FAJER, J., FELTON, R. H. & DOLPHIN, D. (1970) The π-cation radical of chlorophyll *a*. *Proc. Natl. Acad. Sci. U.S.A. 67,* 813–820

BOXER, S. G. & CLOSS, G. L. (1976) A covalently bound dimeric derivative of pyrochlorophyllide *a*. A possible model for reaction center chlorophyll. *J. Am. Chem. Soc. 98*, 5406–5408

BRODY, S. S. (1958) New excited state of chlorophyll. *Science (Wash. D.C) 128*, 838–839

BRODY, S. S. & BROYDE, S. B. (1968) Low temperature absorption spectra of chlorophyll *a* in polar and nonpolar solvents. *Biophys. J. 8*, 1511–1533

CHOW, H. C., SERLIN, R. & STROUSE, C. E. (1975) The crystal and molecular structure of ethyl chlorophyllide·2H₂O. *J. Am. Chem. Soc. 97*, 7230–7237

CLAYTON, R. K. (1963) Toward the isolation of a photochemical reaction center in *Rhodopseudomonas spheroides*. *Biochim. Biophys. Acta 75*, 312–323

CLAYTON, R. K. & WANG, R. T. (1971) Photochemical reaction centres from *Rhodopseudomonas spheroides*. *Methods Enzymol. 23*, 696

COMMONER, B., HEISE, J. & TOWNSEND, J. (1956) Light-induced paramagnetism in chloroplasts. *Proc. Natl. Acad. Sci. U.S.A. 42*, 710–718

COTTON, T. M., LOACH, P. A., KATZ, J. J. & BALLSCHMITER, K. (1978) Studies of chlorophyll–chlorophyll and chlorophyll–ligand interactions by visible absorption and infrared spectroscopy at low temperatures. *Photochem. Photobiol. 27*, 735–750

DUTTON, P. L., LEIGH, J. S. & SIEBERT, M. (1972) Primary processes in photosynthesis: *in situ* ESR studies on the light induced oxidized and triplet state of reaction center bacteriochlorophyll. *Biochem. Biophys. Res. Commun. 46*, 406–413

DUTTON, P. L., LEIGH, J. S. & REED, D. W. (1973) Primary events in the photosynthetic reaction center from *Rhodopseudomonas spheroides* strain R-26: triplet and oxidized states of bacteriochlorophyll and the identification of the primary electron acceptor. *Biochim. Biophys. Acta 292*, 654–664

DUYSENS, L. N. M. (1952) *Transfer of Excitation Energy in Photosynthesis*, Thesis, University of Utrecht

DUYSENS, L. N. M., HUISKAMP, W. J., VOS, J. J. & VAN DER HART, J. M. (1956) Reversible changes in bacteriochlorophyll in purple bacteria on illumination. *Biochim. Biophys. Acta 19*, 188–190

EMERSON, R. & ARNOLD, W. (1931–1932) The separation of the reactions in photosynthesis by means of intermittent light. *J. Gen. Physiol. 15*, 391–420

EMERSON, R. & ARNOLD, W. (1932–1933) The photochemical reaction in photosynthesis. *J. Gen. Physiol. 16*, 191–205

FEHER, G. (1956) Observations of nuclear magnetic resonances via the electron spin resonance line. *Phys. Rev. 103*, 834–835

FEHER, G. (1971) Some chemical and physical properties of a bacterial reaction center particle and its primary photochemical reactants. *Photochem. Photobiol. 14*, 373–387

FEHER, G., HOFF, A. J., ISAACSON, R. A. & McELROY, J. D. (1973) Investigation of the electronic structure of the primary electron donor in bacterial photosynthesis by the endor technique. *Biophys. Soc. Abstr. 13*, 61a

FEHER, G., HOFF, A. J., ISAACSON, R. A. & ACKERSON, L. C. (1975) Endor experiments in chlorophyll and bacteriochlorophyll *in vitro* and in the photosynthetic unit. *Ann. N. Y. Acad. Sci. 244*, 239–259

FONG, F. K. (1974a) Molecular basis for the photosynthesis primary process. *Proc. Natl. Acad. Sci. U.S.A. 71*, 3692–3695

FONG, F. K. (1974b) Energy upconversion theory of the primary photochemical reaction in plant photosynthesis. *J. Theor. Biol. 46*, 407–420

FONG, F. K. & KOESTER, V. J. (1976) *In vitro* preparation and characterization of a 700 nm absorbing chlorophyll–water adduct according to the proposed primary molecular unit in photosynthesis. *Biochim. Biophys. Acta 423*, 52–64

FUHRHOP, J. H. & MAUZERALL, D. C. (1969) The one-electron oxidation of metalloporphyrins. *J. Am. Chem. Soc. 91*, 4174–4181

GOVINDJEE & WARDEN, J. T. (1977) Green plant photosynthesis. Upconversion or not? *J. Am. Chem. Soc. 99*, 8088–8090

HINDMAN, J. C., KUGEL, R., SVIRMICKAS, A. & KATZ, J. J. (1977) Chlorophyll lasers: stimulated light emission by chlorophylls and Mg-free chlorophyll derivatives. *Proc. Natl. Acad. Sci. U.S.A.* 74, 5–9

HINDMAN, J. C., KUGEL, R., WASIELEWSKI, M. R. & KATZ, J. J. (1978) Fluorescence and lasing in covalently linked chlorophyll *a* dimer. *Proc. Natl. Acad. Sci. U.S.A.* 75, 2076–2079

KATZ, J. J. & NORRIS, J. R. (1973) Chlorophyll and light-energy transduction in photosynthesis. *Curr. Top. Bioenerg.* 5, 41–75

KATZ, J. J., NORRIS, J. R., SHIPMAN, L. L., THURNAUER, M. C. & WASIELEWSKI, M. R. (1978*a*) Chlorophyll function in the photosynthetic reaction center. *Annu. Rev. Biophys. Bioeng.* 7, 393–434

KATZ, J. J., JANSON, T. R. & WASIELEWSKI, M. R. (1978*b*) A biomimetic approach to solar energy conversion, in *Energy Chemical Sciences (The 1977 Karcher Symposium)* (Christian, S. D. & Zuckermann, J. J., eds.), Pergamon Press, Oxford, in press

KOHL, D. H., TOWNSEND, J., COMMONER, B., CRESPI, H. L., DOUGHERTY, R. C. & KATZ, J. J. 1965) Effect of isotopic substitution on electron spin resonance signals in photosynthetic organisms. *Nature (Lond.) 206*, 1105–1110.

KOK, B. (1956) Preliminary notes on the reversible absorption change at 705 mμ in photosynthetic organisms. *Biochim. Biophys. Acta 22*, 399–401

KOK, B. (1957) Absorption changes induced by the photochemical reaction of photosynthesis. *Nature (Lond.) 179*, 583–584

LESSING, H. E. (1976) Transient absorption and rotational relaxation in the liquid state. *Opt. Quantum Electron.* 8, 309–315

LEVANON, H. & NORRIS, J. R. (1978) The photo-excited triplet state and photosynthesis. *Chem. Rev.* 78, 185–198

LOACH, P. A. & SEKURA, D. L. (1967) Comparison of decay kinetics of photo-produced absorbance, epr, and luminescence changes in chromatophores of *Rhodospirillum rubrum. Photochem. Photobiol.* 6, 381–393

MCELROY, J. D., FEHER, G. & MAUZERALL, D. C. (1969) On the nature of the free radical formed during the primary process of bacterial photosynthesis. *Biochim. Biophys. Acta 172*, 180–183

MCELROY, J. D., MAUZERALL, D. C. & FEHER, G. (1974) Characterization of the primary reactants in bacterial photosynthesis. II. Kinetic studies of the light-induced EPR signal ($g = 2.0026$) and the optical absorbance changes at cryogenic temperatures. *Biochim. Biophys. Acta 333*, 261–277

MENZEL, E. R. (1976) On the energy upconversion mechanism of light utilization by chlorophyll in photosynthesis. *J. Theor. Biol. 56*, 401–416

MIMS, W. B. (1972) Electron spin echoes, in *Electron Paramagnetic Resonance* (Geschwind, S., ed.), pp. 263–351, Plenum Press, New York

MORY, S., LEUPOLD, D., KÖNIG, R., HOFFMAN, P. & FREGIN, W. (1976) Der Chlorophyll-laser. *Exp. Tech. Phys. 24*, 37–40

NORRIS, J. R., UPHAUS, R. A., CRESPI, H. L. & KATZ, J. J. (1971) Electron spin resonance and the origin of Signal I in photosynthesis. *Proc. Natl. Acad. Sci. U.S.A. 68*, 625–629

NORRIS, J. R., DRUYAN, M. E. & KATZ, J. J. (1973) Electron nuclear double resonance of bacteriochlorophyll free radical *in vitro* and *in vivo. J. Am. Chem. Soc. 95*, 1680–1682

NORRIS, J. R., SCHEER, H., DRUYAN, M. E. & KATZ, J. J. (1974) An electron nuclear double resonance (endor) study of the special pair model for photo-reactive chlorophyll in photosynthesis. *Proc. Natl. Acad. Sci. U.S.A. 71*, 4897–4900

NORRIS, J. R., SCHEER, H. & KATZ, J. J. (1975) Models for antenna and reaction center chlorophyll. *Ann. N. Y. Acad. Sci. 244*, 260–280

SCHEER, H., KATZ, J. J. & NORRIS, J. R. (1977) Proton-electron hyperfine coupling constants of the chlorophyll *a* cation radical by endor spectroscopy. *J. Am. Chem. Soc. 99*, 1372–1381

SHIPMAN, L. L. & KATZ, J. J. (1977) Calculation of the electronic spectra of chlorophyll *a* – and bacteriochlorophyll *a*–water adducts. *J. Phys. Chem. 81*, 577–581

SHIPMAN, L. L., COTTON, T. M., NORRIS, J. R. & KATZ, J. J. (1976) New proposal for structure of special-pair chlorophyll. *Proc. Natl. Acad. Sci. U.S.A. 93*, 1791–1794

SPANGLER, D., MAGGIORA, G. M., SHIPMAN, L. L. & CRISTOFFERSON, R. E. (1977) Stereo-electronic properties of photosynthetic and related systems. 2. *Ab initio* quantum mechanical ground state characterization of Mg porphine, Mg chlorin, and ethyl chlorophyllide *a*. *J. Am. Chem. Soc. 99*, 7478–7489

STENSBY, P. S. & ROSENBERG, J. L. (1961) Fluorescence and absorption studies of reversible aggregation in chlorophyll. *J. Phys. Chem. 65*, 906–909

STROUSE, C. E. (1974) The crystal and molecular structure of ethyl chlorophyllide *a*·2H$_2$O and its relationship to the structure and aggregation of chlorophyll *a*. *Proc. Natl. Acad. Sci. U.S.A. 71*, 325–328

STROUSE, C. E. (1976) Structural studies related to photosynthesis: a model for chlorophyll aggregates in photosynthetic organisms. *Prog. Inorg. Chem. 21*, 159–177

THURNAUER, M. C., KATZ, J. J. & NORRIS, J. R. (1975) The triplet state in bacterial photo-synthesis. Possible mechanisms of the primary photo-act. *Proc. Natl. Acad. Sci. U.S.A. 72*, 3270–3274

WARDEN, J. T. (1976) Experimental examination of the 'energy upconversion' theory for green plant photosynthesis. *Proc. Natl. Acad. Sci. 73*, 2773–2775

WARDEN, J. T. & BOLTON, J. R. (1972) Simultaneous optical and electron spin resonance detection of the primary photoproduct P700 in green plant photosynthesis. *J. Am. Chem. Soc. 94*, 4351–4352

WARDEN, J. T. & BOLTON, J. R. (1973) Simultaneous quantitative comparison of the optical changes at 700 nm (P700) and electron spin resonance signals in System I of green plant photosynthesis. *J. Am. Chem. Soc. 95*, 6435–6436

WASIELEWSKI, M. R., STUDIER, H. M. & KATZ, J. J. (1976) Covalently linked chlorophyll *a* dimer: a biomimetic model of special pair chlorophyll. *Proc. Natl. Acad. Sci. U.S.A. 73*, 4282–4286

WASIELEWSKI, M. R., SMITH, U. H., COPE, B. T. & KATZ, J. J. (1977) A synthetic biomimetic model of special pair bacteriochlorophyll *a*. *J. Am. Chem. Soc. 99*, 4172–4173

WRAIGHT, C. A. & CLAYTON, R. K. (1973) The absolute quantum efficiencies of bacterio-chlorophyll photo-oxidation in reaction centres of *Rhodopseudomonas spheroides*. *Biochim. Biophys. Acta 333*, 246–260

Discussion

Malkin: Has anyone been concerned with the difference in line-width between P865$^+$ and P700$^+$? The line-width is significantly different in the spectra of the bacteriochlorophyll dimer and the chlorophyll *a* dimer *in vivo*.

Katz: The two chlorophylls have significantly different structural formulae and, therefore, will have different e.p.r. line-widths! The largest electron–proton hyperfine coupling constants originate by interaction of the unpaired spin with the additional protons in rings IV of chlorophyll *a* and the additional two in ring II of bacteriochlorophyll *a*. The spin-echo modulation (Fig. 2) results from the interaction of the unpaired spin with the nitrogen atoms and cannot be observed by ordinary endor spectroscopy. The proton–electron interactions account for only about 60–80% of the hyperfine interactions. It is not practical to calculate the intrinsic line-width, but there is nothing about the broader line-width of bacteriochlorophyll *a* relative to chlorophyll *a* that appears unusual.

Junge: The e.p.r. signal from the chlorophyll *a* dimer is narrowed by a factor of $\sqrt{2}$. This narrowing is precise to within less than 10%. Could you translate this into a possible range of probabilities for residence of the unpaired spin on one of the two chlorophylls? What line-width would be expected if the spin resided, say, 40% on one chlorophyll and 60% on the other?

Katz: The $\sqrt{2}$ relationship for spin sharing by $Chl_{sp}^{+\cdot}$ is deduced on the basis of equal sharing at equivalent molecular sites, and thus cannot be used to calculate the line-shape for unequal spin sharing.

Thornber: The line-width of bacteriochlorophyll $b^{+\cdot}$ in oxidized reaction centres does not narrow by $\sqrt{2}$. Why not?

Katz: The photosynthetic bacteria *Rhodopseudomonas viridis* and *Thiocapsa pfenningi* contain bacteriochlorophyll *b*, which has a structural formula quite different from that of bacteriochlorophyll *a*. The line-widths of the primary donor bacteriochlorophyll *b* in the reaction centre of these photosynthetic bacteria has the line-width characteristic of monomeric Bchl $b^{+\cdot}$. A possible explanation for the lack of line-narrowing in these organisms is the absence of a special pair. Or, there may be a special pair but the external environment of one of the bacteriochlorophyll *b* molecules may differ sufficiently to affect the delocalization so that the unpaired spin is effectively localized on one of the bacteriochlorophyll *b* molecules. Another possibility is a distortion in the geometry of a special pair that reduces the rate of spin exchange so that delocalization is prevented.

Clayton: We speak of dimers and infer interactions; three of these involve singlet excitons, triplet excitons and the sharing of an electron deficiency. Are these types of interaction consistent in the 'special pair' bacteriochlorophyll *a in vivo*? Have they been examined sufficiently with bacteriochlorophyll *b* to see if they are consistent for that? Could a pair of molecules be classified as a dimer in terms of one of the mechanisms of interaction but not another?

Katz: Different numbers of chlorophyll molecules may be required to account for spin-sharing and for optical red-shifts, and experiments with the covalently linked bacteriochlorophyll *a* dimer lead to the same conclusion for bacterio-chlorophyll molecules. Thus, two bacteriochlorophyll *a* molecules account satisfactorily for line-narrowing in P865$^{+\cdot}$ but do not account for the red-shift to 865 nm. Evidently interactions with additional bacteriochlorophyll *a* or bacteriopheophytin *a* molecules are necessary for a red-shift to 865 nm in the bacteriochlorophyll *a* special pair.

Duysens: You suggested that *in vitro* photooxidation is associated with formation of a special pair. Van Gorkom *et al.* (1974) found that prolonged exposure of chloroplasts to deoxycholate gave the monomer P680. Nevertheless the monomer was photooxidized. This indicates that the special pair may not be necessary for the photooxidation; the dimer might be needed for capturing excitation energy more efficiently or for transferring electrons more easily across the membrane.

Porter: Does the monomer not photooxidize *in vitro*?

Katz: No, it can be photooxidized.

Porter: Why then do you say that the dimer is necessary for photooxidation?

Katz: I did not say that a Chl_{sp} is necessary for photooxidation. In all cases so far studied where Chl *a* or Bchl *a* is involved, nature uses a special pair as the primary donor in photosynthesis. This may be related to a higher chemical stability for a $Chl_{sp}^{+\cdot}$ as compared to $Chl^{+\cdot}$ and to a greater ease of oxidation of a Chl_{sp} as compared to the monomer. Incidentally no monomeric Chl *a* species has been prepared in the laboratory that is red-shifted to the extent that its absorption maximum occurs at 680 nm.

Porter: Do you mean that the reactivity is greater?

Katz: No, a chlorophyll *a* monomer π-cation free radical may be expected to disproportionate and/or undergo internal oxidation-reduction reactions that lead rapidly to its destruction at room temperature.

Porter: Does it disproportionate into a dication and neutral molecule?

Katz: The experimental evidence available now only indicates that *in vitro* Chl $a^{+\cdot}$ is rapidly converted into diamagnetic species of unknown structure.

Porter: So, experimentally, the life-time of the dimer cation (as judged by e.p.r.) is longer than that of the monomer cation?

Katz: The experimental evidence on the relative stability of Chl $a^{+\cdot}$ and $Chl_{sp}^{+\cdot}$ is still too meagre for a definitive statement. Monomeric chlorophyll *a in vitro* is a species that absorbs between 660 and 670 nm. That is not where the absorption maximum of reaction-centre P700 is observed.

Porter: That is a different point. I am asking about the reactivity with respect to electron transfer of the dimer; is it less reactive or, when it is formed, is it less stable?

Katz: It appears easier to remove an electron from a special pair of chlorophylls than from the corresponding monomer chlorophyll. The π-cation free radical that results when the Chl_{sp} is oxidized appears not to be as reactive a species as that formed from the monomer.

Porter: In other words, the ionization potential is less for the pair.

Katz: The redox behaviour of the covalently linked dimers has been measured: the special pair is significantly easier to oxidize. The covalently linked dimer in the folded configuration can be regarded as a supermolecule in which the highest-occupied (HOMO) and lowest-unoccupied (LUMO) molecular orbitals will be split into higher and lower energy levels. Removal of an electron from the higher-energy HOMO of the supermolecule will be energetically less demanding than from the monomer.

Porter: But isn't this dimer trap going to be in its excited state when it transfers energy? You need to know the redox potential in the excited state, not the ground state.

Katz: The energy of the H_+ orbital of the covalently linked dimer in its folded configuration will be higher than that of the highest-occupied molecular orbital for monomeric chlorophyll *a* and so it seems reasonable to us that the synthetic Chl_{sp} will be easier to oxidize.

Porter: Shouldn't we be more interested in the lowest-unoccupied molecular orbital?

Katz: Experimental observations of the oxidizability of the folded dimer indicate that it is significantly easier to oxidize than the corresponding monomer by several hundred millivolts. How significant this is remains a question for the future.

Butler: The folded chlorophyll dimer fluoresced and lased but the open dimer only fluoresced. Why?

Katz: The fluorescence of the open covalently linked dimer is similar to that of ordinary monomeric chlorophyll (see Table 5). No fluorescence at 730 nm is observed as is the case for the folded dimer at room temperature. The fluorescence yield from the open linked dimer is relatively high (but not quite as high as that of monomeric chlorophyll *a*).

Robinson: What information do the lasing experiments give?

Katz: Lasing behaviour reflects the photophysics; it is based on a considerable theoretical background in terms of the various cross-sections and intersystem crossings. The fact that we detect a population inversion in one case but not in the other is significant to photophysicists.

Robinson: What was the optical lay-out for the laser experiments?

Katz: The excitation beam irradiated a cell with parallel, polished windows and lasing was observed in a direction perpendicular to the excitation beam.

Robinson: A fluorescence quantum yield of 0.2 means that the excitation beam must be extremely powerful to stimulate lasing in the perpendicular direction.

Katz: The power requirement is high but not extraordinarily so.

Robinson: But there are difficulties: DODCI, which has a 99.9% quantum yield, will not lase even when pumped with a picosecond pulse.

Katz: Such a pulse is not high-powered.

Robinson: It has a high peak power but its energy content is low. What are you pumping the cell with?

Katz: A nitrogen laser, at power densities from about 3 to 35 MW/cm^2

depending on the particular chlorophyll or chlorophyll derivative.

Porter: If I may summarize, you are saying that in the green plant the P700 is a folded dimer. Is the light-harvesting chlorophyll a monomer?

Katz: The spectral properties of green plant antenna chlorophyll are better accounted for by chlorophyll oligomers, [Chl]$_n$, than by monomer chlorophyll.

Porter: What do you think is the trap, the P690, in photosystem II?

Katz: The only chlorophyll *a* species that absorbs at 680 nm so far prepared in the laboratory is the oligomer [Chl *a*]$_n$, which has been used as a model for antenna chlorophyll. Hydrated Chl *b* in hydrocarbon solvents absorbs near 680 nm. Whether either of these species can be identified with photosystem II chlorophyll absorbing at 680 nm is still under investigation.

Butler: You assume that P700 is analogous to the folded dimer and that the 730 nm fluorescence band which appears in the emission spectrum of chloroplasts at low temperatures is due to P700. However, I showed (Butler 1961) that the 730 nm emission from chloroplasts at -196 °C is fluorescence from a form of chlorophyll that absorbs at 705 nm, C-705, which traps excitation energy in the antenna chlorophyll of photosystem I. That the 730 nm fluorescence is due to C-705 rather than to P700 can be shown readily since P700 can be oxidized, either by light or by mild chemical oxidants, but C-705 shows no such redox behaviour. The 730 nm emission band from chloroplasts at low temperature is completely independent of the redox state of P700. It seems to me that your folded dimer is an excellent model for C-705 but perhaps not for P700. Have you examined the redox behaviour of your folded dimer and is the 730 nm fluorescence lost or altered on oxidation?

Katz: First let me mention that a linear stack, Chl·H$_2$O Chl·H$_2$O Chl, is expected to have its red absorption maximum near 705 nm. The fluorescence properties of short linear Chl·H$_2$O stacks have not been studied in detail. The possible formation of such species *in vivo* at low temperatures should not be overlooked.

Oxidation of the covalently linked folded dimer bleaches the red maximum. The π-cation Chl *a*$^{+}$· has an absorption maximum near 860 nm. The covalently linked dimer cation presumably will also be red-shifted relative to the unoxidized species, but there is no reason to suppose that these oxidized species will fluoresce at 730 nm.

Clayton: During the formation of the oxidized bacteriochlorophyll special pair *in vivo*, in the bacterial system, bands are lost at 805 and 865 nm and appear at 795 nm. Do you see any counterpart of this *in vitro*?

Katz: The linked, folded bacteriochlorophyll does not replicate the optical properties of P865. The oxidized covalently linked bacteriochlorophyll dimer in its folded configuration has an absorption maximum at 1150 nm, in the same region of the spectrum where P865$^+$· has an absorption maximum.

Clayton: It is intuitively plausible that if a positive charge is hopping between two bacteriochlorophyll molecules a monomer-like band might remain in the absorption spectrum. We observe this in the *in vivo* system; previously we attributed it entirely to a band shift of an 800 nm component, but polarization studies refute this.

Katz: The acetyl groups are not used in the folding of the synthetic covalently linked bacteriochlorophyll *a* dimer and are thus available for coordination interactions with additional bacteriochlorophyll or bacteriopheophytin molecules. Additional interactions of this kind with the synthetic bacterio-chlorophyll special pair might provide a red-shift to about 865 nm.

Thornber: It is pleasing to have proof that the cationic form of the dimeric pair absorbs at about 1200 nm—we had expected this (cf. Thornber *et al.* 1978) but lacked proof. Your *in vitro* spectrum of the oxidized form of the dimer of bacteriochlorophyll showed two peaks at about 1200 nm, but *in vivo* only one peak is observed in this region.

Clayton: *In vivo* there are three!

Thornber: But not with the same relative intensities as were seen in Dr Katz's figures.

Clayton: No; the bands are 980, 1140 and 1250 nm. The 1250 nm one is the most intense one for bacteriochlorophyll *a in vivo*.

Butler: Recently at the Royal Institution we have examined that question (Butler *et al.* 1978) from measurements of the lifetime and the relative yield of the 730 nm fluorescence from chloroplasts over a range of low temperatures (—60 to —196 °C) where the yield changes markedly. We conclude (see pp. 248–249 for the rationale of the experiment) from the experimental finding that the lifetime of fluorescence is directly proportional to its yield that C-705 is present over the entire temperature range but is less fluorescent at the higher temperatures because more of the excitation energy of C-705 is transferred on to P700.

Malkin: With regard to whether the antenna chlorophyll is monomeric or dimeric, a few years ago we looked for the e.p.r. signal of P680$^+$ in chloroplasts and found what we thought was the signal from this species (Malkin & Bearden 1973). We saw a signal for the free radical with a line-width of a normal chlorophyll dimer radical (about 8 G). Now we know that this signal does not originate from P680$^+$ but that it may arise from an oxidized form of antenna chlorophyll. The species has the same line-width as for P700$^+$ (and for P680$^+$ when we subsequently found it). So the question again comes up as to whether the antenna contains monomers or dimers.

Katz: A free radical generated in a chlorophyll *a* dimer or oligomer gives an e.p.r. signal with the line-width of the monomer free radical because the dimer or oligomer does not have the ability to delocalize an unpaired electron. This is possibly a consequence of the orthogonal orientation of the chlorophyll molecules in $[Chl]_2$ or $[Chl]_n$. Thus, line-width measurements in this situation do not easily allow an unambiguous differentiation between a monomer free radical Chl $a^{+\cdot}$ and $[Chl\ a]_n^{+\cdot}$.

Porter: So if it were dimer, one could not distinguish it easily by e.p.r. from a monomer.

Amesz: The size of the light-induced e.p.r. signal due to oxidized antenna chlorophyll and also of the corresponding absorbance changes (Visser *et al.* 1977) is about the same as that of oxidized P680 and P700. So it represents only a small amount of oxidized chlorophyll, which may be a special dimer close to the reaction centre, not necessarily representative of the bulk antenna chlorophyll *a*. (For further discussion, see pp. 53–59 and 344–351.)

References

BUTLER, W. L. (1961) A far-red absorbing form of chlorophyll *in vivo. Arch. Biochem. Biophys. 93*, 3–25

BUTLER, W. L., TREDWELL, C. J., MALKIN, R. & BARBER, J. (1978) in press

MALKIN, R. & BEARDEN, A. J. (1973) Detection of a free radical in the primary reaction of chloroplast photosystem II. *Proc. Natl. Acad. Sci. U.S.A. 70*, 294–297

THORNBER, J. P., DUTTON, P. L., FAJER, J., FORMAN, A., HOLTEN, D., OLSON, J. M., PARSON, W. W., PRINCE, R. C., TIEDE, D. M. & WINDSOR, M. W. (1978) in *Proceedings of the Fourth International Congress of Photosynthesis Research* (Reading, England), in press

VAN GORKOM, H. J., TAMMINGA, J. J., HAVEMAN, J. & VAN DER LINDEN, I. K. (1974) Primary reactions, plastoquinone and fluorescence yield in subchloroplast fragments prepared with deoxycholate. *Biochim. Biophys. Acta 347*, 417–438

VISSER, J. W. M., RIJGERSBERG, C. P. & GAST, P. (1977) Photooxidation of chlorophyll in spinach chloroplasts between 10 and 180 K. *Biochim. Biophys. Acta 460*, 36–46

Properties of chlorophyll on plasticized polyethylene particles

G. R. SEELY

C. F. Kettering Research Laboratory, Yellow Springs, Ohio

Abstract   There are several reasons for suspecting that there is a specific inter-action between chlorophyll and galactolipids in the chloroplast.   The model system described is intended to detect association of chlorophyll with polar lipids and other surfactants at a hydrocarbon–water interface. It consists of chlorophyll and other lipids or surfactants absorbed to the surface of poly-ethylene particles, which have been swelled with undecane to allow the lipophilic parts of these molecules to be anchored firmly in the hydrocarbon substrate. The absorption spectrum of adsorbed chlorophyll is usually modified by the presence of surfactant, and usually in the direction of decreased order of aggre-gation. Spectra in the presence of glycolipids in particular seem peculiar to the surfactant. The particles are strongly fluorescent, at room temperature as well as at 77 K, and emission bands from aggregated chlorophyll species are observed along with fluorescence of monomeric chlorophyll.

Photosynthesis appears to be unique among photobiological systems in that it uses an efficient multimolecular transport system for delivering excited state energy to centres of photochemical reaction.   How chlorophylls and other components of the photosynthetic unit (PSU) are arranged to accomplish this is a perennial subject for speculation. Dissection of chloroplasts with detergents has led to the conclusion that there are only a few distinct kinds of chlorophyll–protein complexes, each molecule of which contains perhaps six or seven chlorophyll molecules (Thornber *et al.* 1976).   The individual proteins are associated with lipids, perhaps in a mosaic matrix such as has been proposed by Anderson (1975), in which chlorophylls serve as interfacial lipids connecting the protein and lipid bilayer components of the membrane.

While preparing a recent review of the properties of chlorophyll in model systems (Seely 1977), I noted that there were several reasons for postulating a special role for galactosyldiglycerides in the ordering of the chlorophyll–

41

FIG. 1. Courtauld model of postulated complex between chlorophyll and monogalactosyl-dilinolenyldiglyceride. Near the centre of the picture are the hydrogen bonds between the C-10a (upper) and C-9 (lower) carbonyls of chlorophyll and the 2- and 3-hydroxy groups of galactose.

protein complexes in the lipid bilayer. For example, Rosenberg (1967) pointed out that galactolipids were formed at the same time as chlorophyll and were degraded at the same rate in the dark. He also noted a good fit between the linolenyl chains of galactolipids and the phytyl group of chlorophylls, and suggested that these lipids were responsible for anchoring chlorophyll in the lamellar membrane. Trosper & Sauer (1968) reported a strong interaction between chlorophyll and glycolipids in CCl_4 solution and in mixed monolayers. With the aid of Courtauld models of chlorophyll and monogalactosyldilino-lenyldiglyceride, I noted that when the linolenyl and phytyl groups were packed together as proposed by Rosenberg, it was possible to form hydrogen bonds simultaneously between the 2- and 3-hydroxy groups of the galactose and the C-10a and C-9 carbonyl oxygen atoms of chlorophyll, respectively, as is shown in Fig. 1. If this is the way in which chlorophyll is bound *in vivo*, it provides a means for locking chlorophylls more firmly into positions oriented favourably for energy transfer. An incidental feature of this hypothesis is that it supplies a rationale for the existence of the β-ketoester function of chlorophyll, and for the predominance of galactolipids in the photosynthetic unit.

There is as yet no direct evidence that chlorophyll *in vivo* is bound in this way. However, Lutz (1977) has found that the C-9 carbonyl vibrations in the resonance Raman spectra of chloroplasts lie in a group centred about 1680 cm^{-1}, which does not resemble the carbonyl spectra of anhydrous chlorophyll oligomers or the microcrystalline hydrate as much as the spectra of chlorophyll *a* and methyl chlorophyllide *a* in CCl$_4$ and CHCl$_3$ containing ethanol, in which the carbonyls are known to be hydrogen bonded to hydroxy groups (Katz *et al.* 1963).

Because a direct inquisition into the detailed environment of chlorophyll in chloroplasts is difficult, more might be learned from a systematic investigation of the interaction of chlorophyll with other lipids in a suitable model system. The system I shall describe is intended to model the interface between one side of a lipid membrane and the aqueous stroma phase, and to allow examination of the association of amphiphilic lipids in equilibrium.

EVOLUTION OF A MODEL

The numerous model systems for the study of chlorophyll may be conveniently grouped into two categories: those, like micelles and mixed mono-layers, which are essentially liquid and in which molecules are free to diffuse to some extent, and those, like microcrystals and adsorbates, which are essentially solid and in which the molecules, once located, are locked in place.

It is typical of systems which are essentially liquid that the absorption spectra, fluorescence and photochemistry of chlorophyll at low concentration is like that of chlorophyll in dilute homogeneous solution, but at high concentration the fluorescence and the photochemistry are quenched, though there is little change in the absorption spectrum. Beddard & Porter (1976) associated the quenching of fluorescence with transfer of excitation energy to dimeric associations of chlorophyll, in which the interaction between the molecules is much weaker than in the better-known dimers of chlorophyll in dry non-polar solvents, for example. I have suggested that the mechanism for quenching in these associations is collapse of an ion pair produced by electron transfer from one chlorophyll to the other within the dimer (Seely 1978). The mechanism is effective in these systems because the chlorophylls are free to move in the force field of their own making, and the consequence of it is that chlorophyll loses its capacity for photochemistry at high concentration.

In the essentially-solid systems chlorophyll is usually much aggregated even when its overall concentration is low. Absorption spectra of chlorophyll in these systems are usually varied (often the main reason for studying these systems), but fluorescence and photochemistry are often weak, absent or of uncertain origin.

An important exception to this generalization exists. When chlorophyll or related pigments are adsorbed to the surface of particles of plastic (polystyrene, polycaprolactam or polyacrylonitrile), sensitization of photochemical re-actions persists or increases at pigment densities at which aggregated forms are evident and fluorescence is quenched (Cellarius & Mauzerall 1966; Komissarov *et al.* 1963; Kapler & Nekrasov 1966). It is possible that the rigid attachment of pigment in these systems prevents quenching of the ion pair by the mechanism discussed and allows conversion into the photochemically-active triplet state.

However, chlorophyll adsorbed on plastic particles, as in most essentially-solid systems, is not a particularly good model of chlorophyll *in vivo*, because the phytyl groups are probably unable to penetrate the polymer matrix so as to be anchored there as in a membrane. Essentially-liquid systems may allow chlorophyll to associate with other lipids and possess the good fluorescence and energy transfer properties desirable in a model, but lose both fluorescence and photochemistry if chlorophylls associate very much with each other. Perhaps the most faithful models in common use are bilayer lipid membranes, but they are hard to work with because of their fragility.

The present model system attempts to retain the desirable properties of liquid and solid systems by a compromise which makes use of the property of plastics to swell by uptake of low-molecular-weight liquids. Plasticized particles are soft solids, rather than viscous liquids, into the matrix of which the phytyl chain of chlorophyll and the fatty portions of other lipids would be expected to work their way. The structure of the system could provide at the same time anchorage for these groups and a degree of rigidity that would retard or prevent quenching of fluorescence. Such particles, if suspended in an aqueous medium by a suitable surfactant, could serve as useful models of the interface between lipid and aqueous phases in the photosynthetic unit.

EXPERIMENTAL SECTION

The particles used in this study are composed of polyethylene plasticized with undecane. They were prepared by suspending polyethylene powder (5 g) in undecane (60 ml), dissolving it by heating to 90 °C, and cooling through the cloud point (about 77 °C) with vigorous stirring to precipitate the swollen particles. The particles were filtered, resuspended in methanol (50 ml) to dissolve adhering undecane, filtered and air-dried. The weight recovered (22.5 g) corresponded to a 350% increase. The particles are angular or prismatic in appearance, with dimensions typically in the range 2–5 μm.

The particles are coated with chlorophyll and other lipids or surfactants by procedures such as the following. A weighed quantity of particles is ground in a mortar with lipid or surfactant dissolved in methanol–water (9 : 1), and the suspension is diluted to 20 ml with 90% methanol containing dissolved chlorophyll. Most of the chlorophyll remains in solution at this point. Chlorophyll is precipitated onto the particles by addition of 60% methanol (20 ml) in portions. Progress of the adsorption is followed by spectra of the supernatant. An equilibrium appears to exist between chlorophyll in solution and adsorbed chlorophyll, which shifts to at least 95% adsorbed at 75% methanol. The green particles are filtered and air-dried briefly.

Absorption spectra of the particles are best measured on optically-dense (absorbance > 2) mulls in aqueous suspending media made viscous with highly polymeric solutes, such as sodium alginate or poly(vinyl alcohol). The spectra were recorded between 800 and 300 nm on a Cary 14R spectrophotometer, with a special cuvette to minimize the distance between the strongly-scattering sample and the photomultiplier.

All samples prepared so far are moderately to strongly fluorescent at room temperature as well as at 77 K. Their fluorescence spectra were recorded from 800 to 640 nm on an apparatus assembled by Dr B. Mayne. Because fluorescence transmitted by the sample is measured in this instrument, spectra of all but the most dilute mulls are subject to significant reabsorption.

RESULTS

When water is added to a solution of chlorophyll a in methanol, there is little change in the absorptivity of the 665 nm band until the water content exceeds 17% by volume, whereupon increasing absorbance in the near infrared indicates the separation of chlorophyll aggregates. When polyethylene–undecane particles are present, chlorophyll is adsorbed to them progressively as the water content increases. At 10% water, little chlorophyll is adsorbed; at 15% water, about half of the chlorophyll is taken up by 0.5 g of particles in 24 ml of solution, and at 25% water, about 97.5%.

The green particles do not disperse well in aqueous media, but absorption spectra of dense samples show bands of nearly equal intensity near 680 and 715 nm, probably representing chlorophyll hydrates aggregated to different degrees (Ballschmiter & Katz 1968). The fluorescence spectrum appears weak, compared with those that follow, but at 77 K it reflects the independent emissions of the two principal aggregated forms of chlorophyll (Table 1).

Although the specific surface area of the particles is not known yet, the amount of chlorophyll used in these preparations probably approaches the

TABLE 1

Absorption and fluorescence bands of chlorophyll *a* on polyethylene–undecane particles with various surfactants[a]

Surfactant	Absorption at 295 K (nm)	Fluorescence	
		77 K	295 K
None	679, 714	682, 715.5	664, 701
Zephiran (alkyl-dimethylbenzyl-ammonium chloride)	663, 741	n.d.	n.d.
Potassium linolenate	*662*, 742	(R) 682, 708	n.d.
N,N-Dimethyl-myristamide	661–667[b]	676.5, *722*	667, 722(sh)
Digitonin	680(sh), *719–724*,[b] 742(sh)	680.5, 707, 759	n.d.
Phosphatidyl-ethanolamine	682, 715, 737(sh)	n.d.	n.d.
Phosphatidyl-inositol ('50%')	*672.5*, 712, 740(sh)	*678*, ~725(sh) (R) 682, 702(sh), 737(?), *758.5*	n.d. (R) 683, *703*, 733(?), 752
Monogalactosyl-diglyceride fraction	670	683.5, *721* (R) 691(sh), *732*	n.d. (R) 706 (broad)
Chloroplasts	n.d.	686, 694(sh), *732*	n.d.

[a]Positions of absorption bands listed for the red and near infrared regions only. An italic band is distinctly stronger than the others, after correction for photomultiplier sensitivity in the case of fluorescence. Symbols: n.d., not determined; (sh), shoulder; (R), strongly re-absorbing samples, for which the shorter wavelength bands are displaced to the red, and longer wavelength bands are enhanced.
[b]Position of the band appears to be sensitive to the manner of preparation or to the suspending medium.

amount which would cover them with a monolayer, and some degree of aggregation is to be expected.

When a surfactant, biological lipid or not, is present, better dispersion of the particles in aqueous methanol is expected and is in all cases obtained. Although the surfactant may solubilize chlorophyll, its presence does not seem to delay adsorption of the pigment to the particles when water is added, and in some cases may accelerate it. Complications arise when the surfactant (e.g., dipalmitoylphosphatidylethanolamine) is insoluble in 10% aqueous methanol, or when undecane is solubilized, as by N,N-dimethylmyristamide, producing turbid solutions. However, addition of water to 25% causes at least 95% of the chlorophyll and at least some of the surfactant to adhere to the particles.

The simplest evidence for an interaction between a surfactant and chlorophyll would be a tendency to form unique aggregates or at least aggregates of lower order, including monomers, in the presence of the surfactant. In the absence of interaction, the surfactant might compete with chlorophyll for the surface of the particles, and encourage the formation of higher chlorophyll aggregates of the sorts found in dense films or monolayers. In fact, most surfactants exert a distinct 'monomerizing effect' on chlorophyll, and give rise to absorption and fluorescence spectra that are characteristic of the surfactant.

Results with several surfactants are summarized in Table 1. The particles for which these results were obtained were prepared as described above. Particles prepared by adding plasticized polyethylene to suspensions of chlorophyll and surfactant in 75% methanol had similar characteristic spectra but usually with stronger absorption in the near infrared region by chlorophyll hydrate oligomers.

Zephiran and potassium linolenate represent simple cationic and anionic surfactants expected to have little specific interaction between their polar groups and chlorophyll. The chlorophyll appears mainly in monomeric form (662 nm) or as the 'microcrystalline' dihydrate (742 nm); the latter is suppressed by an excess of linolenate. The fluorescence is mainly from the monomer, though strongly reabsorbed in the sample reported.

N,N-Dimethylmyristamide possesses a $CO \cdot NR_2$ function that might interact with chlorophyll as would the peptide groups of proteins. The spectrum of chlorophyll shows mainly monomer, but there is evidence for the presence of dimers or oligomers in the absorption tail extending into the infrared (Fig. 2). The sample is as strongly fluorescent at room temperature (295 K) as at 77 K, but the monomer fluorescence (667 nm) dominates at the former and the 722 nm fluorescence at the latter temperature.

In contrast to the myristamide, the complex glycolipid digitonin seems to promote aggregation. Curiously, the strongest fluorescence (Fig. 3) seems to come from a minor form of probably-monomeric chlorophyll which is scarcely distinguishable in the absorption spectrum. In view of the important role of

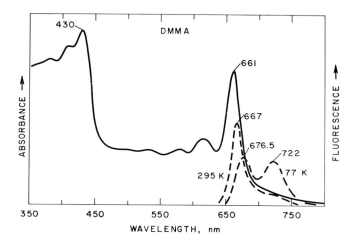

FIG. 2. Absorption and fluorescence spectra of chlorophyll *a* adsorbed to polyethylene/ undecane particles with *N,N*-dimethylmyristamide (DMMA): ——, absorption spectrum in a carbowax–glycerol–water mull; –––, fluorescence spectra in a mull with 2.5% sodium alginate, recorded at 295 K and at 77 K under 435 nm excitation. The fluorescence spectra were recorded at the same detector sensitivity setting but are not corrected for the wavelength dependence of photomultiplier sensitivity.

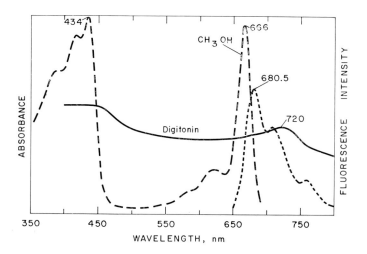

FIG. 3. Absorption and fluorescence spectra of chlorophyll adsorbed to particles with digitonin: ——, absorption spectrum of a mull with sodium alginate solution; -------, uncorrected fluorescence spectrum in alginate mull at 77 K, excited at 435 nm; –––, absorption spectrum of chlorophyll in 83.3% methanol solution.

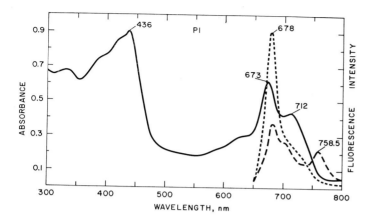

FIG. 4. Absorption spectrum of chlorophyll absorbed to polyethylene/undecane particles with crude phosphatidylinositol (——), and fluorescence spectra of optically thin (-----) and dense (– – –) samples, excited at 435 nm and 77 K, and recorded at the same detector sensitivity setting.

digitonin in isolation of chlorophyll–protein complexes, a strong interaction with the pigment is suspected.

The zwitterionic surfactant dipalmitoylphosphatidylethanolamine, although it dispersed the particles well, showed no indication of interaction with chlorophyll. The spectrum when adsorbed on these particles was scarcely distinguishable from that in the absence of surfactant.

A commercial soybean phosphatidylinositol fraction, assayed at 50% by the supplier (Sigma), gave the absorption spectrum shown in Fig. 4. The strongest band is that of a probably monomeric form at the unusually long wavelength of 672.5 nm. The fluorescence spectra in Fig. 4 show the effects of reabsorption. At low particle density, the fluorescence band at 678 nm corresponds to the absorption at 672.5 nm; at high particle density, one of the more prominent features is the fluorescence band at 758.5 nm, which probably emanates from the form of chlorophyll responsible for the absorption shoulder at 740 nm.

Since the development of this model system was instigated by an hypothesized interaction of chlorophyll with monogalactosyldilinolenyldiglyceride, it is appropriate to conclude with an examination of the effect of this lipid on the spectrum of adsorbed chlorophyll. The monogalactosyldiglyceride fraction was isolated from soybean leaf lipids by chromatography on a sugar column in chloroform–methanol. In the presence of this diglyceride, the spectra take on the apparently simple forms shown in Fig. 5. There is one prominent ab-

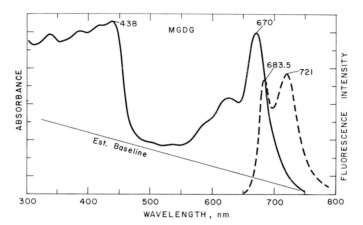

FIG. 5. Absorption spectrum (———) of chlorophyll adsorbed to polyethylene/undecane particles with monogalactosyldiglyceride (MGDG) fraction from soybean leaves, and (uncorrected) fluorescence spectrum (------) excited at 435 nm and at 77 K of a very thin sample, both in alginate mulls.

sorption band in the red, at 670 nm, although the tail of the band extends some distance into the near infrared. The fluorescence spectrum of a dilute dispersion consists of a band at 683.5 nm, evidently corresponding to the main absorption band, and a stronger band at 721 nm.

Some features of these spectra are worth noting. The 670 nm band appears to be about twice as broad as a band for monomeric chlorophyll (cf. the spectrum in methanol, Fig. 3, and the spectra in Figs. 2 and 4). The structure of the spectrum in the remainder of the visible region resembles that in methanol, but broadened and shifted 5 nm to the red. The ratio of the absorbance of the Soret band at 438 nm to that of the red band is less than 1, and the Stokes shift, 13.5 nm, is greater than that usually associated with monomeric chlorophyll (about 8 nm). The band width, the stronger absorption in the red region than in the blue, and the large Stokes shift seem more characteristic of known chlorophyll dimers than of monomers. It is not known if spectra like that in methanol are diagnostic of hydrogen bonding to the C-9 carbonyl.

DISCUSSION OF RESULTS

Most lipids and other surfactants have the disaggregating effect on the spectrum of adsorbed chlorophyll that would be expected if they interact with the pigment. The spectroscopic forms evoked appear characteristic of the surfactant. The three glycolipids in particular produced unique spectra that merit closer examination. The results with monogalactosyldiglyceride suggest that the hypothesis of a special interaction with chlorophyll remains viable.

An interesting and useful feature of these systems is the strong and complex fluorescence that they show at room temperature as well as at 77 K, probably in consequence of firm attachment of chlorophyll to the particle surface. Since the rate of intersystem crossing to the chlorophyll triplet state is about twice that of fluorescence (Bowers & Porter 1967), the population of triplet states and the retention of their photochemical activity are predictable. The often poor mirror-image relation between absorption and fluorescence spectra suggests that non-trivial energy transfer may be extensive, but this process is at present difficult to disentangle from that of reabsorption and re-emission.

Fong *et al.* (1976) have established in one case that fluorescence at 720 nm comes from a chlorophyll dimer with a band near 700 nm. Fluorescence bands near 722 nm in several of the samples of Table 1 suggest that dimers with absorption bands near 700 nm may be prevalent in these particulate systems also, although this is not obvious from the spectra.

ACKNOWLEDGEMENT

I thank Dr B. Mayne for his assistance in obtaining and interpreting the fluorescence spectra.

References

ANDERSON, J. M. (1975) Possible location of chlorophyll within chloroplast membranes. *Nature (Lond.) 253*, 536–537

BALLSCHMITER, K. & KATZ, J. J. (1968) Long wavelength forms of chlorophyll. *Nature (Lond.) 220*, 1231–1233

BEDDARD, G. S. & PORTER, G. (1976) Concentration quenching in chlorophyll. *Nature (Lond.) 260*, 366–367

BOWERS, P. G. & PORTER, G. (1967) Quantum yields of triplet formation in solutions of chlorophyll. *Proc. R. Soc. Lond. A 296*, 435–441

CELLARIUS, R. A. & MAUZERALL, D. (1966) A model for the photosynthetic unit — photochemical and spectral studies on pheophytin *a* adsorbed onto small particles. *Biochim. Biophys. Acta 112*, 235–255

FONG, F. K., KOESTER, V. J. & POLLES, J. S. (1976) Optical spectroscopic study of (Chl $a \cdot H_2O)_2$ according to the proposed C_2 symmetrical molecular structure for the P700 photoactive aggregate in photosynthesis. *J. Am. Chem. Soc. 98*, 6406–6408

KAPLER, R. & NEKRASOV, L. I. (1966) Sensitization of the reduction reaction of methyl red by adsorbed chlorophylls *a* and *b*. *Biofizika 11*, 420–426

KATZ, J. J., CLOSS, G. L., PENNINGTON, F. C., THOMAS, M. R. & STRAIN, H. H. (1963) Infrared spectra, molecular weights, and molecular association of chlorophylls *a* and *b*, methyl chlorophyllides, and pheophytins in various solvents. *J. Am. Chem. Soc. 85*, 3801–3809

KOMISSAROV, G. G., GAVRILOVA, V. A., NEKRASOV, L. I., KOBOZEV, N. I. & EVSTIGNEEV, V. B. (1963) The dependence of the photosensitizing activity of chlorophyll, adsorbed on caprone, on the surface concentration. *Dokl. Akad. Nauk S.S.S.R. 150*, 174–175

LUTZ, M. (1977) Antenna chlorophyll in photosynthetic membranes — a study by resonance Raman spectroscopy. *Biochim. Biophys. Acta 460*, 408–430

ROSENBERG, A. (1967) Galactosyl diglycerides: their possible function in *Euglena* chloroplasts. *Science (Wash. D.C.) 157*, 1191–1196

SEELY, G. R. (1977) Chlorophyll in model systems: clues to the role of chlorophyll in photosynthesis, in *Primary Processes of Photosynthesis* (Topics in Photosynthesis; vol. 2) (Barber J., ed.), pp. 1–53, Elsevier, Amsterdam

SEELY, G. R. (1978) Photochemistry of chlorophyll in solution: modeling Photosystem II. *Curr. Top. Bioenerg. 8*, 3–37

THORNBER, J. P., ALBERTE, R. S., HUNTER, F. A., SHIOZAWA, J. A. & KAN, K.-S. (1976) The organization of chlorophyll in the plant photosynthetic unit. *Brookhaven Symp. Biol. 28*, 132–148

TROSPER, T. & SAUER, K. (1968) Chlorophyll *a* interactions with chloroplast lipids *in vitro*. *Biochim. Biophys. Acta 162*, 97–105

Discussion

Porter: Sheena Carlin in our laboratory has been studying the effect of changing the lipid in vesicles and liposomes on concentration quenching (Beddard *et al.* 1976). Vesicles and liposomes made from mixtures of chlorophyll and these lipids behave like ordinary solutions; they fluoresce with a high yield at low concentrations of chlorophyll; as the concentration of chlorophyll with respect to the lipid increases, eventually the fluorescence is almost completely quenched. She has found that, as you have been suggesting, the concentration at which one observes half-quenching is considerably higher (about 50%) with galactolipids than with lecithin. The best we have attained at present is a 3 : 1 ratio of the monogalactosyldiglyceride to the digalactosyldiglyceride lipids, in which the half-quenching concentration of chlorophyll corresponds to one molecule of chlorophyll in 17 of lipid—an extremely high ratio. If the concentration quenching is, as we believe, due to chlorophyll molecules being close enough to interact when one of them is excited, then statistically this chlorophyll should not fluoresce and so the chlorophyll molecules must be held apart by some complexing with the lipid. Did you study the effect of chlorophyll concentration on the phenomena which you described?

Seely: Not seriously; we cannot yet record good absorption spectra of these pastes.

Porter: In the spectrum with digitonin (Fig. 3) was the high absorption due to scattering?

Seely: The spectra are recorded in conditions of multiple scattering, where little or no light is transmitted. Spectra with other lipids are sharp but do vary from one experiment to another—at least in details. I have kept the chlorophyll concentration constant and varied the lipid in this preliminary survey; one certainly could vary the chlorophyll concentration to see how that changes the spectrum.

Cogdell: You illustrated that this interaction greatly distorts the absorption spectrum (compared to that without lipid). When one strips all the lipid from a chlorophyll-protein complex isolated from a chloroplast by leaving it in a strong detergent such as SDS, is there a big shift in the absorption spectrum?

Clayton: Treatment of a bacterial antenna complex with a high concentration of lauryldimethylamine oxide (LDAO) (which, we believe, removes most of the lipid) does not change the spectrum except to shift the 850 nm peak to 845 nm; this shift is reversed on removal of the detergent by dilution.

Seely: In those preparations protein holds the chlorophyll molecules in place; in ours, we have no protein, just lipid.

Butler: With the proper use of digitonin one can fractionate chloroplasts and isolate pure fractions of photosystem I particles, photosystem II particles and the light-harvesting chlorophyll *a/b*–protein complex (Wessels *et al.* 1973; Satoh & Butler 1978). The low-temperature absorption and fluorescence emission spectra are characteristic for each of the three fractions and the wavelength maxima of the spectral components are the same as those observed in the low-temperature spectra of the intact chloroplasts. That is, the photochemical apparatus of chloroplasts can be fractionated by digitonin without altering the spectral characteristics of the various forms of chlorophyll which are distinguishable by low-temperature spectroscopy. That is not true with other detergents. If the chloroplasts are fractionated with Triton X-100 or if the digitonin-purified fractions are incubated with 0.1% Triton a short time before freezing to $-196\,^{\circ}$C, the 730 nm fluorescence of photosystem I shifts to 679 nm with a strong increase in the yield, the fluorescence of photosystem II particles also shifts to 679 nm and the forms of chlorophyll absorbing at longer wavelengths shift their absorption maxima to shorter wavelength.

Porter: Does that mean that the chlorophyll in the complete system interacts with lipid?

Butler: Yes; it also means that each particle in a particular digitonin-purified fraction has a complete set of pigments which is representative of the pigments of that fraction and that these pigments are held together on the particles in the same relation to one another as in intact chloroplasts. The fluorescence emanates only from the longest-wavelength form of chlorophyll of that particular fraction (e.g. from C-705 of photosystem I particles) and all the excitation energy flows from the shorter-wavelength forms to the longest-wavelength form in the particle. With the digitonin-purified particles no fluorescence appears from the shorter-wavelength forms. However, after treatment with Triton, that tight energy coupling becomes loosened and chlorophyll behaves more like monomeric chlorophyll (Satoh & Butler 1978).

Knox: Dr Seely, your tentative model of the 670 nm bands in the spectra with monogalactosyldiglyceride implies the existence of a fairly specific dimer because, if there are two bands with oscillator strength and you are only seeing one in emission, then there must be strong excitonic interaction ($\gtrsim kT$) to prevent the upper one from being populated and emitting. For both states to absorb, the transition moments of the two monomers must be more nearly perpendicular to each other than parallel. This considerably reduces the 'phase space' of possible dimers.

Seely: I cannot comment further until we do more experiments but I tend to agree with you.

Beddard: With the galactolipid vesicles and liposomes we saw no difference in the shape of the absorption band from when we used lecithin; the absorption maximum was shifted in wavelength but it was not broader. The emission was at the wavelength we expected (unlike the example in Fig. 5).

Barber: Do the digitonin particles have lipids in them?

Thornber: Both the digitonin and the Triton particles retain lipid (Allen *et al.* 1971) but the SDS particles are depleted of almost all lipid (Thornber *et al.* 1967).

Wessels: Digitonin-photosystem I particles contain only 8% non-pigment lipids. The fact that treatment of the particles with phospholipase or galactolipase had no effect on either the absorption spectrum or the photochemical activity suggests that digitonin can substitute for lipids in providing the reaction centre with the appropriate medium and structure.

Seely: If the chlorophyll were bound to a galactolipid, one would expect the digitonin to replace the galactolipid.

Porter: Is the difference between absorption and emission due to the difference in temperature at which the spectra are recorded?

Seely: In systems other than the galactolipid one we do not see this much difference between the absorption and emission band-widths.

Wessels: Did the absorption spectra differ above and below the phase-transition temperature of the lipid?

Seely: That presumably would not be important because the lipid is in the surface of the particle but I do not know whether it could be indirectly responsible for the difference between the spectra at room temperature and at 77 K.

Katz: The measurement of fluorescence at a temperature different from that at which the absorption was measured bothers me. I see no reason to assume that the same chlorophyll species are present at low temperature as at room temperature. Chlorophyll–chlorophyll interactions and coordination interactions between the chlorophylls and ligands such as lipids ought to be considered in equilibrium terms. Competition for coordination at the central magnesium atom of chlorophyll depends on temperature. A nucleophile such as a lipid or a nucleophilic group in a protein side-chain might not compete successfully with the keto carbonyl group of another chlorophyll molecule for coordination at magnesium at room temperature but it might be more effective at low temperature. The fluorescence of almost any chlorophyll system *in vitro* changes at low temperature, presumably owing to the formation of new chlorophyll species often present in such small amounts that they are not easily detected in a visible absorbance spectrum.

Seely: More data are needed before we can evaluate the relative importance of the factors you have mentioned. The equilibrium between chlorophyll and the other lipids is a specific objective of our studies.

Katz: The visible and fluorescence spectra of your systems imply to me a complicated mixture of different chlorophyll species. With the possible exception of the absorption at 673 nm, all the other absorption maxima in your spectra can be replicated in the laboratory without any lipid at all.

Beddard: The concentration of chlorophyll in the particles seems uncertain. Have you observed any energy transfer between different chlorophyll molecules?

Seely: The appearance of the fluorescence spectra indicates some energy transfer. How much, I don't know. There ought to be quite a bit because the chlorophyll density approaches that of a monolayer on the particles in some cases.

Beddard: Is there any concentration quenching associated with the chlorophyll?

Seely: The fluorescence is bright and is not decreased—even in the strongly-reabsorbing examples, although the emission spectra are greatly distorted and displaced to longer wavelengths. So there must be strong reabsorption and re-emission.

Porter: But everything depends on concentration: at very low concentration one would expect a strong fluorescence.

Seely: Yes; in these systems the chlorophyll concentration is probably high.

Robinson: What is the mechanism of concentration quenching? Is it better intersystem crossing, energy transfer to some trap, or what?

Porter: The mechanism that Dr Beddard and I visualize is as follows. The chlorophyll molecules are fixed—there is no diffusion—and are close enough together, on average, to transfer energy efficiently. As the excitation, or exciton, moves there is a chance that it will land on one chlorophyll which has another chlorophyll molecule close to it. I do not mean a dimer but refer to a purely statistical distribution. If they are close enough, when one is excited there is a chance of interaction in the excited state to form an excimer and that quenches.

Robinson: Is that because the lower state of the excimer does not emit or because it loses energy by intersystem crossing?

Porter: Either the chlorophyll excimer does not emit (it is non-radiative and crosses to the ground state) or it emits so far in the red that we have not seen it (and we have looked for it). It is not due to intersystem crossing because we have shown that the triplet yield is not increased by quenching.

Katz: How then do you account for the fluorescence of the folded linked dimer in which the chlorophyll molecules are about as close as they can be?

Porter: I don't know. On the whole, one expects dimers of chlorophyll not to fluoresce.

Katz: But a dimer formed by a carbonyl–magnesium interaction is, essentially, a non-fluorescent species. Its fluorescence yield may be zero.

Robinson: Is that the open dimer?

Katz: No; the open dimer is the equivalent of monomeric chlorophyll and always fluoresces with high yield except when the concentration is high. But the disappearance of fluorescence at high concentration seems to be due to more than just proximity alone.

Porter: I do not subscribe to that. Dimers, like molecules, may differ in their properties of intersystem crossing and radiationless conversion into the ground state. If they are able rapidly to convert radiationlessly, they will not fluoresce. Excimers, to which I am referring, have no bonding in the ground state— bonding is only in the excited state. There are excimers that fluoresce and excimers that do not—just as there are molecules that do or do not fluoresce, depending on the Frank–Condon factors for radiationless crossing. With two neighbouring chlorophyll molecules an excimer may be formed (or, rather, several excimers in the random distributions) in which the fluorescence yield is almost zero.

Seely: I have suggested elsewhere (1978*a,b*) a possible mechanism for what seems to be a general phenomenon of concentration quenching. This is particularly relevant because, according to Huppert *et al.* (1976), chlorophyll singlet states are quenched by quinones in a few picoseconds. The general mechanism of heteroquenching and concentration quenching for molecules such as chlorophyll may be as follows. If two chlorophyll molecules are close together and one is excited into the singlet state, it is energetically possible for an electron to transfer from one to the other producing an ion pair. The creation of this ion pair is sudden and the solvent medium is not prepared for it.

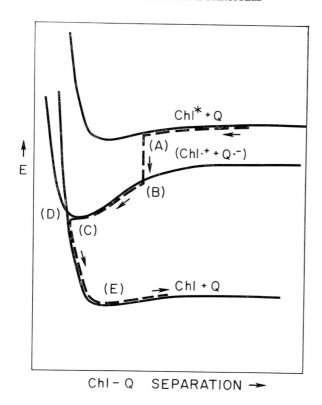

FIG. 1 (Seely). Energy surfaces as a function of the distance of separation for concentration quenching of an excited chlorophyll molecule (Chl*) through formation of an ion pair (A→B) and radiationless collapse to the ground state (C→D→E). (From Seely 1978a.)

As we assume that the two chlorophylls are close together, probably in contact, the attractive charge suddenly created between them will draw them together; the electrostatic charge is uncompensated by rearrangement of the solvent dipoles. The time required for this movement ought to be comparable with the usual vibrational times—that is, in the picosecond range. So, within a few picoseconds, the two molecules collapse towards each other and, if the energy surface for the ion pair crosses the energy surface for the ground state near the energy minimum of the former, there will be an adiabatic transition to the ground state in the time of one vibration (see Fig. 1). This could be general for molecules which form ion pairs.

Katz: How much closer can two chlorophyll molecules be than they are in the linked dimer reaction?

Seely: The linkage may prevent the collapse by holding them apart.

Katz: That is closer than any electrostatic force will bring them.

Porter: It depends on what is meant by the distance of separation.

Seely: The equilibrium separation in the ion pair is less than that in the weakly associated ground state because of the electrostatic attraction between the two molecules.

Porter: I agree. There is also the interaction due to 'resonance' between the two excimer configurations without any charge transfer. The importance of charge transfer in the excimer is an open question. Practically all pairs of molecules form better bonding in the excited state of two molecules together than in the ground state.

Katz: I take exception to that remark! In carbon tetrachloride (in which no extraneous nucleophile is present) nothing prevents the two parts of the dimer from coming into contact perpendicularly. In this configuration they are not fluorescent. No matter what the concentration of the solution is, every chlorophyll has a nearest neighbour. It all depends on whether those energy surfaces in Fig. 1 cross; there is no operational definition by which we know in advance whether they will cross. I maintain that there is something exceptional about chlorophyll.

Seely: I agree that whether the surfaces cross seems to be a property of the molecule. We need some explanation for concentration quenching that will apply uniformly to the many instances where it is observed.

Porter: The explanation of concentration quenching being due to interaction of one molecule with an excited close neighbour to form an excimer explains the facts and there seems to be no alternative for chlorophyll in solution. It is only when we try to describe this interaction in terms of the detailed structure of the potential energy curves that we are uncertain.

Seely: Do you identify the excimer with an ion pair?

Porter: Not necessarily; it may be so and is likely to contribute to the interaction terms.

Junge: I cannot understand what Dr Seely means by the distance of separation. If there is no solvent molecule between the two chlorophylls, they can be as close as their van der Waals' radius allows, but as the potential is very steep not much closer. Obviously there will be no major movement: the separation will change by a small amount. If, however, solvent molecules do separate the chlorophylls, viscosity will prevent any movement on the picosecond scale.

Seely: The molecules can move closer than their van der Waals' radius when no solvent molecule separates them; the suddenly developed charge is an additional force that tends to pull them together. If solvent molecules are present between them, some may be squeezed out. During this compression the molecules can cross over to the ground state.

Junge: Do you mean that the separation might change by as little as 0.01 nm?
Seely: Yes.

Robinson: Is it necessary to talk so specifically about crossing? When any energy is lost, the molecule drops below the excited state and you cannot get it back again.

Butler: The interesting question is not why there is concentration quenching at high concentrations of chlorophyll *in vitro* but rather why there is so little quenching at equal or higher concentrations *in vivo*. Also, I suspect that the effective chlorophyll concentration in Dr Katz's dimers which fluoresce is higher than the concentration at which Dr Seely finds strong concentration quenching.

Katz: They fluoresce *in vitro* because there is no electron acceptor. With charge transfer an easy mechanism is available to get back to the ground state. Presumably if it were engaged in forming charge-transfer complexes it would not fluoresce.

Beddard: Dr Katz, if you made a concentrated solution of the open dimer forms, you would not see concentration quenching between two of these dimer molecules, would you? Each chlorophyll would behave as an isolated molecule and there should then be little or no fluorescence in solution.

Katz: When our experiments with the linked pheophytin are complete, we shall able to discuss the concentration quenching better.

References

ALLEN, C. F. *et al.* (1971) in *Methods Enzymol. Vol. 21*

BEDDARD, G. S., CARLIN, S. E. & PORTER, G. (1976) Concentration quenching of chlorophyll fluorescence in bilayer lipid vesicles and liposomes. *Chem. Phys. Lett. 43*, 27–32

HUPPERT, D., RENTZEPIS, P. M. & TOLLIN, G. (1976) Picosecond kinetics of chlorophyll and chlorophyll/quinone solutions in ethanol. *Biochim. Biophys. Acta 440*, 356–364

SATOH, K. & BUTLER, W. L. (1978) Low temperature spectral properties of subchloroplast fractions purified from spinach. *Plant Physiol. 61*, 373–380

SEELY, G. R. (1978a) Photochemistry of chlorophyll in solution: modeling Photosystem II. *Curr. Top. Bioenerg. 8*, 3–37

SEELY, G. R. (1978b) The energetics of electron-transfer reactions of chlorophyll and other compounds. *Photochem. Photobiol. 27*, 638–654

THORNBER, J. P., STEWART, J. C., HATTON, M. W. C. & BAILEY, J. L. (1967) Antibodies reactive with specific folic acid determinants. *Biochemistry 6*, 2006–2014

WESSELS, J. S. C., VAN ALPHEN-VAN WAVEREN, O. & VOORN, O. (1973) Isolation and properties of particles containing the reaction center complex of photosystem II from spinach chloroplasts. *Biochim. Biophys. Acta 292*, 741–752

The preparation and characterization of different types of light-harvesting pigment–protein complexes from some purple bacteria

RICHARD J. COGDELL and J. PHILIP THORNBER*

Department of Botany, The University, Glasgow and Department of Botany, Imperial College of Science and Technology, London

Abstract A general strategy, with some specific examples, is given for the isolation and purification of detergent-soluble, antenna pigment–protein complexes from the photosynthetic membranes. Absorption, fluorescence and circular dichroism spectra, and the pigment and protein composition of B800–B850–protein and B890–protein complexes of some purple bacteria (*Rhodospirillum rubrum*, *Rhodopseudomonas sphaeroides* and *Rps. capsulata* and *Chromatium vinosum*) are discussed. We conclude that there are probably two major classes of antenna carotenochlorophyll-proteins in purple bacteria containing bacteriochlorophyll *a*: a B890 complex which has one carotenoid and two bacteriochlorophyll molecules in the minimal unit (probable molecular weight around 20 000), and a B800 + B850 complex which has one carotenoid and three bacteriochlorophyll molecules in a similar-sized minimal unit. The whole cell spectrum of any purple bacterium can be reconstituted by combining different proportions of the spectra of these two complexes with that of the photochemical reaction centre.

For most photosynthetic bacteria the biochemical organization and function of the few photosynthetic pigment molecules, which together with proteins and other compounds comprise the energy-converting photochemical reaction centre, is far better understood than that of the antenna pigment–protein complexes (e.g., see reviews by Parson & Cogdell 1975; Thornber *et al.* 1978). In this paper we shall begin to redress the balance in the case of purple bacteria containing bacteriochlorophyll *a*.

In these bacteria carotenoids as well as bacteriochlorophyll *a* function as antenna pigments: bacteriochlorophyll/carotenoid ratios lie typically between 1.5–3/1 compared with 4–5/1 in most other photosynthetic organisms. The near-infrared absorption spectrum of purple bacteria allows us to categorize

---

Permanent address: Department of Biology and Molecular Biology Institute, University of California, Los Angeles, California 90024, USA

them into three classes (cf. Thornber *et al.* 1978): (*a*) those like *Rhodospirillum rubrum* which have all their antenna chlorophyll absorbing at 870–890 nm; (*b*) those like *Rhodopseudomonas sphaeroides* and *Rhodopseudomonas capsulata* which have their antenna chlorophylls absorbing maximally at 800, 850 and 870–890 nm; (*c*) those like *Rhodopseudomonas acidophila* and *Chromatium vinosum* which can absorb like class (*b*) but which can also have antenna bacteriochlorophylls absorbing at 800 and 820 nm in certain growth conditions. A nomenclature has been given for these different spectral forms (Vredenberg & Amesz 1967): B800, B820, B850 and B890 are used to refer to antenna chlorophylls absorbing at about 800, 820, 850 and around 870–890 nm, respectively. Thornber *et al.* (1978) proposed that the antenna of all known purple bacteria is composed of two biochemically-different types of caroteno-bacteriochlorophyll–protein complexes—a B890 complex and a B800 + B850 complex; and some molecules of the latter can exist in a different state in some organisms (e.g. *Chr. vinosum*) so that they absorb maximally at about 800 and 820 nm. Thus, *Rsp. rubrum* would be expected to contain only one of the two types, the B890 complex, *Rps. sphaeroides* should contain both, and *Chr. vinosum* should contain both with the B800 + B850 complex existing some-times in two different states. The isolation and biochemical and spectral characterization of these complexes obtained from *Rps. sphaeroides*, *Rsp. rubrum*, *Rps. capsulata* and *Chr. vinosum* described here go a long way towards substantiating this notion. We expect that further studies will lead to knowledge of how each antenna pigment is arranged with respect to the other so that we shall be able to understand at the molecular level how the antenna pigments function so efficiently in directing excitation energy to the reaction centre.

ISOLATION OF ANTENNA PIGMENT–PROTEIN COMPLEXES

General principles

   These components are membrane-bound, water-insoluble entities. For their isolation and for ease of subsequent investigation of the organization of antenna pigments within them, the membrane structure must be dissociated so that the pigments are released as the smallest possible homogeneous entity that is still representative of their *in situ* state. This is most readily done by dissolving photosynthetic membranes in a detergent solution and applying standard protein fractionation techniques to the resulting extract (see later). The near-infrared spectrum is a sensitive monitor of whether the *in vivo* intramolecular organization has been disturbed during isolation and purification: the B800,

B850 etc. spectral forms absorb at wavelengths far removed from that of monomeric bacteriochlorophyll *a* in organic solvents or in detergent solutions owing to the specific environment of each pigment molecule within its protein framework (cf. Fenna & Matthews 1977). A change in a pigment's interaction with its immediate environment will cause a spectral shift; for example, in the extreme case complete disruption of a particle containing B890 into detergent-complexed bacteriochlorophyll and denatured protein will shift the 890 nm peak to about 760 nm. Lesser shifts can and do occur with only a slight change in interactions (cf. Thornber *et al.* 1978).

Isolation of hydrophobic proteins is fraught with difficulty compared with isolation of water-soluble proteins. But, nevertheless, certain advantages of working with antenna pigment–protein complexes largely offset these difficulties: as we just described a rapid and sensitive assay for the desired component exists; secondly, the antenna complexes are *major* components of the membrane; and thirdly, there is no difficulty in locating them during purification because of their colour. However, unlike water-soluble proteins for which a universally-applicable recipe can be given for their isolation, detergent-soluble proteins need a flexible approach to a described isolation procedure because the physicochemical behaviour of detergent-soluble components during fractionation is far less well understood than that of enzymes, for example. Thus, slight differences in the condition of a chromatographic medium used for isolation of a detergent-soluble component can result in a method working *exactly* as published in one laboratory but not in another for reasons that remain to be explained. Secondly, the ratio of detergent to amount of the required component is almost certainly critical if we are to explain its behaviour on any particular fractionation technique. However, it is almost impossible to obtain the same value for the ratio on each occasion since the proportion of the required component varies with respect to other detergent-adsorbing compounds not only in a solution at any particular stage in purification but also in different batches of whole cells. As a result an empirical approach must be used and alterations in procedure may be required each time a preparation of the desired component is made. To surmount these difficulties we advise starting with a large batch of cells so that one can vary the ratio of detergent to bacteriochlorophyll (in this case) on aliquot portions until a procedure that works is obtained, then it can be repeated *exactly* several times before a new batch of cells must be used and the procedure modified slightly. Finally we advise that as it is all too easy to continue to pursue any one attempt at a preparation long after it has become obvious that it is not working, one should not hesitate to stop a preparation that is obviously not going correctly and make a fresh start.

We propose a general strategy that can be adopted for attempting to isolate some spectral form of chlorophyll from an organism that has not been previously studied:

(1) Select a detergent that dissociates the membrane structure into the smallest possible bits but does not cause any large shift in the absorbance of the bulk pigment(s). We prefer the anionic detergent sodium dodecyl sulphate (SDS) to the weakly zwitterionic detergent lauryldimethylamine oxide (LDAO), and the latter to the non-ionic detergent, Triton X-100, for obtaining a product which is the smallest representative form of the pigment(s) and which contains the least extraneous material. However, SDS is more likely than LDAO (which is more likely than Triton X-100) completely to dissociate and denature any pigment–protein into its component parts. Mixing two or more detergents or changing detergents during the isolation process can be advantageous (see later section on *Chr. vinosum*).

(2) Once the required entity is solubilized, the following fractionation techniques (in order of usefulness in our experience) can be applied: (*a*) DEAE-cellulose chromatography, (*b*) hydroxyapatite chromatography, (*c*) fractionation by ammonium sulphate precipitation, (*d*) polyacrylamide gel electrophoresis, (*e*) density gradient centrifugation, (*f*) gel filtration. In general one must include the solubilizing detergent in buffers used with most (*a, b, d–f*) of these techniques to prevent reaggregation of the required entity and to allow it to be eluted from the columns etc. However, the longer the required component is in contact with the detergent the more likely its spectra will be altered. Thus, the faster the procedure can be completed the better the product. Note that SDS extracts cannot be chromatographed on DEAE-cellulose since the detergent binds too tightly to the charged group on the supporting material and that SDS must be used in electrophoretic separations not only to keep the desired component in solution but also to provide it with a charge.

Specific examples

Isolation of the B890-complex of Rsp. rubrum. A 30% solution of LDAO (0.5 ml) was added to the chromatophores (10 ml; $A_{890} = 136$ cm^{-1}), and the mixture was diluted with water to 25 ml. (High concentrations of detergent can prevent components binding to a chromatographic support.) This solution was loaded onto a hydroxyapatite column (5 × 2.5 cm) equilibrated in 10mM-sodium phosphate, pH 7.0. The column was washed with first 10mM- and then 150mM-phosphate buffer; the latter eluted the reaction centre plus a small fraction of antenna (B890) pigment. (The reaction centre can be further purified, if desired, on DEAE-cellulose columns.) The hydroxyapatite column was washed with one column-volume of 0.5% LDAO and again eluted with

10mM- and 150mM-phosphate buffer; the eluate (free bacteriochlorophyll) was discarded. Elution with 250mM-phosphate–0.1% LDAO removed the B890 complex which was further purified on a DEAE-cellulose column (3 × 1.5 cm) equilibrated in 50mM-Tris–HCl, pH 8.0. After the absorbed material on the column had been washed with Tris buffer containing 50mM-NaCl, the B890 complex was removed with 50mM-Tris–135mM-NaCl–0.1% LDAO, pH 8.0.

Isolation of the B800–B850 and the B890 complexes of Chr. vinosum. An earlier method (Thornber 1970) using SDS had provided the B800–B850 complex, but the B890 material obtained (Fraction A) had the reaction-centre component tightly associated with it. We have found that, by using LDAO and later SDS, all three pigment–proteins can be separated. In brief, we added ammonium sulphate to an LDAO extract (made as for *Rsp. rubrum*) until precipitation was evident. The precipitate was isolated, redissolved and chromatographed on DEAE-cellulose. Elution with 50mM-Tris–100mM-NaCl–0.1% LDAO yielded the reaction centre, and increasing the NaCl concentration to 250 mmol/l eluted the B890 complex free of the reaction centre. The B800–B850 complex was obtained from the dialysed supernatant ammonium sulphate solution. The dialysate was made 0.5% with respect to SDS and adsorbed to a hydroxyapatite column. Solutions of steadily greater concentrations of sodium phosphate, pH 7.0, were passed through the column and the required complex was removed with 0.3M-phosphate. We have purposely omitted many details specific for our material so that someone who wants to repeat the procedures will not view the specific detailed steps as essential and will develop such steps independently on the basis of what happens on their columns, etc.

SPECTRAL AND BIOCHEMICAL CHARACTERIZATION OF ANTENNA COMPLEXES

Absorbance spectra

Fig. 1 shows the room-temperature spectra of complexes obtained from *Chr. vinosum*. Low-temperature spectra of whole cells reveal five spectral forms: B800 (2), B820, B850 and B890 (Thornber 1970; Vredenberg & Amesz 1967). We obtained the B890 form in a carotenobacteriochlorophyll–protein complex (B890 complex) that lacked all the other forms as well as the photochemical reaction centre. A second isolated component contained the B800 and B850 forms and this, when treated with LDAO, could be converted into a B800 + B820 complex; this shift can be reversed on dialysis and readdition of SDS. Other treatments have been reported to cause these same shifts (cf. Thornber *et al.* 1978). This phenomenon needs further investigation because it could

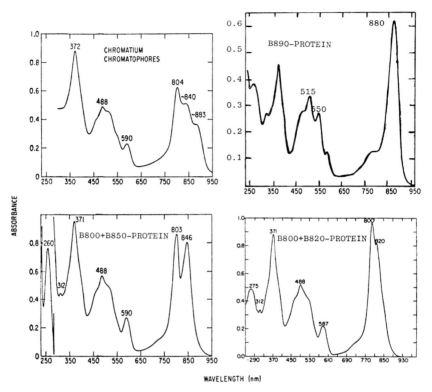

FIG. 1. The room-temperature absorption spectra of *Chromatium vinosum* chromatophores, and of the B890–protein, B800 + B850–protein and B800 + B820–protein (Thornber 1970) isolated from *Chr. vinosum* chromatophores. All samples are in a Tris buffer, pH 8.0.

lead to a greater understanding of some spectral forms in purple bacteria.

The B890 complex isolated from *Rsp. rubrum* had an identical near-infrared spectrum (Fig. 2) to that of the *Chr. vinosum* complex, and a *Rps. sphaeroides* B800 + B850 complex prepared by us (Fig. 3) but first isolated by Clayton & Clayton (1972), had a two-peaked near-infrared spectrum like that of the B800 + B850 complex of *Chr. vinosum* (Fig. 1) but with the 850 nm peak higher than that at 800 nm. A B800 + B850 complex obtained from *Rps. capsulata* had an identical spectrum (data not shown) to that of the *Rps. sphaeroides* component. The 850 and 890 nm bands in all our preparations shift to longer wavelengths when the temperature is lowered.

The fourth derivative of an absorbance spectrum can resolve the number

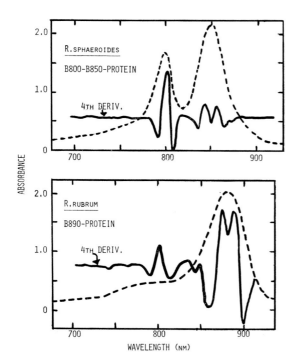

FIG. 2. Room-temperature near-infrared absorption spectrum (---) of B800+B850 complex of *Rps. sphaeroides* (2.4.1) and of B890–protein of *Rsp. rubrum* in 50mM-Tris–HCl–1 % LDAO, pH 8. The fourth derivative of each spectrum (——) is shown; the derivative intervals were 4.0, 3.5, 3.0 and 2.5 nm.

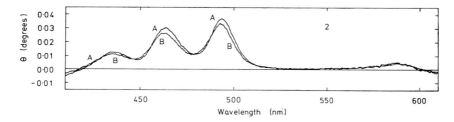

FIG. 3. Circular-dichroism spectra of *Rps. sphaeroides* (2.4.1) B800+B850 complex (A) and chromatophores (B) in 50mM-Tris–HCl, pH 8.0; 1 % LDAO was present in the sample of the isolated complex. The concentration of each was adjusted so that $A_{500\ nm} \approx 1.0$

and wavelength maximum of the spectral forms in a single peak in the spectrum if it is composed of more than one transition. With the B800+B850 complex of *Rps. sphaeroides* (Fig. 2), two forms were observed in the 850 nm peak but only one was present in the 800 nm band (Cogdell & Crofts 1978). Thus two bacteriochlorophyll molecules (possibly more; however, see below) must give rise to the 850 nm absorbance but just one could contribute to the 800 nm peak. Similar analysis of the 890 nm peak in the *Rsp. rubrum* B890 complex shows that it also contains two spectral forms (Fig. 2). Since some forms may not be detected by fourth-derivative analysis, other techniques must be applied to resolve the situation unequivocally.

Circular-dichroism spectra

We used circular-dichroism spectra of whole chromatophores and of the isolated complexes to analyse the spectral forms present and to detect any change during isolation in any pigment–pigment or pigment–protein interactions. C.d. spectra are even more sensitive than absorption spectra in detecting slight changes in the environment of a pigment (Sauer 1975).

In the visible region of the c.d. spectra (Fig. 3) carotenoids give rise to an induced circular dichroism due either to their asymmetrical binding to protein or to exciton coupling with other pigments (Cogdell & Crofts 1978). The induced c.d. spectrum is essentially the same for the carotenoids within the intact chromatophore membrane and within the isolated antenna complexes. This is strong evidence to support the view that the isolated complexes retain their native conformation. In the near-infrared region the spectra reveal that two bacteriochlorophyll molecules in an exciton interaction give rise to the 850 nm as well as to the 890 nm peaks in whole chromatophores and in complexes isolated from *Rps. sphaeroides* and *Rsp. rubrum* (Sauer & Austin 1978). No exciton interaction was evident in the 800 nm peaks.

Fluorescence spectra

The emission spectra of *Chromatium* pigment-proteins have been published previously (Thornber 1970). The spectra indicated that efficient energy transfer occurs between the 800 and 850 nm forms in the isolated B800+B850 complex.

Summary of the spectral data

All the data indicate that the B800+B850 complexes of *Rps. sphaeroides*,

Rps. capsulata and *Chr. vinosum* and the B890 complexes of *Rsp. rubrum* and *Chr. vinosum* have been fractionated with little or no alteration from their state *in vivo*. It appears that the B800+B850 complexes obtained from different organisms are homologous, as are the B890 complexes. A minimum of four pigment molecules (three chlorophylls and one carotenoid) are required in the basic unit of the B800+B850 complexes, and three (two chlorophylls and one carotenoid) in the B890 complexes to explain the spectral data (see also below).

Biochemical composition of isolated complexes

By determining the pigment and polypeptide composition and the bacterio-chlorophyll/protein ratio in the two different types of complexes we aimed to describe the composition of the basic building blocks of the antenna portion of the photosynthetic apparatus.

(*a*) *Pigment composition.* Table 1 gives the carotenoid and chlorophyll content of complexes isolated from four bacteria. Several points are striking: first, if one postulates only one carotenoid molecule in the minimal unit, the number of chlorophyll molecules present in each unit would be close to a whole number—two in the B890 unit and three in the B800+B850 unit—and would be in agreement with the interpretation of the spectral data (see above). Second, there is little specificity in the nature of the carotenoid present: no one carotenoid type is present as the sole carotenoid in only one complex in an organism; furthermore, mutations affecting carotenoid synthesis (*Rps. sphaeroides*, Table 1) change the carotenoid type in the B800+B850 complex without altering radically the near-infrared spectrum and, although the B890 complexes obtained apparently show a preference for containing spirilloxanthin, it is most probable that homologous complexes from other bacteria and from mutants with altered carotenoid composition will show that the B890 complex has no absolute specificity for the carotenoid type present. Some change, however, in the wavelength of the near-infrared maximum might occur with different carotenoids present (cf. Thornber *et al.* 1978).

(*b*) *Polypeptide composition and size of the isolated particles.* Electrophoresis of the isolated complexes on polyacrylamide gels shows that all the pigments co-migrate with protein; i.e., they are pigment–protein complexes. Further-more, after electrophoresis their near-infrared absorption spectrum has not changed. The B800+B850 complex of *Chromatium* migrates in a buffer con-taining 0.13% SDS at the rate expected for a component of molecular weight about 100 000 (Thornber 1970). Clayton & Clayton (1972) reported a molecular

TABLE 1

Carotenoid and bacteriochlorophyll content of B800+B850 and B890 complexes isolated from *Chromatium vinosum, Rhodopseudomonas capsulata* and *sphaeroides* and *Rhodospirillum rubrum*

Organism	Bacteriochlorophyll/ carotenoid ratio	Carotenoid composition
B800+B850 complexes		
Chr. vinosum[a]	2.83/1	Rhodopin (42–67%)
	3.21/1	Spirilloxanthin (20–25%)
	3.20/1	Rhodovibrin[e] (11–28%)
	2.82/1	Lycopene (1–4%)
Rps. capsulata[b] n 22	2.80/1	Neurosporene (65%)
	2.91/1	Chloroxanthin (20%)
		Dihydroxyneurosporene (15%)
Rps. sphaeroides 2.4.1	2.92/1[d]	Spheroidene (91 %)
		Spheroidenone (9 %)
Rps. sphaeroides GA	3.07/1[d]	Neurosporene (60 %)
		Dihydroxyneurosporene (27%)
		Chloroxanthin (13 %)
Rps. sphaeroides GlC	2.91/1[d]	Neurosporene (100 %)
B890 complexes		
Chr. vinosum[c]	2.12/1	Spirilloxanthin (76–80 %)
	2.16/1	Rhodopin (10–15 %)
	1.97/1	Rhodovibrin[e] (1–4 %)
	1.96/1	Lycopene (1–4 %)
Rsp. rubrum[c]	1.93/1	Spirilloxanthin (99 %)
	1.94/1	Unknown (1 %)
	1.94/1	

[a]Three different preparations; [b]one preparation; [c]two preparations; [d]averages of several determinations on different preparations. [e]Rhodovibrin may be P481 or P481–OH (these were not distinguished).

weight of $>$ 100 000 for the B800+B850 complex of *Rps. sphaeroides*. However, when the isolated complexes we have described are electrophoresed in 1 % SDS with no LDAO in the buffer, we observed that all complexes exhibit one or two pigmented protein zones in the 40 000–60 000 region. Complete denaturation of the complexes (i.e. heating the complexes in SDS and breaking all disulphide bonds so that the pigments are no longer associated with protein) reveals that they are composed of a small (7000–12 000 molecular weight) polypeptide or polypeptides (cf. also Thornber *et al.* 1978). Sauer & Austin (1978) have recently concluded that there are two polypeptides both of about 10 000 molecular weight in the B800+B850 complex isolated from *Rps. sphaeroides*, and Tonn *et al.* (1977) have reported that, although the protein part associated with B890 in *Rsp. rubrum* electrophoresed as a 12 000 molecular weight component, its most probable molecular weight was 18 800.

(*c*) *Bacteriochlorophyll/protein ratios.* Sauer & Austin (1978) have determined this ratio for a B860 complex (two bacteriochlorophylls per 20 000 molecular weight) and the B800+B850 complex (three bacteriochlorophylls per 20 000 molecular weight) of *Rps. sphaeroides*. For the B890 complex of *Rsp. rubrum* Tonn *et al.* (1977) have calculated that 3–7 bacteriochlorophyll and 1–2 carotenoid molecules are associated with the 18 800 polypeptide.

CONCLUSIONS

Previous studies on chlorophyll–protein complexes of other classes of photosynthetic organisms (Fenna & Matthews 1977; Thornber *et al.* 1977) have indicated that in general it requires some 5000–7000 g protein to accommodate one pigment molecule. Thus on that basis it would require about 15 000–21 000 and 20 000–28 000 g protein to accommodate the smallest number of pigment molecules (three in the B890 and four in the B800+B850 complexes, respectively) that spectral and pigment composition data indicate are the most likely numbers of pigment molecules contained in the basic building blocks. It would be most unlikely that a single polypeptide chain of the size generally reported for the denatured complexes (molecular weight about 10 000) would suffice to accommodate them and, therefore, it is most probable that the basic unit of the B800 + B850 complex contains *two* 10 000 polypeptide chains and four pigment molecules, and the B890 complex contains three pigment molecules and a polypeptide of molecular weight about 19 000. The electrophoretically measured molecular weight of the isolated material (40 000–60 000 and 100 000) indicates that some oligomeric form(s) of such basic units is particularly stable. Whether one can further dissociate the isolated material to a molecular weight expected for the basic unit (about 20 000) with the pigments still remaining conjugated with the protein requires further study.

FUTURE INVESTIGATIONS

We now have a good idea of the composition of the basic building blocks used to make up the energy-collecting portion of the photosynthetic apparatus of purple bacteria containing bacteriochlorophyll *a*. We can account for just about every photosynthetic pigment molecule in the photosynthetic apparatus of several purple bacteria in the B890, B800+B850 and reaction-centre complexes. It is not possible to do this for other classes of photosynthetic organisms in which the location of most of the carotenoids and some of the chlorophylls *in vivo* (cf. Thornber *et al.* 1977) is not yet known. In the case of the purple bacteria such knowledge is just a beginning for an overall understanding of energy migration through the antenna at the molecular level. We now need to determine the relative orientations of the pigment molecules within the building blocks and to know how many blocks of each complex there are in a photosynthetic unit (i.e. their *in vivo* oligomeric structure) and what are the relative orientations of the blocks with respect to each other. We must study reconstitution of each complex from its component parts as well as reconstitution of the photosynthetic apparatus from the available complexes. A more detailed examination must be made of the protein chemistry of each complex, and the control mechanisms for their synthesis requires further investigation.

Future data will enable us to explain why there is such a difference in the efficiency of energy transfer from carotenoids to the photochemically different purple bacteria (e.g. 30% efficiency in *Rsp. rubrum*, 50–60% in *Chr. vinosum*, 80–90% in *Rps. sphaeroides*); is it the nature of the carotenoid or is it differences in the line-up of their transition moments with respect to bacteriochlorophyll that determines these efficiencies? The data should also allow us (*a*) to suggest why most purple bacteria make two spectrally and biochemically different antenna components when it would seem that one type would suffice, (*b*) to learn what controls how much of each type is synthesized, and (*c*) to know why some organisms (e.g. *Rps. acidophila* and others; Cogdell & Schmidt, unpublished data; see also Thornber *et al.* 1978) are better adapted than others for varying the proportions of the complexes with respect to each other, and for altering the near-infrared spectrum of the B800+B850 complex. Some insight should be obtained about the optimum ratio of antenna to reaction centre pigments: for reasons which are not understood plants have more than twice the ratio that is found for most purple bacteria, and it appears that the ratio in plants results in a lower overall efficiency of energy conversion (cf. Radmer & Kok 1977). Is there something that can be learnt from studying purple bacteria that may be beneficial in improving plant photosynthetic efficiency by manipulation of their genetic machinery?

ACKNOWLEDGEMENTS

Grants from the Science Research Council, the National Science Foundation (PCM 75–20252) and the Guggenheim Memorial Foundation supported the research. We are greatly indebted to Drs J. Barber and A. R. Crofts for providing some of the facilities used in this study, and to Dr K. Sauer who supplied preprints of pertinent articles.

References

CLAYTON, R. K. & CLAYTON, B. J. (1972) Relations between pigments and proteins in photosynthetic membranes of *Rhodopseudomonas sphaeroides*. *Biochim. Biophys. Acta 283*, 492–504

COGDELL, R. J. & CROFTS, A. R. (1978) Analysis of the pigment content of an antenna pigment/protein complex from three strains of *Rhodopseudomonas sphaeroides*. *Biochim. Biophys. Acta 502*, 409–416

FENNA, R. E. & MATTHEWS, B. W. (1977) Structure of a bacteriochlorophyll *a*–protein from *Prosthecochloris aesturii*. *Brookhaven Symp. Biol. 28*, 170–182

PARSON, W. W. & COGDELL, R. J. (1975) The primary photochemical reaction of bacterial photosynthesis. *Biochim. Biophys. Acta 416*, 105–149

RADMER, R. & KOK, B. (1977). Photosynthesis: limited yields, unlimited dreams. *BioScience 27*, 599–605

SAUER, K. (1975) Primary events and the trapping of energy, in *Bioenergetics of Photosynthesis* (Govindjee, ed.), pp. 115–181, Academic Press, New York

SAUER, K. & AUSTIN, L. A. (1978) Bacteriochlorophyll–protein complexes from the light-harvesting antenna of photosynthetic bacteria. *Biochemistry 17*, 2011–2019

THORNBER, J. P. (1970) Photochemical reactions of purple bacteria as revealed by studies of three spectrally different carotenobacteriochlorophyll–protein complexes isolated from *Chromatium*, strain D. *Biochemistry 9*, 2688–2698

THORNBER, J. P., ALBERTE, R. S., HUNTER, F. A., SHIOZAWA, J. A. & KAN, K.-S. (1977) The organization of chlorophyll in the plant photosynthetic unit. *Brookhaven Symp. Biol. 28*, 132–148

THORNBER, J. P., TROSPER, T. L. & STROUSE, C. E. (1978) Bacteriochlorophyll *in vivo*: relation of spectral forms to specific membrane components, in *The Photosynthetic Bacteria* (Clayton, R. K. & Sistrom, W. R., eds.), Plenum Press, New York, in press

TONN, S. J., GOGEL, G. E. & LOACH, P. A. (1977) Isolation and characterization of an organic solvent soluble polypeptide component from photoreceptor complexes of *Rhodospirillum rubrum*. *Biochemistry 16*, 877–885

VREDENBERG, W. J. & AMESZ, J. (1967) Absorption characteristics of bacteriochlorophyll types in purple bacteria and efficiency of energy transfer between them. *Brookhaven Symp. Biol. 19*, 49–61

Discussion

Duysens: When a fraction of the reaction centres is bleached the energy absorbed by the B890 associated with these reaction centres is transferred to other reaction centres. For that reason I do not believe that a barrier of B800 molecules separates two reaction centres.

Cogdell: That is why we suggest that the reaction centres are joined by B890

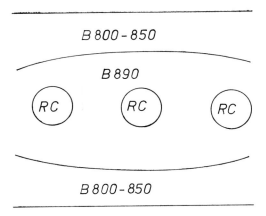

FIG. 1 (Cogdell). A model for a possible arrangement of the pigment–protein complexes in the photosynthetic membrane of *Chromatium vinosum*.

complexes. According to Monger & Parson (1977) about 10 reaction centres must be in close communication. Fig. 1 depicts the proposed organization with the B800–850 complex surrounding a cluster of B890 complexes around the central reaction centres.

Porter: Would that arrangement satisfy your experimental observations, Dr Duysens?

Duysens: Possibly, but I am not sure.

Staehelin: Would you not expect a more random organization of reaction centres? Assuming a relatively high density of such complexes in the membrane one could imagine that at any one moment an average of 10 reaction centres could be in a position to communicate with each other.

Cogdell: I am sure that some sort of statistical distribution centred on about 10 reaction centres in a unit is more realistic.

Seely: Nothing prevents energy transfer between non-adjacent complexes except the distance. If that is not too great, the excitation could easily 'skip' from one complex to another.

Duysens: Yes, but if the distance is appreciably larger than between the individual molecules in the units, the relation between the fluorescence yield and the fraction of closed reaction centres will not be one of the Stern–Volmer type, which is in general observed.

Seely: What is the diameter of this complex?

Thornber: The complex is a small monomeric unit (molecular weight about 10 000) and, therefore, the diameter is likely to be smaller than 2 nm.

Barber: Are they the whole units or part of a bigger unit?

Cogdell: When the complexes are not treated with great excesses of detergent they occur as aggregates (about 50 000–100 000 molecular weight) which are probably about 3.0 nm in diameter.

Clayton: A randomly constituted antenna that makes no special effort to focus the exicitons may transfer energy with 92% efficiency. A four-fold improvement (which is the most we may expect) by an ingenious arrangement of pigments in a sequence of descending excitation energies would give about 98% efficiency. What does this improvement mean in terms of survival?

Cogdell: The biosynthesis of these complexes depends on the light intensity. When these organisms grow under high light intensity they produce the B800–850 complexes in relatively small amounts. Only when they are stressed with low light intensities do they synthesize this extra variable amount of the B800–850 form to increase the antenna size and, so, the probability of capturing the photon. The amount of B890 formed bears roughly a fixed relation to the amount of reaction centres.

Clayton: Rhodopseudomonas sphaeroides behaves similarly. The main result of this adaptation to weak light is a larger antenna, so that the dark reactions are driven at rates close to their limits. A lesser consequence might concern the focusing of excitons in order of decreasing energy.

Anderson: What about the lipid content of the membrane? Are these complexes fixed or mobile?

Breton: We have no information yet about the mobility of the pigment–protein complexes (antenna or reaction centres) within the membrane. Several approaches are possible: fluorescence polarization, photodichroism, saturation transfer e.s.r. spectroscopy.

Cogdell: The native complexes do not need lipid to retain their normal spectrum. On SDS gels the phospholipid is often revealed at the ion front on staining.

Anderson: Is the lipid in the membrane predominantly glycolipid or phospholipid?

Cogdell: It contains no galactolipid, nor any glucose. The carbohydrate is probably present as surface glycoproteins.

Anderson: Are the fatty-acid chains largely unsaturated? In other words, is it a fluid membrane?

Cogdell: These membranes have a high protein/lipid ratio.

Barber: As Dr Anderson implied, the system may be more dynamic with the pigment–protein complexes able to diffuse freely so as to change their interrelationship in specific conditions.

Junge: Maybe ordered domains are present in a statistical (fluctuating) sense. With the free energy of a particle in an ordered domain being lower by a small amount, say 1 kcal/mol ($\sim$4 kJ/mol), ordered structures will prevail in 85% of the cases.

Thornber: We do not envisage the B890 complexes moving in the membrane as separate entities because they are tightly bound to the reaction centre and other electron-transport components, but rather the whole conglomerate of reaction centre with its associated B890 and electron-transport components may move through the B800–850 complex to associate with another similar conglomerate containing B890 elsewhere in the membrane; i.e. we envisage the complexes moving like corks in a sea, as in the Singer–Nicholson membrane model.

Porter: It seems that nature is trying to package the chlorophyll molecules so as to keep them separate, because if they come together there is concentration quenching. Otherwise what is the purpose of the protein envelope?

Clayton: There is another constraint on fluidity: the reaction centre is also the core of the vectorial electron- and proton-transport system. If the membrane were too fluid, the reaction centre would not be coupled in an orderly way to cytochromes and quinones which have to interact with specific surfaces of the membrane.

Staehelin: Freeze-fracture data appear to support the notion that the association between the reaction centres and the B890 complexes is probably tighter than with the B800–850 complexes on the outside, since the particles seen on freeze-fractured photosynthetic membranes of bacteria are between 6.0 and 10.0 nm in diameter. In chloroplast membranes whose systems are better understood, the system II reaction centres are surrounded by and form a tight complex with the light-harvesting complexes.

Porter: That 6.0–10.0 nm dimension implies about 20 or more of these in the freeze-fracture particles.

Katz: How do you visualize the nature of the interaction between the bacteriochlorophyll and the protein that gives rise to the red shift?

Thornber: The major interaction is a chlorophyll–chlorophyll one but chlorophyll–protein interactions have some effect on the exact absorption maximum of the chromophore.

Cogdell: Maybe the chlorophyll–protein interaction is the equivalent of lowering the temperature and holding the chromophore in a fixed orientation.

Robinson: What causes the 591 nm peak in the absorption spectrum of the B890 complex?

Katz: The absorption maximum near 590 nm can be assigned to the Q_x band. In bacteriochlorophyll $a \cdot L_1$, in which the magnesium is pentacoordinate, this band shifts to 603 nm. The change in coordination does not affect the Q_y band. Thus the absorption band in the orange is a rough indicator of the average coordination number of the magnesium in bacteriochlorophyll *a*.

Breton: What is the magnitude of the splitting in the spectrum of the B890 protein of *R. rubrum* and does it correspond to a strong or weak coupling?

Cogdell: The fourth derivative of the B800–850 spectrum (Fig. 2) was measured at room temperature. The 800 nm band sharpens as the temperature is lowered and the 850 nm band shifts to longer wavelengths. At 77 K we can no longer see the two transitions in the 850 nm band and we have not determined whether this is weak or strong coupling.

Duysens: The absorption band at 590 nm is more or less the same in both the B800–850 and B890 complexes (see Fig. 1) and, since it does not shift appreciably, the coordination number of magnesium in the three types—B800, B850 and B890—is probably the same. The shifts in absorption in the infrared may have something to do with another effect of the π-system.

Thornber: I should stress at this point that in the photosynthetic bacteria that we have been studying almost all the carotenoids (about 99%) present in the cell are contained in pigment–protein complexes, whereas in higher plants little of the total carotenoid can be accounted for in any protein complexes.

Anderson: It is probably fair to say that the P700–chlorophyll *a* complexes from all species contain β-carotene.

Thornber: But, in terms of the total percentage of carotenoid in the chloroplast, the β-carotene associated with P700 makes up only about 10% of the total β-carotene and, therefore, even less of the total carotenoid.

Duysens: According to Table 1 99% of the carotenoid in *Rhodospirillum rubrum* is spirilloxanthin. I found that *old* bacteria contain a lot of spirilloxanthin but young bacteria contain a different carotenoid which has a shoulder not at 570 nm but at a shorter wavelength. The carotenoid composition changes with age; this is also apparent in energy transfer (Duysens 1952). Do the particles you extract have different properties, depending on the age of the bacteria?

Thornber: We always use 'old' bacteria. What is the carotenoid in the young bacteria?

Duysens: It may be rhodopin.

Robinson: When the carotenoid acts as a light-harvesting pigment, is it coupled closely to the bacteriochlorophyll molecule or to some other intermediate pigment? If not, surely it cannot transfer the energy because of non-radiative processes on the picosecond time-scale?

Cogdell: The efficiency of energy transfer depends on the carotenoid type. It is not clear whether with the same complex (and, therefore, the same binding site) but with different carotenoids the efficiency of energy transfer would be controlled by the binding and the orientation factor or whether it would just be due to the carotenoid type itself. We have derived from *Rps. sphaeroides* a battery of mutants for carotenoid biosynthesis: they can make proteins that are red, brown, green etc. But we still have to determine whether the carotenoid type or the binding site controls the efficiency of energy transfer.

Clayton: This question should be accessible to experiment, as the efficiencies of transfer from carotenoid to bacteriochlorophyll range from about 10 to 90% in different organisms.

Robinson: What happens to the rest of the energy?

Duysens: H. Rademaker (unpublished work) found that it created triplet carotenoids.

Cogdell: That must be the first demonstration of direct excitation of the carotenoid to its triplet state because normally the carotenoids in solution neither fluoresce nor form triplets on direct excitation.

Porter: Does this quantitatively account for the loss of transfer to chlorophyll?

Duysens: I am not sure that it does.

Porter: Could there be singlet energy transfer to the chlorophyll and then back-transfer to form the triplet carotene?

Katz: As the energy of the carotene triplet is lower than that of the chlorophyll *a* triplet, the latter will populate the carotene triplet state.

Robinson: It does not go *via* bacteriochlorophyll?

Duysens: It does not go *via* fluorescent bacteriochlorophyll—that is evident from the fluorescence action spectrum—but it is possible that it goes *via* a small fraction of the bacteriochlorophyll that is not fluorescent.

Cogdell: It is energetically impossible for it to go from the carotenoid to the major bacteriochlorophyll and back again with such a high degree of efficiency. If it were, then exciting the bacteriochlorophyll should have the same result as exciting the carotenoid itself. The photochemistry proceeds with a high efficiency if the bacteriochlorophyll is directly excited. This means that all that energy cannot be lost to the carotenoids otherwise photochemistry would be a very inefficient process.

Clayton: Does the formation of the carotenoid triplet depend on whether the reaction centres are open or closed?

Duysens: Apparently not.

Katz: In green plants chlorophyll triplets do not seem to form in the presence of carotenoids.

Seely: What are the energy levels of the bacterial carotenoids?

Cogdell: Nobody knows. The major function of the carotenoids is not light-harvesting but protection; light-harvesting is a bonus that may derive from evolutionary pressures. That might explain the great variation in efficiency of energy transfer. But at the same time, they protect the chlorophyll pigments extremely well from the harmful photodynamic reaction.

Clayton: I would reverse the statement: the carotenoids developed as light-harvesting pigments before the atmosphere became aerobic and then those photosynthetic bacteria that contained carotenoids were amazed and delighted to discover that they were immune to the poison, oxygen!

Katz: Do any of these organisms contain bacteriopheophytin?

Cogdell: We have not looked for it but it is not obviously prominent in the spectra.

References

DUYSENS, L. N. M. (1952) *Transfer of Excitation Energy in Photosynthesis,* pp. 1–96, Thesis, University of Utrecht

MONGER, T. G. & PARSON, W. W. (1977) Singlet-triplet fusion in *Rhodopseudomonas sphaeroides* chromatophores: probe of organization of photosynthetic apparatus. *Biochim. Biophys. Acta 460,* 393–407

Chlorophyll–protein complexes of brown algae: P700 reaction centre and light-harvesting complexes

JAN M. ANDERSON and JACK BARRETT*

CSIRO, Division of Plant Industry, Canberra, Australia

Abstract   Thylakoid membranes from several brown algae have been fragmented with the non-ionic detergent, Triton X-100. Three intrinsic chlorophyll–protein complexes with different pigment compositions have been isolated by sucrose density gradient centrifugation. Brown algae contain the photosystem 1 reaction-centre complex, a P700–chlorophyll *a*–protein which has similar spectroscopic and chemical properties to those of higher plants. This complex represents about 10–20 % of the total chlorophyll in all species; the *Acrocarpia paniculata* complex has a chlorophyll/P700 ratio of 38. Two main light-harvesting complexes have also been isolated, which have properties unique to brown algae. The heavier of these, an orange fraction, is a fucoxanthin–chlorophyll *a/c*–protein; this complex contains most of the fucoxanthin and has only chlorophyll c_2. The other, a green fraction, is a chlorophyll *a/c*–protein enriched in violaxanthin. Neither of these complexes possesses detectable photosystem 1 or photosystem 2 activities. Both of these complexes efficiently transfer light energy to chlorophyll *a*, indicating that the molecular arrangement of their pigments is similar to that *in vivo*. Differential extraction of thylakoid membranes indicates that the P700–chlorophyll *a*–protein is the complex most firmly embedded in the membrane, but the fucoxanthin–chlorophyll *a/c*–protein is the least firmly bound. We suggest that the fucoxanthin complex is the most variable component of the photosynthetic unit of brown algal chloroplasts.

In the blue-green light of coastal waters veritable forests of marine seaweeds flourish, adding to our sources of foodstuffs, fibre materials and potential liquid fuels. As well as these large marine plants, there are the myriads of phytoplankton that dwell in the vast areas of surface oceanic waters. The exact contribution of marine seaweeds and phytoplankton to the global fixation of

*ARGC Senior Research Fellow on leave from School of Biological Sciences, Macquarie University, North Ryde, New South Wales 2113, Australia

81

carbon dioxide is difficult to estimate but it is thought to be at least 30%
(Fogg 1972), of which the greater part is accomplished by those seaweeds and
phytoplankton which contain chlorophyll c as a photoaccessory pigment.

Apart from descriptions of the pigment composition of brown algae, little
is known about photosynthesis in this important class of marine algae and
nothing was known about their chlorophyll–protein complexes. By contrast,
the main intrinsic chlorophyll–protein complexes of higher plants and green
algae, the P700–chlorophyll a–protein and the light-harvesting chlorophyll
a/b–protein, are well characterized (Boardman et al. 1978; Thornber 1975;
Anderson 1975). Hence the aim of our study has been to isolate and characterize
the chlorophyll–protein complexes of brown algae.

PIGMENTS OF BROWN ALGAE

The habitat of brown algae ranges from the shallow rock pools of coastal
waters to ocean depths of 100 m. As the depth of water increases, there is a
marked decrease of light intensity and the spectral quality of light is altered
owing to the rapid attenuation of red and then blue light. Fig. 1 illustrates this
effect; the light available even at 5 m depth is predominantly blue-green, and
chlorophylls absorb weakly in this region. Hence brown algae depend largely
on the light absorbed by pigments other than chlorophyll a (Haxo 1960),
namely chlorophyll c and fucoxanthin; the pigment composition of the chloroplasts reflects the spectral quality and intensity of light available.

In addition to chlorophyll a, brown algae contain chlorophyll c which replaces
the chlorophyll b of higher plants and green algae. Chlorophyll c (Fig. 2) is a
porphyrin rather than a chlorin derivative with an unsaturated ring IV which
has an acrylic acid instead of the usual propionic acid side-chain at C-7
(Dougherty et al. 1966). As the acrylic side-chain is not esterified either by
phytyl or other long-chain alkyl groups, chlorophyll c is more polar than either
chlorophyll a or b. The absorption maximum in the red of chlorophyll c in vivo
is much less intense and at lower wavelength (633 nm) than those of chlorophylls a and b, but the Soret band in the blue at 465 nm is relatively more
intense; this is a characteristic feature of porphyrins.

Chlorophyll c isolated from many algal classes is a mixture of two closely
related pigments, chlorophylls c_1 and c_2, so named because chlorophyll c_1 has
one vinyl group (as has chlorophyll a), whereas chlorophyll c_2 has two vinyl
groups (like protoporphyrin). Chlorophyll c_2 is universally distributed in all
chlorophyll c-containing algae examined. Brown seaweeds (Phaeophyceae),
diatoms (Bacillariophyceae), golden-brown flagellates (Chrysophyceae and
Haptophyceae), and fucoxanthin-containing dinoflagellates (Dinophyceae)

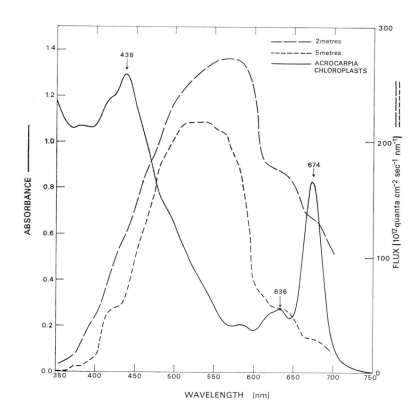

FIG. 1. Comparison of the absorption spectrum of *Acrocarpia paniculata* chloroplasts and the calculated spectroscopic distribution of the downward quantum flux assuming a solar altitude of 45° and no light scattering at a depth of 2 and 5 m in coastal water (quantum flux from Fig. 3 in Kirk 1976): ——————, absorbance; — — —, quantum flux at 2 m; – – –, quantum flux at 5 m.

also contain chlorophyll c_1; in contrast, peridinin-containing dinoflagellates and cryptomonads (Cryptophyceae) contain only chlorophyll c_2 (Jeffrey 1976).

Fucoxanthin (Fig. 3) is the main carotenoid of brown seaweeds and diatoms. It is an allenic carotenoid with several oxygen functions whose unusual structure was established by Bonnett *et al.* (1969). Although its absorption spectrum in organic solvents (λ_{max} 470, 444 and 424 nm, in light petroleum (b.p. 60–80 °C)) is similar to that of other carotenoids, its principal absorption *in vivo* lies between 500 and 590 nm (Fig. 1), which partly accounts for the brown colour

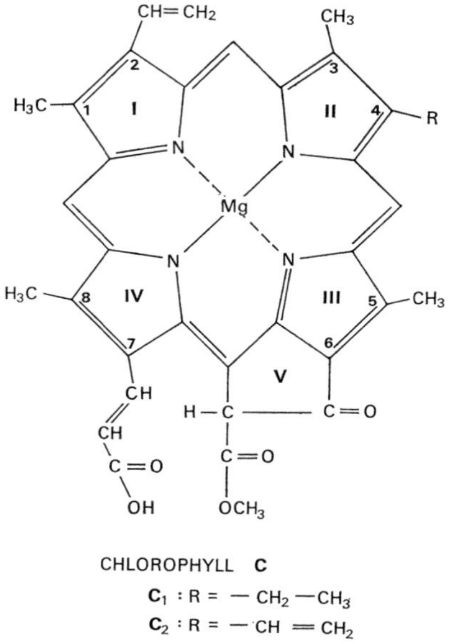

CHLOROPHYLL C

$C_1 : R = -CH_2-CH_3$

$C_2 : R = -CH = CH_2$

FIG. 2. Chlorophylls c_1 and c_2

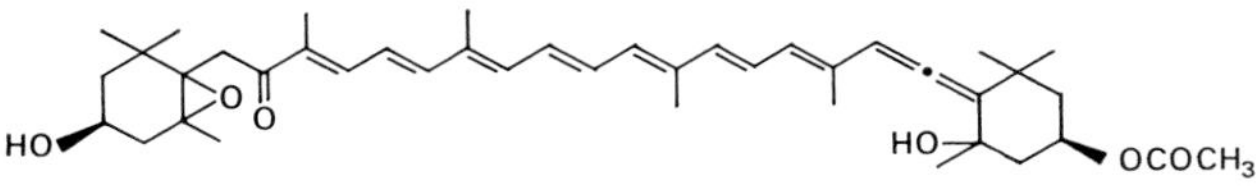

FIG. 3. Fucoxanthin.

of this algal class. This large red shift *in vivo* far exceeds the usual shift of 20 nm observed with most carotenoids, and allows brown algae to use the prevailing blue-green light. The other principal marine carotenoid, peridinin, present in dinoflagellates also has a large red shift *in vivo* (Haxo 1960). Apart from fucoxanthin, brown algae contain two other carotenoids, β-carotene and violaxanthin, which are also found in higher plants, and possibly traces of other carotenoids (Jensen 1966).

ISOLATION AND CHARACTERIZATION OF PIGMENT–PROTEIN COMPLEXES

We have isolated chloroplasts from several brown algal species including *Ecklonia radiata* (Laminariales), *Cladostephus spongiosus* (Sphacelariales),

Scytothamnus australis (Dictyosiphonales), *Phyllospora comosa* (Fucales), *Sargassum sp.* (Fucales) and *Acrocarpia paniculata* (Fucales). Seaweeds were collected from the south-eastern coast of New South Wales and transported to Canberra in chilled, aerated seawater. Chloroplast isolation is very difficult primarily owing to the extensive, tough cell walls and to the secretion of copious mucilage by injured cells. Furthermore, there are only one or two layers of cells containing chloroplasts and each cell has few chloroplasts. The current use of 1M-sorbitol in seawater and a differential washing procedure has greatly increased the yields (Barrett & Anderson, unpublished work); nonetheless, the yields of chlorophyll are low. The chlorophyll *a*/chlorophyll *c* molar ratio varies from 3–10 and is not necessarily the same for a particular species throughout the year.

Extensively washed thylakoid membranes were fragmented with 1% Triton X-100 in 50mM-Tricine (*N*-tris(hydroxymethyl)methylglycine) buffer, pH 8.0, at 0 °C for 30 min with stirring; the Triton X-100/chlorophyll ratio was 60. The solubilized extract was centrifuged at 15 000 **g** for 20 min, and the supernatant liquid generally contained 75–90% of the chlorophyll. Chloroplasts stored in liquid nitrogen for some time, however, often required multiple extractions to release all the chlorophyll. The solubilized extracts were separated into fractions with different pigment compositions by hydroxyapatite chromatography or by sucrose density gradient centrifugation (Barrett & Anderson 1977); the latter method has the better resolution and reproducibility. Apel (1977) has also used sucrose density gradient centrifugation for the successful isolation of the chlorophyll–proteins from Triton X-100 extracts of the green alga, *Acetabularia mediterranea*. Methods for estimations of pigments, absorption and fluorescence spectroscopy are given in Barrett & Anderson (1977).

A typical sucrose density gradient separation with Triton X-100 solubilized thylakoid membranes of *Acrocarpia* gives five main bands (Fig. 4): an applegreen band towards the bottom of the tube, an orange band and a stronger orange band at a greater buoyancy, a main green band and a yellow-green band situated towards the top of the tube. Instead of two orange zones, only a single orange band is obtained with some brown algal species, *e.g. Sargassum sp.* and *Ecklonia radiata* (Barrett & Anderson 1977).

The heavy, apple-green fraction has an absorption spectrum, with a λ_{max} in the red at 674 nm, which clearly shows the predominance of chlorophyll *a* (Fig. 5). This spectrum is similar to that of the P700–chlorophyll *a*–protein complex isolated from *Ecklonia radiata* (Barrett & Anderson 1977). The *Acrocarpia* complex has no chlorophyll *c*. It has a chlorophyll/P700 ratio of 38. This complex was isolated from fresh chloroplasts of all the brown algal species examined; the yields varied from 10–20% of the total chlorophyll. It

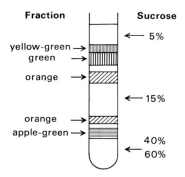

FIG. 4. Schematic representation of the sucrose density gradient separation of fractions from Triton X-100-solubilized thylakoids of *Acrocarpia paniculata*: gradient, 60% sucrose (1 ml), 40% sucrose (3 ml), 5–15% sucrose (7 ml), Triton X-100 extract (1 ml) all in 50mM-Tricine buffer, pH 8.0; centrifugation in a Beckman Spinco SW 41 rotor for 48 h at 270000 **g**.

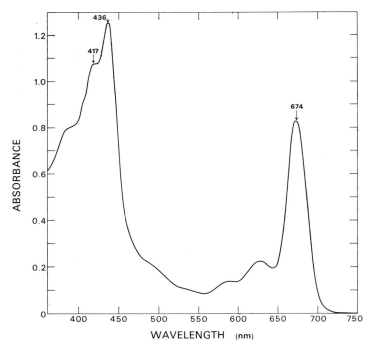

FIG. 5. Absorption spectrum of *Acrocarpia paniculata* P700–chlorophyll *a*–protein complex.

was difficult to extract the P700-complex from some of the thylakoid membranes which had been stored in liquid nitrogen for long periods. The fluorescence of the P700-complex is extremely low, as it is with higher plant P700-complexes (Boardman *et al.* 1978). Its excitation spectrum has bands at 418 and 436 nm in the blue, with a small contribution from a carotenoid identified as β-carotene. The isolation of a P700–chlorophyll *a*–protein from several brown algal species points to the universality of its distribution in both aquatic and terrestrial plants. Further, the P700–chlorophyll–protein complex of brown algae has about 40 mol of chlorophyll *a* for each mol of P700 as is found with the P700–chlorophyll *a*–protein complexes of higher plants, green and blue-green algae (Thornber *et al.* 1977).

The orange and green fractions are enriched in chlorophyll *c* compared to chloroplasts. The pigment composition and spectral properties of the two

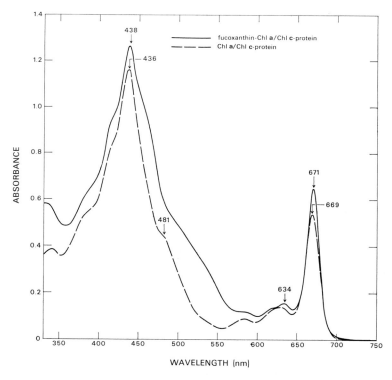

FIG. 6. Absorption spectra of *Acrocarpia paniculata* light-harvesting protein complexes: ————, fucoxanthin–chlorophyll *a/c*–protein complex; — — —, chlorophyll *a/c*–protein.

TABLE 1

Molar ratios of different chlorophylls (chl) in *Acrocarpia* chlorophyll–protein complexes

Chlorophyll–protein complex	$\dfrac{[chl\ a]}{[chl\ c_1 + c_2]}$	$\dfrac{[chl\ c_2]}{[chl\ c_1]}$	$\dfrac{[chl\ a]}{[P700]}$
P700–chl *a*–protein	>66		38
Light-harvesting fucoxanthin–chl *a/c*–protein	2.0	chl c_2 only	0
Light-harvesting chl *a/c*–protein	3.0	1.0	0

Acrocarpia orange fractions are identical. The absorption spectrum of the main *Acrocarpia* orange fraction (Fig. 6) shows enhanced absorption at 465 and 634 nm due to chlorophyll *c* and in the 500–550 nm region due to fucoxanthin. These orange fractions contain most of the fucoxanthin. The chlorophyll *a/c* molar ratio is 2.0 (average value); significantly, no chlorophyll c_1 is present at all (Table 1). The fluorescence emission spectrum (Fig. 7) of this fraction has a major peak at 684 nm and a minor one at 735 nm, and its excitation spectrum clearly shows peaks due to chlorophyll *a* (418 and 438 nm), chlorophyll *c* (465 nm) and fucoxanthin at 510 and 540 nm. This excitation spectrum resembles those of intact *Acrocarpia* chloroplasts or other brown algal chloroplasts (Goedheer 1970) except that the excitation bands due to chlorophyll *c* and fucoxanthin are enhanced compared to those of the chloroplast spectra. Thus the light absorbed by both chlorophyll *c* and fucoxanthin can be efficiently transferred to chlorophyll *a*, indicating that the molecular arrangement of the pigments in this complex has not been greatly altered by isolation from the thylakoid membrane.

By contrast, the absorption of the green fraction (Fig. 6) shows enhanced absorption at 460 and 633 nm (due to chlorophyll *c*) compared to that of chloroplasts, and there is no absorption in the region of 500–550 nm for fucoxanthin. The main carotenoid of this complex is violaxanthin. The chlorophyll *a*/chlorophyll *c* molar ratio is 3.0 (average value) and it contains about equimolar amounts of chlorophyll c_2 and chlorophyll c_1 in *Acrocarpia* (Table 1). The fluorescence emission spectrum (Fig. 7) of this fraction is similar to that of the orange fraction; however, the excitation spectrum is different showing the transfer of light energy from the Soret bands of chlorophylls *a* and *c*, but little transfer occurs in the region of 500 to 560 nm, as expected, since this complex does not contain fucoxanthin. It is important to note that all the fluorescence

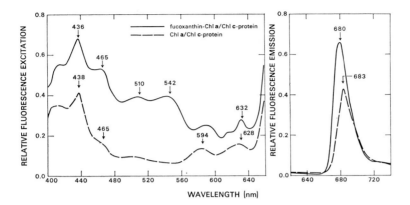

FIG. 7. Fluorescence excitation and emission spectra of *Acrocarpia paniculata* light-harvesting protein complexes at 77 K. The emission wavelength for the excitation spectrum was 681 nm and the excitation wavelength for the emission spectrum was 440 nm. The absorbance was 0.1 at 671 nm: ——————, fucoxanthin–chlorophyll *a*/*c*–protein complex; — — —, chlorophyll *a*/*c*–protein complex.

is emitted by chlorophyll *a* and there is no separate fluorescence emission from chlorophyll *c* at 636 nm with either the green or orange complex (Figs. 6 and 7). This also demonstrates that the molecular organization of the pigments of both of these complexes has not been altered during isolation.

In contrast to the pigment–protein complexes just described, the absorption and fluorescence spectra of the yellow-green fraction at the top of the sucrose density gradient indicate that this fraction contains some free chlorophyll *a*.

Sucrose density gradient centrifugation separated chlorophyll–protein complexes from Triton X-100-solubilized thylakoid membranes of all the brown algae examined. In all cases, three distinct complexes were obtained with similar pigment composition and spectral properties to those illustrated here with *Acrocarpia*. The similarity of the absorption maxima of the isolated complexes to the absorption maxima present in the chloroplasts (Fig. 1) and those observed in various brown algal chloroplasts by Goedheer (1970) demonstrates that we have isolated pigment–protein complexes and not merely free pigments. Moreover, the fluorescence excitation spectra reinforce this conclusion.

The yield of the P700–chlorophyll *a*–protein varied between 10–20% of the total chlorophyll; the yields of the green complex and the free chlorophyll towards the top of the tube were also somewhat variable, but the fucoxanthin–chlorophyll *a*/*c*₂-protein was the most variable component. The yield of this complex was rather low in *Phyllospora comosa*, variable in *Ecklonia radiata* depending on the site of collection and the season, and represented about 40% of the total chlorophyll in *Acrocarpia paniculata* and *Sargassum sp.*

Photosystem 1 activity was detected only in the P700–chlorophyll *a*–protein. In all the brown algae examined, P700 was detected both by oxidized *minus* reduced difference spectra and by light-induced absorbance changes. No photosystem 2 activity was detected in any of the isolated chlorophyll–proteins, with either (1) water as the electron donor and ferricyanide as the electron acceptor or (2) diphenylcarbazide as the electron donor and dichlorophenylindophenol as the electron acceptor (a method which allows photosystem 2 to be detected even though the more-sensitive water oxidation has been destroyed). We have no means of measuring P680, the reaction-centre chlorophyll of photosystem 2, and cannot, therefore, state unequivocally that either the orange or the green complexes do not contain P680. We are naming both the orange and green fractions light-harvesting chlorophyll–proteins, since neither photosystem 1 nor photosystem 2 activities have been detected in these fractions, but it will be necessary to locate P680. In higher plants, the P680–chlorophyll *a*–protein is labile and not well characterized biochemically (Boardman *et al.* 1978).

Although the storage of chloroplasts in liquid nitrogen is logistically convenient, the extractability of the chlorophyll–protein complexes is diminished with some brown algal chloroplasts. We have taken advantage of this to extract differentially the pigment complexes with 1% Triton X-100 from several brown algal thylakoids. We find that the fucoxanthin–chlorophyll *a/c*–protein complex is the most readily extracted from thylakoid membranes, then the green chlorophyll *a/c*–protein complex, and the most difficult complex to remove is the P700–chlorophyll *a*–protein. Thus, although the complexes of brown algae are all intrinsic proteins, as is the case for the complexes of higher plants and green algae (Anderson 1975), it is apparent that the P700–chlorophyll *a*–protein complex is the least accessible to Triton X-100 in the brown algal thylakoid membrane.

We have also used polyacrylamide gel electrophoresis to fractionate the pigment–protein complexes of brown algae. We have had some success in preliminary experiments using either 0.1% sodium dodecyl sulphate or 0.02% Triton X-100. It is difficult to find conditions in which the polypeptides of the brown algal complexes retain their pigments during sodium dodecyl sulphate polyacrylamide gel electrophoresis. Our results confirm the existence of several discrete pigment–protein complexes and prove unequivocally that the pigments are associated with polypeptides of different molecular weights. The P700–chlorophyll *a*–protein is found towards the top of the gel with a high apparent molecular weight and the orange fucoxanthin–chlorophyll *a/c*–protein runs nearer the gel front with an apparent molecular weight of 24 000; several green bands are observed between these two complexes.

In summary, we have isolated three distinct intrinsic chlorophyll–protein complexes by sucrose density gradient centrifugation of Triton X-100 extracts of six brown algae comprising four orders. In each case, these include a P700–chlorophyll a–protein complex whose spectral properties and pigment composition are indistinguishable from those of the P700–chlorophyll a–protein complex isolated from higher plants, green and blue-green algae. There are also two light-harvesting complexes which have no detectable photosystem 1 or 2 activities. One is a fucoxanthin–chlorophyll a/c_2–protein, which is present in variable amounts and in some cases is found at two different positions in the sucrose density gradient, and the other is a green chlorophyll a/c–protein, enriched in violaxanthin. Some free chlorophyll which is not attached to protein is found at the top of the density gradient tube. We have not established how many polypeptides are present in each of these fractions, but we are confident that the three complexes isolated are the major complexes of brown algae.

PROPOSED BINDING OF PIGMENT MOLECULES WITH PROTEIN

Structural studies of chlorophyll–protein complexes, with one exception, are not so far advanced as those of the haemoglobins and cytochromes. The arrangement of chlorophyll and carotenoid molecules in the chloroplast pigment complexes and the mode of interaction of pigment and protein is unknown, except for a water-soluble bacteriochlorophyll–protein complex, whose three-dimensional structure has been determined by X-ray at 0.28 nm resolution (Fenna & Matthews 1977). The structure thus revealed may serve as a model for the brown algal complexes. The bacteriochlorophyll-complex consists of three subunits each of 42 000 molecular weight; each subunit contains seven bacteriochlorophyll molecules arranged in an ellipsoid, which is completely enfolded by the polypeptide chains. The Mg atoms of six of the bacteriochlorophyll molecules are ligated to amino acid residues; liganding is restricted to one side of the ring and no ligand to the sixth position of magnesium has been detected. This position is shielded, however, by the phytyl chain folded back over the ring. The arrangement of the bacteriochlorophyll molecules is determined by their interaction with specific amino acids of the polypeptide chain.

Thornber et al. (1977) pointed out similarities in composition between the P700–chlorophyll a–protein complexes of higher plants and blue-green algae and the bacteriochlorophyll-complex, and they suggest that the chlorophyll molecules will be buried in the interior of the P700–chlorophyll a–protein as found with the bacteriochlorophyll-complex. However, the latter is water-soluble and an arrangement such as this with *all* the chlorophyll molecules surrounded by polypeptide chains may not be ideal for all intrinsic protein complexes. One of us (Anderson 1975) suggested a model in which some of the phytyl chains of chlorophylls *a* and *b* are associated with the hydrophobic exterior of their intrinsic proteins; in this model, chlorophyll forms part of the boundary lipid of these proteins. The hydrophilic side of the chlorin macrocycle containing the isocyclic ring is postulated to interact with the exposed hydrophilic segment of the protein at the membrane surface and the more hydrophobic part of the ring would be buried in the protein. Such an organization might account for the chlorophyll which is readily released from protein by either Triton X-100 or sodium dodecyl sulphate. Obviously chlorophyll *c* with no phytyl chain could not be located as a boundary lipid. Significantly, chlorophyll *c* is not readily accessible to Triton X-100, because we do not find any free chlorophyll *c*.

It is probable that the main attachment of chlorophyll *c* to the protein is through ligation of amino acids (such as histidine) to at least one side of the central Mg atom. Further, it is likely that the sixth position of the Mg will also be occupied with a coordinating amino-nitrogen atom, since there is no phytyl chain to shield the Mg in chlorophyll *c*, as is the case in the bacteriochlorophyll complex. Cytochrome b_5, although the tetrapyrrole is coordinated to iron rather than Mg, may serve as a better model for the linkage of chlorophyll *c* to protein. Other linkages to the protein could occur through the free carboxy group of the acrylic acid side-chain and the carbonyl group of the isocyclic ring; the aliphatic side-chains would provide extra, but rather weak links.

The presumptive linkage of fucoxanthin to the protein responsible for the large *in vivo* absorption shift to the red is readily modified with heat or high Triton X-100 concentrations (Goedheer 1970; Kirk 1977). Nevertheless, the isolation of the fucoxanthin light-harvesting complex demonstrates that the linkage of fucoxanthin to the protein and the arrangement of fucoxanthin relative to the chlorophyll molecules can be preserved. Our evidence points to fucoxanthin being an integral component of the orange light-harvesting complex of brown algae.

LIGHT-HARVESTING COMPONENTS OF ALGAL CLASSES

We suggest that the fucoxanthin–chlorophyll *a*/*c*–protein is the variable component of the photosynthetic unit of brown algal chloroplasts. Both the light intensity and spectral quality of light vary dramatically in the marine environment (Fig. 1) and, moreover, fucoxanthin is the main pigment which enables brown algae to capture the blue-green light which prevails in the underwater environment. Consequently, it is reasonable to suggest that the fucoxanthin-containing light-harvesting complex is the variable component of the photosynthetic unit. Indeed we have found that the yield of this complex is the most variable of the three complexes isolated from brown algae; further, the yield may vary within a particular species depending on the season and the site of collection. Moreover, our studies with differential extraction of the chlorophyll–protein complexes from brown algal thylakoid membranes demonstrate that the fucoxanthin-containing complex is the least-firmly bound in the membrane, and the green chlorophyll a/c–protein and the P700–chlorophyll a–protein are more·firmly bound within thylakoid membranes. This suggests that the P700-complex, and possibly the chlorophyll a/c–protein complex, are fundamental to the organization of the photosynthetic unit but that it is the fucoxanthin-complex which is synthesized and assembled in response to the special light environment. The light-harvesting chlorophyll a/b–protein of higher plants, which contains all the chlorophyll b, is present is greater amounts in shade plants and varies with light and other environmental growth conditions (Boardman *et al.* 1978). Thus, this complex is not only the major light-harvesting component of higher plants but also is the variable component of the photosynthetic unit of these plants. We suggest that the fucoxanthin–chlorophyll a/c–protein of brown algae has an analogous function to that of the light-harvesting chlorophyll a/b–protein of higher plants.

A comparison can now be made of the light-harvesting complexes of brown algae with those of other algal classes. The only other chlorophyll c-containing complex yet isolated is from the diatom, *Phaeodactylum tricornutum* (Holdsworth & Arshad 1977). This complex has a molecular weight of about 850 000. It is probably made up of 40 protein subunits (of 25 000 molecular weight), 40 mol of chlorophyll a, 20 mol of chlorophyll c, 20 mol of fucoxanthin, 8 g atoms of copper, and 0.6–2 g atoms of manganese. Moreover, this complex is probably an integral part of photosystem 2 since it has photosystem 2 activity, and its fluorescence properties and electron paramagnetic resonance spectra are also somewhat similar to those of photosystem 2 in higher plant chloroplasts.

Photosynthetic light-harvesting complexes of an entirely different nature have been isolated from several marine dinoflagellates. These water-soluble complexes are peridinin–chlorophyll a–proteins (Haxo *et al.* 1976; Prézelin & Haxo 1976; Song *et al.* 1976) which contain peridinin and chlorophyll a in the molar ratio of 4:1 or 9:2. These complexes have no chlorophyll c_2. The proteins of particular species have different isoelectric points and molecular weights; the usual range for the molecular weights of the complexes is between 35 000 and 39 000.

It is generally agreed that the two complexes containing the reaction centres of photosystem 1 and photosystem 2 and a postulated third chlorophyll *a*–protein complex are common to all plants, and that during the course of evolution the light-harvesting component varied to meet demands of specialization (Thornber *et al.* 1977; Boardman *et al.* 1978). The light-harvesting components which have now been demonstrated to occur in the various algal classes are listed in Table 2. These include the chlorophyll *a*/*b*–protein of green algae and *Euglena*, and the fucoxanthin–chlorophyll *a*/*c*–protein, and the chlorophyll *a*/*c*–protein of brown algae. These light-harvesting chlorophyll–proteins are all intrinsic proteins. However, it is remarkable that the more primitive aquatic plants have water-soluble, light-harvesting complexes. These include the fucoxanthin–chlorophyll *a*/*c*–protein of a diatom and the peridinin–chlorophyll *a*–proteins of peridinin-containing dinoflagellates. Finally, the red algae and the prokaryotic blue-green algae, which do not have chlorophylls *b* or *c*, have water-soluble phycobiliproteins as their light-harvesting complexes. The light-harvesting complexes of the golden-brown flagellates, the fucoxanthin-containing dinoflagellates and the cryptomonads have not been isolated or characterized as yet. It will be of great evolutionary interest to see if those unicellular algae which also contain chlorophylls c_1 and c_2 and fucoxanthin possess chlorophyll *a*/*c*–proteins and fucoxanthin–chlorophyll *a*/c_2–proteins corresponding to the green and orange complexes which we have isolated from brown algae. Finally, it has not escaped our notice that the apparent molecular weights of the fucoxanthin–chlorophyll *a*/c_2–protein of brown algae and a

TABLE 2

Light-harvesting components of algal classes

Algal class	Isolated complexes[a]	Protein
Green algae + *Euglena*	Chl *a*/*b*–protein	Intrinsic
Brown algae	Fucoxanthin–chl *a*/c_2–protein	Intrinsic
	Chl *a*/c_1 + c_2–protein	Intrinsic
Diatoms	Fucoxanthin–chl *a*/*c*–protein	Water-soluble
Dinoflagellates	Peridinin–chl *a*–protein	Water-soluble
Cryptomonads	Phycobiliproteins	Water-soluble
Red algae	Phycobiliproteins	Water-soluble
Blue-green algae	Phycobiliproteins	Water-soluble

[a]chl. chlorophyll.

diatom are in the same range as that of the light-harvesting chlorophyll a/b–protein of higher plants.

ACKNOWLEDGEMENTS

We thank Dr S. W. Thorne for the fluorescence spectra, Dr H. B. S. Womersley (University of Adelaide) for identification of the algae and our many colleagues who assisted in the collection of seaweed. One of us (J.B.) is indebted to the Australian Research Grants Committee for a Senior Fellowship and a maintenance grant.

References

ANDERSON, J. M. (1975) The molecular organization of chloroplast thylakoids. *Biochim. Biophys. Acta 416*, 191–235

APEL, K. (1977) Chlorophyll–proteins from *Acetabularia mediterranea. Brookhaven Symp. Biol. 28*, 149–161

BARRETT, J. & ANDERSON, J. M. (1977) Thylakoid membrane fragments with different chlorophyll *a*, chlorophyll *c* and fucoxanthin compositions isolated from the brown seaweed, *Ecklonia radiata. Plant Sci. Lett. 9*, 275–283

BOARDMAN, N. K., ANDERSON, J. M. & GOODCHILD, D. J. (1978) Chlorophyll–protein complexes and structure of mature and developing chloroplasts. *Curr. Top. Bioenerg. 8*, 35–109

BONNETT, R., MALLAMS, A. K., SPARK, A. A., TEE, J. L., WEEDON, B. C. L. & McCORMICK, A. (1969) Carotenoids and related compounds. Part XX. Structure and reactions of fucoxanthin. *J. Chem. Soc. C*, 429–454

DOUGHERTY, R. C., STRAIN, H. H., SVEC, W. A., UPHAUS, R. A. & KATZ, J. J. (1966) Structure of chlorophyll *c. J. Am. Chem. Soc. 88*, 5037–5038

FENNA, R. E. & MATTHEWS, B. W. (1977) Structure of a bacteriochlorophyll *a*–protein from *Prosthecochloris aestuarii. Brookhaven Symp. Biol. 28*, 170–182

FOGG, G. E. (1972) *Photosynthesis*, 2nd edn., English Universities Press, London

GOEDHEER, J. C. (1970) On the pigment system of brown algae. *Photosynthetica 4*, 97–106

HAXO, F. T. (1960) The wavelength dependence of photosynthesis and the role of accessory pigments, in *Comparative Biochemistry of Photoreactive Systems* (Allen, M. B., ed.), pp. 339–360, Academic Press, New York

HAXO, F. T., KYCIA, J. H., SOMERS, G. F., BENNETT, A. & SIEGELMAN, H. W. (1976) Peridinin–chlorophyll *a* proteins of the dinoflagellate *Amphidinium carterae (Plymouth 450). Plant Physiol. 57*, 297–303

HOLDSWORTH, E. S. & ARSHAD, J. H. (1977) A manganese–copper–pigment–protein complex isolated from the photosystem II of *Phaeodactylum tricornutum. Arch. Biochem. Biophys. 183*, 361–373

JEFFREY, S. W. (1976) The occurrence of chlorophyll c_1 and c_2 in algae. *J. Phycol. 12*, 349–354

JENSEN, A. (1966) *Carotenoids of Norwegian Brown Seaweeds and of Seaweed Meals*, Norwegian Institute of Seaweed Research, Report No. 31, Tapir

KIRK, J. T. O. (1976) Yellow substance (gelbstoff) and its contribution to the attenuation of photosynthetically active radiation in some inland and coastal south-eastern australian waters. *Aust. J. Mar. Freshwater Res. 27*, 61–71

KIRK, J. T. O. (1977) Thermal dissociation of fucoxanthin–protein binding in pigment complexes from chloroplasts of *Hormosira* (Phaeophyta). *Plant Sci. Lett. 9*, 373–380

PRÉZELIN, B. B. & HAXO, F. T. (1976) Purification and characterization of peridinin–chlorophyll *a*–proteins from the marine dinoflagellates *Glenodinium sp.* and *Gonyaulax polyedra. Planta 128*, 133–141

SONG, P. S., KOKA, P., PRÉZELIN, B. B. & HAXO, F. T. (1976) Molecular topology of the photosynthetic light-harvesting pigment complex, peridinin–chlorophyll *a*–protein, from marine dinoflagellates. *Biochemistry 15*, 4422–4427

THORNBER, J. P. (1975) Chlorophyll–proteins: light-harvesting and reaction center components of plants. *Annu. Rev. Plant Physiol. 26,* 127–158
THORNBER, J. P., ALBERTE, R. S., HUNTER, F. A., SHIOZAWA, J. A. & KAN, K.-S. (1977) The organization of chlorophyll in the plant photosynthetic unit. *Brookhaven Symp. Biol. 28,* 132–148

Discussion

Wessels: I am not surprised that you detected no photochemical activity in the chlorophyll–protein complexes since you use a high concentration of Triton.

Anderson: We detected a little activity in the initial extracts but that quickly disappeared. We had to use high concentrations of Triton X-100 and that will interfere when one is using diphenylcarbazide to assay photosystem II. The brown algal membranes are extremely resistant to attack by detergents and, since they were not fragmented by digitonin, we were forced to use Triton X-100 and high Triton X-100/chlorophyll ratios. The membranes may be rich in glycoproteins; the outer surface is different from that of higher plants.

Staehelin: Doesn't that suggest that the lipid composition differs in this system?

Anderson: Yes, but we know nothing about the lipid composition.

Thornber: In higher plants the P700–chlorophyll a–protein is the more easily-solubilized pigmented complex (cf Brown & Duranton 1964). In the brown algae you used you find the reverse. Is this due to the lack of phytyl chains on the outside of the protein, making the complex less hydrophobic and, therefore, more soluble?

Anderson: No, I don't think so. The fucoxanthin complex which is most readily extracted with brown algal membranes contains chlorophyll a as well as chlorophyll c. Certainly, the porphyrin ring of chlorophyll c has to be buried deep in the protein.

Thornber: Have you ever treated any isolated chlorophyll–protein or any photosynthetic membrane fraction (e.g. digitonin particles) with chlorophyllase to see whether their solubility properties change?

Anderson: No.

Katz: It is not impossible that an active chlorophyllase removes the phytyl chain from chlorophylls c_1 and c_2 during the isolation process (and diatoms take a long time to harvest by centrifugation).

Anderson: And one would not detect that spectroscopically anyway.

Katz: Whether chlorophylls c_1 and c_2 are esterified *in vivo* ought to be checked before one speculates about the relative polarities of the chlorophyll environments.

Staehelin: What is the protein composition of the light-harvesting complex?

Anderson: We have not analysed the polypeptides on slab-gels in great detail. Unfortunately, with these slab-gels, although we apparently have high protein concentrations initially, when we stain with Coomassie blue we see barely any protein. The only band we see from the fucoxanthin complex, as isolated from the density gradient, is one at a molecular weight of 24 000. As well as the polypeptide of the P700–chlorophyll *a*–complex we see several green bands of lower molecular weight. However, these results are still preliminary.

Thornber: 'Amido-black' sometimes stains detergent-solubilized proteins better than Coomassie blue does.

Anderson: Yes, but the staining is still not as good as we want it to be. We are trying to do fluorescence scanning.

Cogdell: Have you stained for carbohydrate? Coomassie blue staining can be poor when glycoprotein is plentiful.

Anderson: A lot of glycoprotein is present.

Katz: I might point out that one has to differentiate between carotenes that may have no donor properties and xanthophylls that contain hydroxy or carbonyl donor groups. N.m.r. spectroscopy indicates that the hydroxy group of lutein can compete successfully with the carbonyl group of chlorophyll for coordination to magnesium. In such a coordination complex the lutein protons that are situated just above the plane of the chlorophyll macrocycle are shifted up-field. β-Carotene cannot act in this fashion as it lacks any donor function, so its association with chlorophyll is presumably mediated mainly by interaction with the chlorophyll phytyl group. Fucoxanthin, to the contrary, could coordinate directly to magnesium.

Anderson: Yes, it could.

Junge: Sewe & Reich (1977) studied electrochromism of chlorophyll mono-layers in contact with carotenoid-doped monolayers in a microcapacitor set-up. They interpret the major peak of electrochromic signals in chloroplasts (at 520 nm; see Emrich *et al.* 1969) as being due to an aggregate between lutein and chlorophyll *b*. Interaction of one OH group on lutein with the central Mg in chlorophyll *b* creates a strong pre-polarization of the originally-symmetrical lutein. This could be the reason why the electrochromism of symmetrical carotenoids, which should be second order if these pigments were unperturbed, is linearly related to the electric-field strength both in chloroplasts and in bacterial chromatophores.

Clayton: This could explain the shifting and non-shifting pools of carotenoids in photosynthetic bacteria; the shifting pool is about 10% of the total. But the coordination is not necessarily to the reaction centres; there might not be enough room for each 'shifting' carotenoid to be coordinated with a magnesium atom in the reaction centre.

Cogdell: Unfortunately for that idea, a mutant of *Rhodopseudomonas sphaeroides* (G1C) contains the hydrocarbon neurosporene as the only carotenoid (see Table 1, p. 70) yet it shows a large carotenoid band-shift.

Amesz: As in the wild strain, there seem to be two pools of carotenoid in this mutant, of which only one (comprising 20–30% of the total) shows electro-chromism (De Grooth & Amesz 1977).

Thornber: Do you envisage that *all* your isolated complexes containing chlorophyll *c* feed their absorbed light energy primarily to the trap of photosystem II, or does one feed photosystem I and the other photosystem II?

Anderson: We have no evidence one way or the other. I think that the fucoxanthin complex is the main light-harvesting complex.

Joliot: Have you looked at the action spectrum of the fluorescence *in vivo* of this type of algae to try to determine the pigment which sensitizes photosystem II? This could easily be done *in vivo*.

Anderson: No.

Porter: What was the molecular weight of the main light-harvesting protein? Was it the same as that of the chlorophyll *a/b*–protein?

Anderson: Yes; its polypeptide has an apparent molecular weight of 24 000.

Porter: How many chlorophyll molecules did it have on it?

Anderson: We cannot yet say for the fucoxanthin complex.

Thornber: For the chlorophyll *a/b*–protein purified from a green alga we find six chlorophyll molecules per molecular weight of 29 000. Dr Knox confirms this number of pigments in the minimal unit of the higher-plant complex as evidenced by his student's recent work (Van Metter 1977*a*, *b*) on the spectral properties of the complex.

Knox: Yes.

Anderson: We do not yet know the ratio of chlorophyll to protein. The protein is probably a glycoprotein and we have only estimated it by Lowry's method, which is not good enough.

Staehelin: Does the proposal of six chlorophylls for each protein of molecular weight 24 000 take into account the fact that all the supposed light-harvesting complexes that can be isolated contain two proteins—one pigmented and one non-pigmented? When the electrophoretic gels of pea or spinach thylakoids are run at 4 °C, one can isolate a complex with a molecular weight of about 70 000. When that is separated and re-run at room temperature, one non-pigmented glycoprotein and one pigmented protein are obtained. The latter is presumably the equivalent of the protein with molecular weight 24 000.

Thornber: Does the glycoprotein also have a molecular-weight equivalency of 24 000?

Staehelin: Depending on the conditions, it can run ahead, with, or behind the pigment protein; for example, at 12.5% acrylamide it runs ahead but at 15% behind. That behaviour is typical of glycoproteins.

Anderson: This happens with the algal pigment–protein complexes also (Chua *et al.* 1975; Bar-Nun *et al.* 1977).

Staehelin: The pigmented one runs more-or-less consistently with other proteins at different concentrations of the gels, whereas the glycoproteins behave quite differently.

Thornber: The material on which we made the chlorophyll/protein estimation was isolated chromatographically on hydroxyapatite rather than by gel electrophoresis in the presence of SDS. We do not know whether this glycoprotein was present. Nevertheless estimates of the chlorophyll and nitrogen content— by Kjeldahl estimation and by measuring the nitrogen from an amino acid analysis of samples of known chlorophyll content—gave a figure close to six chlorophyll molecules per 24 000 protein.

Wessels: We have prepared the light-harvesting chlorophyll *a/b*–protein from spinach chloroplasts treated with digitonin by density-gradient centrifugation and ion-exchange chromatography. This complex (F_{III}) produced a green band at 24 000 molecular weight on SDS gels. When the chlorophyll–protein complex was dissolved in SDS at a temperature higher than about 60 °C, the pigment was released and two distinct polypeptides were observed at 24 000 and 27 000, the latter being much more intense than the band at molecular weight 24 000. The F_{III} particles were found to contain 5–8 chlorophyll molecules per unit of molecular weight 24 000. If we assume that only the 27 000 polypeptide was associated with chlorophyll, enhancing its electrophoretic mobility to 24 000 so that it co-migrates with the 24 000 polypeptide, the figure for the proper chlorophyll–protein may increase to, say, 6–9 per 24 000 unit.

Staehelin: We believe that it is a true component of the complex for the following reasons (A. McDonnel, unpublished work, 1978): when one makes a Triton or digitonin preparation of the light-harvesting complex, the colourless polypeptide always remains associated with the pigment protein. At 4 °C about half the chlorophyll runs at 70 000 and half at about 25 000–30 000 but, based on studies of concanavalin A binding, all the glycoprotein is at 70 000. At room temperature all the chlorophyll–proteins run at about 25 000–30 000 and so does the glycoprotein.

Thornber: What do you suppose is the function of this glycoprotein?

Staehelin: It might be involved in membrane stacking; it could be the membrane-adhesion component that complexes with the pigment protein to form the larger complex surrounding the reaction centre core of photosystem II.

Thornber: Do you envisage an interaction between the carbohydrates in glycoprotein groups causing stacking of thylakoid membranes?

Staehelin: At first, that is what I thought but all the evidence suggests that membrane adhesion is an hydrophobic effect and not mediated by ions. Ions shield the charges so that the membranes can come together. In all animal membrane systems so far studied the glycoproteins have their sugar groups exposed on the external or luminal surface. Translated to the chloroplast membrane system, this suggests that the sugar groups are associated with the lumina of the thylakoids, where they could be involved in membrane adhesion, and not with the stromal surface.

Porter: What is the protein–protein interaction between these particles containing 6–8 chlorophyll molecules that makes them up into a photosynthetic unit? For energy transfer these particles must be close with the chlorophyll molecules only about 3 nm apart.

Staehelin: Some recent results may be relevant to that (D. P. Carter, unpublished work, 1978). When we unstack chloroplast membranes and treat the membranes mildly with pronase (to remove only, say, three out of the 30 proteins seen on polyacrylamide gels), the light-harvesting complex (which at 4 °C runs at 70 000) loses about 500 in molecular weight. Once that has happened the membranes cannot be experimentally restacked with ions. However, examination of the membranes after freeze-fracture, once they have been mildly treated with pronase and kept in a low salt medium, shows that the number of EF-face particles (see p. 151) has increased by about 30%. On addition of salts, which normally causes re-stacking of membranes, the density of the particles returns to normal but the particles are larger than in the controls. We conclude that there is some free light-harvesting complex in the membrane which probably gets bound to the larger aggregation of photosystem II core and light-harvesting elements when ions are around—by an equilibrium-type reaction—and that this larger complex is usually involved in stacking. The pronase-digestion experiments allow us to distinguish the stacking process from the formation of the large complexes within the membrane. We assume that the lateral association is hydrophobic—maybe with some salt binding.

Knox: We have discovered, or actually rediscovered, the denaturation of the light-harvesting complexes (see remarks by Thornber *et al.* 1967 and Ogawa *et al.* 1966); we had to keep them at about 2 °C to prevent this process. Apparently the complexes 'unwind' (I adopt a hypothetical image for convenience). The first sign of this is that the fluorescence polarization increases but the absorption spectrum stays the same; that means that the chlorophylls are coming apart (less energy transfer leading to depolarization occurs) but still remain attached to their local protein environment. After some time the chlorophyll *a* starts to detach and the Chl *b*/Chl *a* ratio goes up.

Staehelin: Is that process reversible?

Knox: No.

Thornber: With regard to Sir George Porter's question on the nature of the interaction between chlorophyll–protein particles in the photosynthetic unit, I believe that this is probably a largely hydrophobic one because the chlorophyll *a/b*–proteins can be observed to occur in oligomeric forms on polyacrylamide gel electrophoresis if the detergent concentration in the gel is lowered from that normally used when only the monomeric form is observed. This observation came about because we and others had been concerned that in all eukaryotic organisms one finds a lot of chlorophyll not associated with protein after SDS polyacrylamide gel electrophoresis. We (Markwell *et al.* 1978) have been trying to develop an electrophoretic system which will retain more of the chlorophyll with its associated protein after dissociation of the membrane in detergent and subsequent electrophoresis. Fig. 1 shows two systems. In the

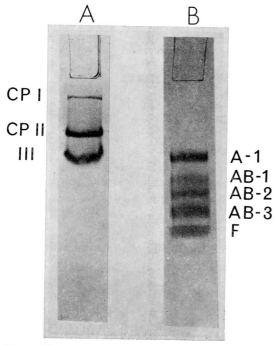

FIG. 1 (Thornber). Two gel electrophoretic systems for the separation of chlorophyll–protein complexes: (A) SDS; (B) results on a system developed by Markwell *et al.* (1978).

old electrophoretic system (A) we saw the SDS-altered form of the P700–chlorophyll *a*–protein (CP I), the chlorophyll *a/b*–protein (CP II) and the free pigment zone (III). The proportion of the total chlorophyll in each was about 10%, 45–55% and 35%, respectively. In the new system (B) we see more bands. The absorption spectrum of A-1 is identical to that of CP I, but the amount of chlorophyll in A-1 is now about 25–30% compared to 10% in system A. Below A-1 are three bands, all of which contain chlorophylls *a* and *b*. At present we cannot say how they are related to CP II but, because of their pigment content, they must all contain material which runs at the same position as CP II in system A. They might contain additional components because there are slight differences in the spectra of AB-1, AB-2 and AB-3. Furthermore AB-3, which is the fastest zone containing chlorophyll *b*, electrophoreses more slowly than CP II. It has a molecular-weight equivalency of about 40 000 (cf. 30 000 for CP II); that of AB-2 is about 60 000 and that of AB-1 about 80 000. One also wonders how these last two components might be related to the 70 000 complex that Dr Staehelin mentioned. The bands AB-1–AB-3 may represent *in vivo* oligomeric forms of the chlorophyll *a/b*–protein. Finally, band F contains about 10% of the total chlorophyll loaded onto the gel; we think the chlorophyll in band F is not bound to protein after electrophoresis. There is a sharp pigmented band located on the top edge of band F which stains strongly with Coomassie blue—however, that does not necessarily mean that it contains protein; lipid also stains with Coomassie blue.

Porter: Are you implying a monomer–dimer–trimer relation of a unit for the AB-1, AB-2 and AB-3 bands?

Thornber: There may be such a relationship between AB-3 and AB-1; AB-2 apparently does not behave as an oligomer.

Anderson: We have also a new gel procedure (Anderson *et al.* 1978). Some 28% of the chlorophyll is in CP I and CP Ia and less than 10% is free chlorophyll. In addition to the three bands containing chlorophylls *a* and *b*, which appear to be aggregates of the light-harvesting complex, we find a weak band which is mainly a chlorophyll *a*–protein complex. Does your AB-2 band contain little chlorophyll *b*?

Thornber: No. They all have lower chlorophyll *a/b* ratios than the intact leaf, as judged by their absorption spectra; however, we have not yet quantitated the ratios. From the absorption spectrum we deduce that AB-3 contains the highest *a/b* ratio of the three AB bands.

Barber: Was there any chlorophyll *b* in band III?

Thornber: No. Most of that chlorophyll originates from dissociation of the CP I complex. The new electrophoretic system now enables some 30% of the total chlorophyll (compared with 10% previously) to be contained in the A-1 band. We had previously suggested that the free pigment (zone III) arose from dissociation of much of the P700–chlorophyll a–protein, CP I (Thornber *et al.* 1977).

Barber: Which band contains the photosystem II reaction centre?

Thornber: I don't know for certain, but I guess that it electrophoreses close to the AB-3 band—such a conclusion would fit Dr Wessels' recent results and those of Hayden & Hopkins (1977) in their studies on the location of chlorophyll a_2 on SDS gel electrophoresis. We hope to relate the bands in the two electrophoretic systems by running a two-dimensional gel using system B in one direction and system A in the second.

Wessels: What is the difference between the two systems? Is it the SDS concentration?

Thornber: This work is currently being done by J. Markwell and Sally Reinman in my laboratory. Since further improvement of the system is still likely I cannot yet give you the final recipe.

Staehelin: How do the results relate to the work of Machold (1978) who, by adding bivalent cations, obtained more stable chlorophyll complexes?

Thornber: I don't know. However, I should like to electrophorese solubilized membranes after they have been extracted with EDTA solutions to see whether some complexes are held to each other by magnesium bridges. (For further discussion, see pp. 351–357.)

References

ANDERSON, J. M., WALDRON, J. C. & THORNE, S. W. (1978) Chlorophyll–protein complexes of spinach and barley thylakoids. Spectral characteristics of six complexes resolved by an improved electrophoretic technique. *FEBS (Fed. Eur. Biochem. Soc.) Lett. 92*, 227–233

BAR-NUN, S., SCHANTZ, R. & OHAD, I. (1977) Appearance and composition of chlorophyll–protein complexes I and II during chloroplast membrane biogenesis in *Chlamydomonas reinhardi. Biochim. Biophys. Acta 459*, 451–467

BROWN, J. S. & DURANTON, J. G. (1964) Partial separation of the forms of chlorophyll a by sodium dodecyl sulfate. *Biochim. Biophys. Acta 79*, 209–211

CHUA, N. H., MATLIN, K. & BENNOUN, P. (1975) A chlorophyll–protein complex lacking in photosystem I mutants of *Chlamydomonas reinhardi. J. Cell Biol. 67*, 361–377

DE GROOTH, B. G. & AMESZ, J. (1977) Electrochromic absorbance changes of photosynthetic pigments in *Rhodopseudonomas sphaeroides* II. Analysis of the band shifts of carotenoid and bacteriochlorophyll. *Biochim. Biophys. Acta 462*, 247–258

EMRICH, H. M., JUNGE, W. & WITT, H. T. (1969) Further evidence for an optical response of chloroplast bulk pigments to a light-induced electrical field in photosynthesis. *Z. Naturforsch. 24b*, 1144–1146

HAYDEN, D. B. & HOPKINS, W. G. (1977) A second distinct chlorophyll a–protein complex in maize mesophyll chloroplasts. *Can. J. Bot. 55*, 2525–2529

MACHOLD, O. (1978) in *Proceedings of the International Congress on Photosynthesis* (Reading, 1977), Biochemical Society, London

MARKWELL, J. P., REINMAN, S. & THORNBER J. P. (1978) *Arch. Biochem. Biophys.*, in press

OGAWA, T., OBATA, F. & SHIBATA, K. (1966) Two pigment proteins in spinach chloroplasts. *Biochim. Biophys. Acta 112*, 223–234

SEWE, K.-U. & REICH, R. (1977) Effect of molecular polarization on electrochromism of carotenoids. 2. Lutein–chlorophyll complexes—origin of field-indicating-absorption change at 520 nm in membranes of photosynthesis. *Z. Naturforsch. 32c*, 161–171

THORNBER, J. P., STEWART, J. C., HATTON, M. W. C. & BAILEY, J. L. (1967) Studies on the nature of chloroplast lamellae I. Chemical composition and further physical properties of two chlorophyll–protein complexes. *Biochemistry 6*, 2006–2014

THORNBER, J. P., ALBERTE, R. S., HUNTER, F. A., SHIOZAWA, J. A. & KAN, K.-S. (1977) The organization of chlorophyll in the plant photosynthetic unit. *Brookhaven Symp. Biol. 28*, 132–148

VAN METTER, R. L. (1977a) *A Study of the Optical Properties of Chlorophyll in Solution and in a Protein Complex*, Ph. D. Thesis, University of Rochester

VAN METTER, R. L. (1977b) Excitation energy transfer in the light-harvesting chlorophyll *a/b*-protein. *Biochim. Biophys. Acta 462*, 642–658

Resonance Raman spectroscopy of chlorophyll-protein complexes

M. LUTZ, J. S. BROWN* and R. RÉMY†

*Service de Biophysique, Département de Biologie, Centre d'Études Nucléaires de Saclay, Gif-sur-Yvette, France, *Carnegie Institution of Washington, Stanford, California, and †Laboratoire de Physiologie Cellulaire Végétale, Université Paris-Sud, Orsay, France*

Abstract Resonance Raman spectra of chlorophyll *a* (Chl *a*) and of Chl *b* were selectively obtained, at low temperature, from chlorophyll–protein complexes prepared from green and blue-green algae and from higher plants.

Antenna Chl *a* in the Chl *a*–P700–protein complexes (CP I) and in the light-harvesting Chl *a/b*–protein complexes (CP II) gives resonance Raman spectra extremely close in all their features to those previously obtained from intact cells and chloroplasts. In particular, the same multiplicity of binding sites for the ketone carbonyl groups of Chl *a* is observed in both CP I and CP II as in intact membranes. These bindings sites are probably the same types as those observed in the intact membranes and are not the magnesium atoms of other chlorophylls. The magnesium atoms of most Chl *a* molecules in both CP I and CP II bind a single external ligand.

Resonance Raman spectra of Chl *b* in CP II preparations, although very similar to those from intact membranes, show partial rearrangement of one of the two environmental subspecies of Chl *b* previously found in intact membranes.

These results provide evidence that chlorophyll–protein complexes closely represent the state of the bulk of antenna chlorophyll *in vivo*.

The relative spatial arrangement of light-harvesting chlorophylls within the photosynthetic membrane determines the conditions of energy transfer from sites of light absorption to reaction centres. This arrangement is not random; light-harvesting chlorophyll molecules assume certain properties of orientation with respect to the membrane plane (Paillotin & Breton 1977). Such orientations must be held by sets of interactions between chlorophyll and surrounding molecules. Determining these interactions and the partner molecules interacting with chlorophyll is thus a matter of considerable current interest.

Resonance Raman (RR) spectroscopy can provide vibrational spectra, with high selectivity, from chlorophyll present in intact chloroplasts or cells (Lutz 1972, 1975). Raman spectra obtained in pre-resonance and resonance conditions with respect to the Soret transitions of the chlorophylls contain bands arising from stretching motions of their 9-keto carbonyl groups as well as of the 3-formyl carbonyl group of chlorophyll b (Chl b) (Lutz 1977; Lutz & Breton 1973). Resonance enhancement of bands involving vibrational modes of the Mg-N_4 grouping was also demonstrated by substitution of ^{26}Mg in Chl a and in Chl b (Lutz et al. 1975). All these functional groups are known to play a predominant role in intermolecular interactions of chlorophyll in vitro (Katz 1973). Ester carbonyl groups in positions 10a and 7c also participate in building certain intermolecular associations of chlorophyll in vitro. They do not give rise to resonance-enhanced Raman bands, being unconjugated to the main π-electron system of the phorbin macrocycle (Lutz 1974). The relative extent of participation of C⋯C and C⋯N bonds in resonating modes of Chl a and Chl b was estimated from ^{15}N substitution (Lutz et al. 1975).

Resonance Raman spectra of antenna chlorophyll present in algae and higher plants show that the Mg atoms as well the 9-keto carbonyl groups of most Chl a molecules and the 3-formyl carbonyl groups of all Chl b molecules are involved in intermolecular bonding. Neither Chl a nor Chl b constitutes a homogeneous population with respect to the interactions assumed by their carbonyl groups. The 9-C=O groups of Chl a take part in at least five different types of interaction and the 3-C=O groups of Chl b in two. These interaction states are most likely identical among higher plants and algae. Multiplicity of interaction states is also observed among Mg atoms of Chl b molecules, and is probable among Mg atoms of Chl a (Lutz 1975, 1977; Lutz & Breton 1973).

Comparison of resonance Raman spectra of antenna chlorophyll with those of various assemblies of chlorophyll in vitro demonstrates that none of the environmental subspecies of antenna chlorophyll may result from the presence of oligomers or polymers of chlorophyll in the membrane. Antenna chlorophyll is thus monomeric and necessarily bound to foreign molecules. Resonance Raman spectra of higher plants and algae are entirely consistent with all antenna chlorophyll molecules being bound to proteins through their Mg atoms and through hydrogen-bonding of carbonyl groups with different amino acids (Lutz 1977). On mild treatment of photosynthetic membranes with diluted detergents, significant amounts of Chl a and of Chl b are extracted as complexes with proteins. Extensive work has substantiated the idea that these complexes represent native states of chlorophyll (Anderson 1975; Thornber 1975). However, in such fractionation experiments significant amounts of chlorophyll are obtained in protein-free states (Brown et al. 1974, 1975). Electronic spectra of protein-bound Chl a differ somewhat from those of antenna chlorophyll a in vivo (Brown et al. 1974; Rémy et al. 1977). The questions thus still arise whether in the chloroplast all chlorophyll molecules are involved in protein complexes and how close the state of chlorophyll in isolated protein complexes is to its native state. It thus appeared useful to compare by resonance Raman spectroscopy the bonding interactions of chlorophyll in these complexes to those previously observed in intact membranes. The two major complexes, P700–Chl *a*–protein (CP I) and light-harvesting Chl *a/b*–protein (CP II), extracted from various plants and algae were investigated at low temperature.

MATERIAL AND METHODS

Chlorophyll–protein complexes

P700–Chl *a*–protein complexes were prepared by treatment of the blue-green algae *Anabaena cylindrica* with Triton-X 100. The samples studied had chlorophyll/P700 ratios of 26 and 51 and contained no cytochromes (Brown 1976). The same procedure was used for spinach (*Spinacia oleracea*). The sample of CP I studied had a chlorophyll/P700 ratio of 38. The photochemically-inactive CP a1 complex from *Euglena gracilis* was prepared according to the method of Brown *et al.* (1974).

Sodium dodecyl sulphate treatment was used to prepare a P700–Chl *a*–protein complex from the blue-green algae *Phormidium curidum* (Malkin *et al.* 1976). This sample was kindly provided by Professor J. P. Thornber and had a chlorophyll/P700 ratio of 50, as determined chemically (P. Mathis & K. Sauer, unpublished results). All the above samples were sent and stored as lyophilizates.

Sodium dodecyl sulphate treatment was used to prepare CP I and CP II complexes from tobacco (*Nicotiana tabacum* L. var. Wisconsin). CP IIa (molecular weight 70 000) and CP IIb (molecular weight 30 000) complexes were studied. The former complex is thought to be a dimer of the latter (Rémy *et al.* 1977).

Spectroscopic methods

The experimental set-up for the resonance Raman spectroscopy as well as techniques for handling and cooling the samples have been previously described (Lutz 1975, 1977). The lyophilized samples were wetted with water or the appropriate buffer to present a smooth surface to the exciting laser beam. Sample temperatures ranged between 25 and 35 K.

RESULTS AND DISCUSSION

Chlorophyll b in light-harvesting Chl a/b–protein complexes

Resonance Raman spectra of CP II from tobacco, when excited at 457.9, 465.8 or 472.7 nm, show bands from Chl *b* and from carotenoids with no apparent contribution from Chl *a*. As in intact chloroplasts, bands from the carotenoids contribute to or obliterate most of the RR bands of Chl *b* in the 1000–1550 cm⁻¹ region. We shall thus restrict our attention here to the 1600–1700 cm⁻¹ region which contains bands from $\nu(C=O)$ modes, and to the 50–1000 cm⁻¹ region which includes bands from $(Mg-N_4)$ modes.

Stretching modes of carbonyl groups. Fig. 1 illustrates the $\nu(C=O)$ regions of Chl *b* in CP IIa from tobacco and in tobacco chloroplasts, at 30 K, on illumination at 465.8 nm. Tobacco chloroplasts have essentially the same spectrum as other green plants and algae (Lutz 1975, 1977). The main band at 1640 cm⁻¹ contains two distinct components at 1634 and 1642 cm⁻¹ arising from bonded 3-carbonyl groups. A shoulder at 1618 cm⁻¹ involves stretching of C⎓C bonds of the phorbin macrocycle—principally methine bridges. A band at 1694 cm⁻¹ is attributed to stretching of the 9-carbonyl groups, either free or weakly interacting with environment. A weak shoulder at 1704 cm⁻¹ could arise from free 9-carbonyl groups in a different environment.

The RR spectrum of CP IIa in the same conditions differs from that of

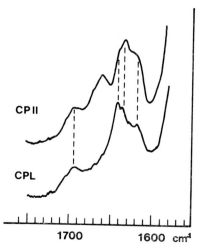

CPII

CPL

1700   1600 cm⁻¹

FIG. 1. Resonance Raman spectra of Chl *b*: the region of carbonyl stretching modes, averaged by summation, at 30 K, excitation 465.8 nm, resolution 7 cm⁻¹: CPL, in thylakoid membranes; CP II, in light-harvesting Chl *a/b*–protein (molecular weight 70 000) from *N. tabacum*.

thylakoid membranes: the major component of the latter at 1642 cm^{-1} markedly decreases with respect to both the 1628 and 1634 cm^{-1} components. The 1634 cm^{-1} band also has a lower relative intensity, with respect to the skeletal component at 1618 cm^{-1}, in spectra of CP IIa than in those of membranes. An additional band occurs at 1661 cm^{-1} in spectra of CP IIa, close to the stretching frequency of free 3-carbonyl groups of Chl *b* (Lutz 1974, 1977).

However, all the frequencies observed for the intact membranes are present in RR spectra of the protein complex. This shows that, most probably, both binding sites for the 3-carbonyl groups of Chl *b* in the membrane are proteins and that CP II includes these sites. Both types of bonds linking 3-carbonyl

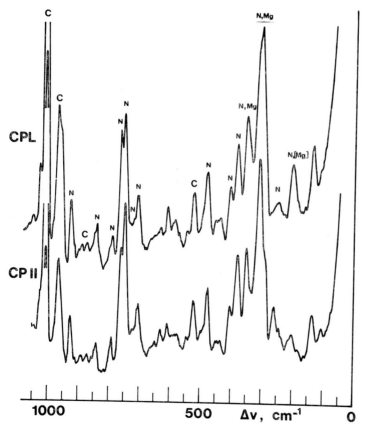

Fig.2. Resonance Raman spectra of Chl *b*, 50–1000 cm^{-1} region, 30 K, excitation 465.8 nm, resolution (at 300 cm^{-1}) 8 cm^{-1}: CPL, in thylakoid membranes; CP II, in light-harvesting Chl *a/b*–protein (molecular weight 70 000) from *N. tabacum*; C, bands with significant contribution from carotenoids; N and/or Mg, bands shifted on corresponding isotope substitution in Chl *b in vitro* and in *Chlorella vulgaris* (Lutz et al. 1975).

groups to the protein are easily broken during the extraction of CP II. The bonds corresponding to the 1642 cm⁻¹ component are more labile than those corresponding to the 1634 cm⁻¹ band. Many molecules whose 3-carbonyl groups are freed during the extraction remain attached by other bonds to the protein.

Lower frequencies (50–1000 cm⁻¹). Several differences may be observed in this region between RR spectra of Chl *b* in tobacco membranes and in CP IIa excited at 465.8 cm (Fig. 2). One of these concerns a band at 300 cm⁻¹, which involves motions of magnesium and of nitrogen (Lutz *et al.* 1975). This band is complex in the spectra of tobacco membranes. As observed for other plants and algae, it contains two components at 312 and 304 cm⁻¹ (Lutz & Breton 1973; Lutz 1975, 1977). The relative intensities of these two components do not follow the same variations with excitation wavelength and must correspond to two different categories of Chl *b* with slightly different positions of their Soret bands. In the first category (hereafter labelled P), corresponding to the 312 cm⁻¹ component are Chl *b* molecules with their magnesium atoms penta-coordinated, i.e. binding a single external ligand, and with their Soret band red-shifted with respect to that of the second category (hereafter labelled H). Chl *b* molecules falling in category H correspond to the 304 cm⁻¹ component and their magnesium atoms are most likely hexacoordinated. RR spectra of CP II from tobacco excited at 465.8 nm do not contain any 304 cm⁻¹ component, and thus CP II appears depleted in Chl *b* molecules of subspecies H.

RR spectra of tobacco membranes excited at 472.7 nm contain a preponderant contribution from subspecies P, and those obtained at 457.9 nm arise mainly from subspecies H. These spectra not only differ in their magnesium-sensitive bands but also in some skeletal bands. Their most conspicuous difference is a relative weakening in intensity, for subspecies P, of the higher frequency component of a doublet at 750–762 cm⁻¹, arising from in-plane bending modes involving motions of nitrogen atoms (Fig. 3). RR spectra of tobacco CP II exhibit this weakening when excited at 465.8 nm as well as at 472.7 nm, thus confirming the depletion of CP II in subspecies H. This category is not completely absent from CP II, however, as a 304 cm⁻¹ band occurs in spectra excited at 457.9 nm.

Differences still occur between spectra of CP II and of membranes excited at 472.7 nm, in both of which subspecies P of Chl *b* is preferentially excited. Discrepancies occur in the structure of a complex, nitrogen-sensitive band at 250 cm⁻¹ and in relative intensities of a band at 173 cm⁻¹ (shoulder for membranes) and of a band at 202 cm⁻¹ (nitrogen- and most likely magnesium-sensitive).

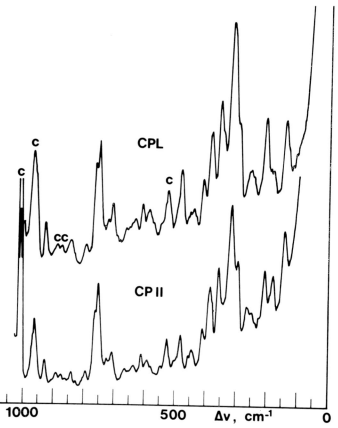

FIG. 3. Resonance Raman spectra of Chl *b* in thylakoid membranes (CPL) and in light-harvesting Chl *a/b*–protein (molecular weight 70000) (CP II) from *N. tabacum*: excitation 472.7 nm; other conditions as in Fig. 2.

These results indicate either that different types of bonding of magnesium atoms to a fifth ligand occur in molecules of subspecies P or that bonding to this ligand(s) is altered in some way in CP II with respect to membranes.

State of chlorophyll b in CP II and in membranes. It is tempting, from the results above, to attribute to the H species of Chl *b*, partly lacking in tobacco CP II, a $\nu(3\text{-C}{=}\text{O})$ frequency at 1642 cm^{-1}, a category also selectively decreased in CP II.

In this hypothesis, most of the Chl *b* molecules belonging to the (P, 1634) subspecies should retain their native network of bonding interactions in CP II.

Most molecules of the (H, 1642) subspecies should either be lost during preparation or undergo environmental rearrangement, by rupture of bonding of their formyl groups or by release of the bond holding a second external ligand on their magnesium, or by both. It may be thought that bonding on both formyl groups and magnesium is altered for these rearranged molecules: it is generally accepted that in non-forcing conditions the magnesium atom of chlorophyll assumes a pentacoordinate rather than hexacoordinate state (Katz *et al.* 1976). Hence, the release of a specific bond on the formyl group

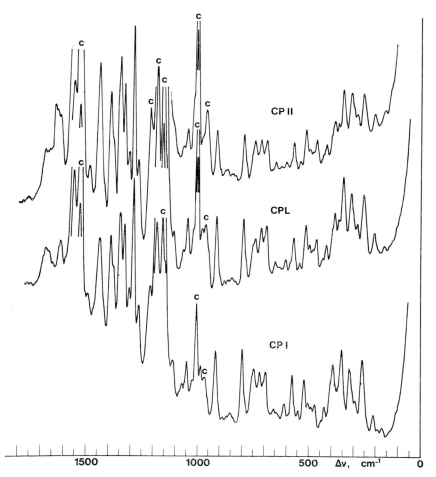

Fig. 4. Resonance Raman spectra of Chl *a*, 30 K, excitation 441.6 nm, resolution (at 1000 cm^{-1}) 8 cm^{-1}: CPL, in thylakoid membranes; CP I, in P700–Chl *a*–protein; CP II, in light-harvesting Chl *a/b*–protein (molecular weight 70 000) from *N. tabacum*; C, bands with significant contribution from carotenoids.

(now vibrating at 1661 cm^{-1}) of a molecule of the (H, 1642) subspecies could well trigger further the release of a second external ligand attached to the magnesium atom.

The multiplicity of subspecies of Chl b in the membrane may, however, be greater than two. Indeed, variations of relative intensities of ν(C=O) and Mg-sensitive bands with excitation wavelength suggested previously that molecules of the P category could have their formyl groups vibrating at 1640 cm^{-1} rather than 1630 cm^{-1} (Lutz 1977).

The present results confirm that at least two *independent* environmental subspecies of Chl b coexist in the membrane, with different bonding networks and electronic spectra. The bonding sites of Chl b most likely are proteic and are preserved in CP II. High lability and rearrangement of the (H, 1642) sub-species was also observed for preparations of CP IIb (molecular weight 30 000) of tobacco, which appeared still more depleted in 3-carbonyl groups vibrating at 1642 cm^{-1}, an observation which suggests that CP IIa (molecular weight 70 000) represents a more native fraction than CP IIb (Rémy *et al.* 1977).

Antenna chlorophyll a in chlorophyll–protein complexes

We used illumination at 441.6 nm to enhance selectively Raman scattering of Chl a present in CP I and CP II complexes from four plants and algae and in CP a1 from *Euglena gracilis*, at 30 K. Fig. 4 shows the RR spectra obtained in these conditions from CP I, CP II and thylakoid membranes from tobacco. Of particular interest in these spectra are the 1600–1750 cm^{-1} and 50–400 cm^{-1} regions which contain information about the states of ketone carbonyl groups and of the magnesium atoms, respectively.

Carbonyl stretching region (1600–1750 cm^{-1}). Bands observed in this region for CP I and CP a1 arise from Chl a only (Fig. 5). A complex, skeletal band near 1615 cm^{-1} has the same origin as in Chl b spectra. A broad cluster of bands at 1650–1700 cm^{-1} arises from ν(9-C=O) modes only (Lutz 1975, 1977). Most of the 9-C=O vibrators have frequencies below 1700 cm^{-1}, a value expected for free groups (Lutz 1974). Hence, as in chloroplasts, most Chl a molecules present in CP I assume intermolecular bonding through their ketone carbonyl groups.

We partially resolved the ν(9-C=O) cluster into five or six components for chloroplasts from spinach and from *Botrydiopsis alpina*, by cooling the samples and averaging the spectra by summation (Lutz 1975, 1977). Such resolution was not achieved for any of the protein complexes studied. However, the ν(9-C=O) clusters show a structure for all the samples. In addition to the main

TABLE 1

Resonance Raman frequencies (cm^{-1}) observed in the carbonyl-stretching region for Chl a and for Chl b in chlorophyll–protein complexes at 30 K

Chl a[a] 441.6[c]	Chl b[b] 465.8[c]	Tobacco CP II 465.8	Tobacco CP I 441.6	Spinach CP I 441.6	Phormidium luridum CP I 441.6	Anabaena cylindrica CP I 441.6	Euglena gracilis CP aI 441.6	Assignments
1616	1618	1615sh	1616m	1614m	1616m	1615m	1615m	Phorbin, $\nu(\mathrm{C_a{\cdots}C_m})$
	1630	1631m	1630vvsh					Chl b, ν(3-C=O)
	1640	1639wsh	1642vw					Chl b, ν(3-C=O)
1653			1653vvsh	1654vvsh	1654vvsh	1654vvsh	1656vvsh	Chl a, ν(9-C=O)
1661		1659w	1664w	1664wsh	1664w	1662wsh	1664vvwsh	Chl a, ν(9-C=O)
		1661w						Chl b, ν(3-C=O)
1670		1670w	1673w	1674w	1674w	~1673w	1673wsh	Chl a, ν(9-C=O)
1681		1679w	1680w	1681wsh	1680w	1681w	1683w	Chl a, ν(9-C=O)
1689		1688w	1691vvsh	1690wsh		1690vvsh		Chl a, ν(9-C=O)
	1694	1694w						Chl b, ν(9-C=O)
1702		1705vvsh	1700vvsh	~1705wsh	~1705vvsh	1702wsh	1703vvsh	Chl a, ν(9-C=O)
		1704vvsh						Chl b, ν(3-C=O)

[a] From seven species. [b] From four species (Lutz 1975, 1977 and this work). [c] 441.6, 465.8: excitation wavelength. $C_a{\cdots}C_m$ are methine bridges.
m: medium relative intensity, w: weak, v: very, sh: shoulder.

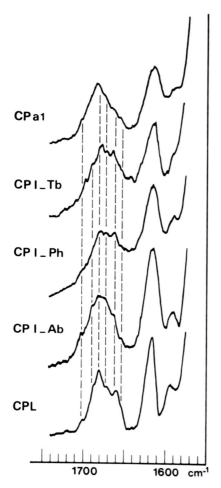

FIG. 5. Resonance Raman spectra of Chl a: the region of carbonyl stretching modes, averaged by summation, 30 K, excitation 441.6 nm, resolution 5 cm^{-1}: CP a1, in CP a1–Chl a–protein from *E. gracilis*; CP I-Tb, in P700–Chl a–protein from *N. tabacum*; CP I-Ph, in P700–Chl a–protein from *P. luridum*; CP I–Ab, in P700–Chl a–protein from *A. cylindrica*; CPL, in whole cells of *Botrydiopsis alpina*.

component at 1680 cm^{-1}, a second band is present at 1662–1664 cm^{-1} for tobacco and for *Phormidium*. Weaker components are observed for all the samples around 1654, 1662, 1672, 1689 and 1704 cm^{-1} (Table 1). The observed frequencies match within 4 cm^{-1} for all samples and, more importantly, they match with the same precision the six average frequencies observed for intact membranes.

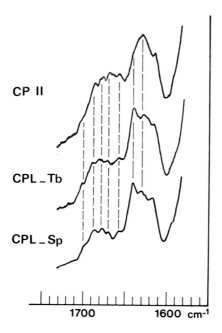

FIG. 6. Resonance Raman spectra of Chl a: the region of carbonyl stretching modes, averaged by summation, 30 K, excitation 441.6 nm, resolution 5 cm^{-1}: CP II, in light-harvesting Chl a/b-protein (molecular weight 30 000); CPL-Tb, in thylakoid membranes from $N.$ $tabacum$; CPL-Sp, in whole chloroplasts of spinach.

The RR spectra of CP II from tobacco obtained at 441.6 mm show an additional contribution from Chl b (Fig. 6). A broad band at 1632 cm^{-1}, together with a weak shoulder near 1640 cm^{-1} in some samples, arise from stretching of formyl groups of these molecules. The 1650–1700 cm^{-1} region primarily arises from 9-C=O groups of Chl a and exhibits some structure, with components at 1659, 1670, 1679, 1688 and 1703 cm^{-1}. These frequencies match closely those observed for CP I and for intact membranes. The 5 cm^{-1} difference found between the 1664 cm^{-1} component of CP I and the 1659 cm^{-1} component of CP II may be significant but may arise from contribution in the latter of the stretching mode of free 3-C=O groups of Chl b.

Resolution of the v(9-C=O) cluster into individual components for certain chloroplasts has shown that antenna Chl a molecules are distributed among not many more than five discrete categories differing by the strengths of bonding interactions on their 9-C=O groups. The closely-matching frequencies of the components strongly suggested that the bonding sites for these groupings are universal. The present results further suggest that all these sites are still present in both CP I and CP II, and that no (or only minor) rearrangement of Chl *a* occurs at this level during extraction of the complexes.

400–1600 cm^{-1} region. Most bands of this region arise from complex modes involving stretching and bending motions of several atoms of the phorbin ring (Lutz 1974; Plus & Lutz 1975; Lutz *et al*. 1975). The formation of (Chl *a*)$_n$ or of (Chl *a*,nH$_2$O)$_m$ oligomers *in vitro* brings characteristic spectral variations to the RR spectra of monomeric Chl *a* in this spectral region. Comparison of RR spectra of Chl *a* in intact chloroplasts with such *in vitro* spectra demonstrated that no sizable pool of antenna Chl *a* might consist of oligomers (Lutz 1977). RR spectra of CP I complexes are still more favourable than those of intact membranes for this comparison, in as much as they contain no, or much lower, contributions from Chl *b* and from carotenoids (Fig. 4). A particularly conspicuous characteristic of the spectra of oligomers is a band at 1228 cm^{-1} instead of 1215 cm^{-1} for the monomers. Samples of CP I from spinach and from *Anabaena* with low carotenoid content and with a consequent decrease in ν_2 satellite usually observed at 1219 cm^{-1} showed the 1215 cm^{-1} band due to monomers but none of the 1228 cm^{-1} band of oligomers. Similarly, positions of bands at 990, 1268 and 1294 cm^{-1} in these spectra were characteristic of monomeric Chl *a*. Thus, the ketone carbonyl groups of antenna Chl *a* in CP I as well as in CP II do not interact with magnesium atoms of other chlorophylls.

50–400 cm^{-1} region. RR spectra of antenna Chl *a* in any of the protein complexes are barely distinguishable from those of Chl *a* in intact chloroplasts in this particular region, which yields information on the state of the magnesium atoms (Fig. 7). As for chloroplasts (Lutz 1977), interaction-sensitive bands of this region may be separated into two categories. Bands of the first category, with frequencies at 390, 290 and 262 cm^{-1}, appear more particularly sensitive to non-directional interactions. For all complexes they occur at frequencies identical to both those of chloroplasts and of monomeric Chl *a in vitro*. In CP I, particularly from tobacco, the 390 cm^{-1} band is enhanced in relative intensity and may indicate a change in environment of Chl *a* molecules. Bands of the second category, at 375, 308–318 and 212 cm^{-1} appeared more specifically sensitive to the state of the magnesium atom. The band at 308–318 cm^{-1} arises from a mode involving motions of nitrogen and the magnesium atoms, and is sensitive to the number of ligands bound to the magnesium (Lutz 1974, 1975; Lutz *et al*. 1975). The increased half-width and the presence of a shoulder at 308 cm^{-1} in spectra of all the complexes with respect to monomeric Chl *a* show that most Chl *a* molecules bind a single external ligand through their magnesium

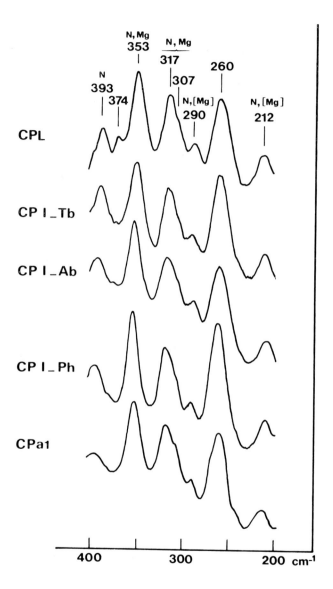

FIG. 7. Resonance Raman spectra of Chl *a*, 200–400 cm⁻¹, averaged by summation, 30 K, excitation 441.6 nm, resolution 8 cm⁻¹: CPL, in thylakoid membranes of *N. tabacum*; CP I-Tb, in P700–Chl *a*-protein from the same; CP I-Ab, in P700–Chl *a*-protein from *A. cylindrica*; CP I-Ph, in P700–Chl a-protein from *P. luridum*; CP a1, in CP a1 Chl *a*-protein from *E. gracilis*; N and/or Mg, bands shifted on corresponding isotope substitution in Chl *a in vitro* and in *Chlorella vulgaris* (cf. Lutz *et al.* 1975).

atoms. This, as well as the frequency and structure of the band at 212 cm^{-1}, confirms that, on average, the state of magnesium is the same as in intact membranes. The interaction-sensitive band at 375 cm^{-1} for Chl a in CP II as well as in chloroplasts appears to bear the same information for CP II (Fig. 4). This band is, however, not observed in RR spectra of Chl a in CP I and CP a1. In as much as it is present in RR spectra of chloroplasts of the Chl b-free mutant barley, which lacks CP II (Thornber & Highkin 1974), its absence in CP I and CP a1 indicates a change in environment of Chl a in these complexes. It now appears likely that this change does not involve the magnesium atoms directly but is due to a change in non-directional interactions, as suggested also by a specific increase in relative intensity of interaction-sensitive bands at 390 cm^{-1} and at 747–755 cm^{-1} in spectra of CP I (Fig. 4).

State of antenna chlorophyll a in protein complexes. A close similarity is thus observed between RR spectra of antenna Chl a bound to the complexes we have examined and of antenna Chl a in chloroplasts. This has several consequences.

These results show that Chl a extracted from chloroplasts as part of these complexes undergoes, at most, minor rearrangement from its native states.

None of the observable, environmental subspecies of antenna Chl a involves Chl–Chl oligomers, either protein-bound or not (Vernon & Klein 1975; Katz *et al.* 1976, 1977). Each component of the ν(C=O) cluster may be considered as resulting from the hydrogen-bonding of 9-C=O groups with a given site of the protein, either a peptidic N-H group or a suitable group on side-chains of amino acids. Similarly, the bulk of antenna Chl a molecules should have their magnesium atoms each bound to one proteic site, either directly or with interposition of water (Lutz 1977).

Only a part of antenna Chl a is presently extractable as protein complexes from chloroplasts (Brown *et al.* 1975; Thornber 1975; Thornber *et al.* 1977). Yet in none of these complexes does the ν(C=O) cluster appear either less complex or more resolved than in membranes. Hence, most if not all the binding sites for ketone groups occurring in the membrane are present in the protein complexes. The bulk of antenna Chl a should thus be complexed to protein in the membrane. However, a possible minor, monomeric fraction of Chl a weakly bound to another type of molecule (e.g. lipid) could not easily be distinguished by RR spectroscopy. CP a1 from *Euglena*, with still the same bonding sites for ketone carbonyl groups of Chl a as CP I and CP II, may represent a third class of polypeptides complexing antenna Chl a *in vivo* (Thornber 1975; Thornber *et al.* 1977). Different proportions of molecules should be bound to each type of site depending on the complex and on the species considered, but these sites should be universal.

The Q_y electronic band of Chl a in protein complexes at low temperature presents the same complexity as in intact chloroplasts and may be decomposed into the same universal components (Brown *et al.* 1975; Rémy *et al.* 1977). RR spectra show that both types of systems contain identical interaction subspecies, a necessary but not sufficient condition for ensuring that identical environmental subspecies, and hence identical absorption forms, occur in all plants and algae and their chlorophyll–proteins. Finally, the fact that we observed no simple correlation between electronic absorption forms and 9-C=O interaction subspecies of Chl a suggests that the optical properties of antenna Chl a are not simply site-determined.

ACKNOWLEDGEMENTS

We thank Professor J. P. Thornber for a sample of CP I from *P. luridum*.

References

ANDERSON, J. M. (1975) The molecular organization of chloroplast thylakoids. *Biochim. Biophys. Acta 416*, 191–235

BROWN, J. S. (1976) P700–Chlorophyll *a*–protein complexes. *Carnegie Inst. Wash. Year Book 75*, 460–465

BROWN, J. S., ALBERTE, R. S., THORNBER, J. P. & FRENCH, C. S. (1974) Comparisons of spectral forms of chlorophyll in protein complexes isolated from diverse groups of plants. *Carnegie Inst. Wash. Year Book 73*, 694–706

BROWN, J. S., ALBERTE, R. S. & THORNBER, J. P. (1975) Comparative studies on the occurrence and spectral composition of chlorophyll-protein complexes in a wide variety of plant material, in *Proceedings of the Third International Congress on Photosynthesis*, vol. 3 (Avron, M., ed.), pp. 1951–1962, Elsevier, Amsterdam

KATZ, J. J. (1973) Chlorophyll, in *Inorganic Biochemistry* (Eichhorn, G. L., ed.), pp. 1022–1066, Elsevier, Amsterdam

KATZ, J. J., OETTMEIER, W. & NORRIS, J. R. (1976) Organization of antenna and photoreaction centre chlorophylls on the molecular level. *Philos. Trans. R. Soc. Lond. B 273*, 227–253

KATZ, J. J., NORRIS, J. R. & SHIPMAN, L. L. (1977) Models for reaction-center and antenna chlorophyll. *Brookhaven Symp. Biol. 28*, 16–55

LUTZ, M. (1972) Spectroscopie Raman de résonance de pigments végétaux en solution et inclus dans les lamelles chloroplastiques. *C. R. Acad. Sci. Paris B 275*, 497–500

LUTZ, M. (1974) Resonance Raman spectra of chlorophyll in solution. *J. Raman Spectrosc. 2*, 497–516

LUTZ, M. (1975) Resonance Raman spectroscopy of the chlorophylls in photosynthetic structures at low temperatures, in *Lasers in Physical Chemistry and Biophysics* (Joussot-Dubien, J., ed.), pp. 451–463, Elsevier, Amsterdam

LUTZ, M. (1977) Antenna chlorophyll in photosynthetic membranes: a study by resonance Raman spectroscopy. *Biochim. Biophys. Acta 460*, 408–430

LUTZ, M. & BRETON, J. (1973) Chlorophyll associations in the chloroplast: resonance Raman spectroscopy. *Biochem. Biophys. Res. Commun. 53*, 413–418

LUTZ, M., KLÉO, J., GILET, R., HENRY, M., PLUS, R. & LEICKNAM, J. P. (1975) Vibrational spectra of chlorophylls a and b labelled with ^{26}Mg and ^{15}N, in *Proceedings of the Second International Conference on Stable Isotopes* (Klein, E. R. & Klein P. D., eds.), pp. 462–469, US Department of Commerce, Springfield, Virginia

MALKIN, R., BEARDEN, A. J., HUNTER, F. A., ALBERTE, R. S. & THORNBER, J. P. (1976) Properties of the low temperature Photosystem I primary reaction in the P700–chlorophyll a–protein. *Biochim. Biophys. Acta 430*, 389–394

PAILLOTIN, G. & BRETON, J. (1977) Orientation of chlorophyll within chloroplasts as shown by optical and electrochromic properties of the photosynthetic membrane. *Biophys. J. 18*, 63–79

PLUS, R. & LUTZ, M. (1975) Resonance Raman scattering of mesoporphyrin IX dimethylester in solution. *Spectrosc. Lett. 8*, 119–139

RÉMY, R., HOARAU, J. & LECLERC, J. C. (1977) Electrophoretic and spectrophotometric studies of chlorophyll-protein complexes from tobacco chloroplasts. Isolation of a light-harvesting pigment protein complex with a molecular weight of 70000. *Photochem. Photobiol. 26*, 151–158

THORNBER, J. P. (1975) Chlorophyll-proteins: light harvesting and reaction center components of plants. *Annu. Rev. Plant. Physiol. 26*, 127–158

THORNBER, J. P. & HIGHKIN, H. R. (1974) Composition of the photosynthetic apparatus of normal barley leaves and a mutant lacking chlorophyll b. *Eur. J. Biochem. 41*, 109–116

THORNBER, J. P., ALBERTE, R. S., HUNTER, F. A., SHIOZAWA, J. A. & KAN, K.-S. (1977) The organization of chlorophyll in the plant photosynthetic unit. *Brookhaven Symp. Biol. 28*, 132–148

VERNON, L. P. & KLEIN, S. M. (1975) Nature of plant chlorophylls *in vivo* and their associated proteins. *Ann. N.Y. Acad. Sci. 244*, 281–296

Discussion

Clayton: Is the region between 800 and 830 nm usable with regard to interference from fluorescence? There you could examine spectra of oxidized P700 and P690.

Lutz: Technical problems of signal detection hinder the application of Raman scattering in this spectral region. Moreover, the fluorescence of chlorophyll would be a major problem as the quantum yield of Raman scattering should be 10^{-8}–10^{-10} in these conditions.

Porter: Doesn't your conclusion that the bonding in the chlorophyll is unchanged by extraction of the complexes disagree with Dr Anderson's model? Perhaps not; all the part of the molecule that you observe by resonance Raman spectroscopy may be in the protein.

Lutz: We concluded that those intermolecular interactions of antenna chlorophyll which involve the phorbin rings and groups conjugated to them remain unaltered for most molecules recovered as part of CP I and CP II. Extracting a chlorophyll–protein complex agreeing with Dr Anderson's (1975) model should only alter, as far as chlorophyll is concerned, the interactions on the phytyl chains. Resonance Raman spectra of chlorophyll do not contain information on the phytyl chains and hence should be insensitive to such alterations anyway. These spectra cannot discriminate between Dr Anderson's model, which locates the phytyl chains on the outer surface of the protein, and other models, e.g. that of Fenna & Matthews (1975), in which they are buried inside the protein.

Porter: So, the chlorophyll that is extracted with the protein was in the protein in the first place.

Lutz: Yes, in so far as it gives the same Raman spectra as in intact chloroplasts.

Katz: If I understood you correctly, some features of your spectra can be better identified with the chlorophyll–water adduct.

Lutz: Only the magnesium-sensitive bands of a certain proportion of antenna chlorophyll *a* and *b* molecules have similar intensities and frequencies to those of homologous bands in Raman spectra of the chlorophyll–water adducts.

Katz: Water could not be inserted into a chlorophyll dimer or oligomer without altering the visible spectrum. Disaggregation would lead to a short-wavelength, fluorescent, monomeric chlorophyll species. It would help in the assignment to have visible spectra corresponding to the Raman spectra.

Porter: What do your results tell us about the coordination of the magnesium atom?

Lutz: The magnesium atoms of most of the chlorophyll *a* are pentacoordinated, both *in vivo* and in protein complexes. The magnesium atoms of a certain fraction of chlorophyll *b* are hexacoordinated *in vivo* (the rest are pentacoordinated). However, most of the hexacoordinated fraction becomes pentacoordinated in CP II. The fifth and sixth ligands are, most likely, donor groups of the peptide chains, although I imagine that water is interposed in some cases.

Porter: Can you eliminate the 5-carbonyl group of another chlorophyll molecule as the ligand?

Lutz: Yes.

Katz: I question that, because the extinction coefficient of the carbonyl vibration may not be the same in the spectrum of a dimer as in that of a monomer. Therefore, in the resonance-enhanced Raman spectrum one cannot necessarily interpret a small absorption peak as indicative of a low concentration.

Lutz: The relevant parameter is the extinction coefficient of the electronic band at the wavelength of excitation, here 441.6 nm. This wavelength falls on the red edge of the Soret band of antenna chlorophyll *a*, and resonance should thus be more favourable for dimers than for monomers. (This has been demonstrated experimentally on Chl *b*: see Fig. 9 in Lutz 1974.) In the terminology of donor and acceptor molecules, the donor molecules, whose Soret band should be slightly red-shifted with respect to those of the acceptor and monomer molecules, would give predominant contributions to the Raman spectra, at equal concentration, and would reveal the presence of dimers.

Katz: The principal differences in the visible spectra resulting from coordination interactions are evident in the red band and not in the Soret band.

Lutz: Yes, but from the behaviour of the Soret band on conversion of monomers into oligomers we can infer that the red-shift in the Q_y band for the donor molecules will be accompanied by a small red-shift in the Soret band (cf. Sauer *et al.* 1966; Shipman *et al.* 1976).

Katz: However, if monomeric chlorophyll *a* is present in your preparations, it should show a short-wavelength absorption and be fluorescent.

I know that technical problems make it difficult to excite in the red band but, nevertheless, that is where it ought to be excited, and in principle there would be less difficulty in interpreting the spectra. However, I realize that it is a considerable technical feat to be able to collect even these spectra. Because many of these spectra have maxima assignable to both chlorophylls *a* and *b* a large absorption at a frequency slightly higher than 1650 cm^{-1} is assigned to a carbonyl group in the chlorophyll *b*. Can one be certain that the 1660 cm^{-1} absorption maximum is not composite and contains a contribution from chlorophyll *a*? It should be worth-while doing this experiment with an organism that does not contain chlorophyll *b*.

Lutz: This has been done (see p. 114 and Lutz 1975, 1977). Bands are present at 1660 cm^{-1} in spectra of chlorophyll *a* in organisms lacking chlorophyll *b*, e.g. *Botrydiopsis alpina*. Moreover, excitation of intact chloroplasts containing chlorophyll *b* between 458 to 478 nm yields Raman contributions from chlorophyll *b* and the carotenoids only. These spectra show that all the formyl carbonyl groups are interacting and vibrate at 1640 and 1630 cm^{-1}. The 1660 cm^{-1} band from free formyl carbonyl groups of chlorophyll *b* was observed for CP II only. In all these cases, the attributions of components around 1660 cm^{-1} either to chlorophyll *a* or to chlorophyll *b* are unambiguous.

Cogdell: What is the carotenoid and what is its isomeric form?

Lutz: Most of the carotenoids in chloroplasts as well as in bacterial chromatophores assume all-*trans* conformations but carotenoids bound to bacterial reaction centres assume *cis* conformations. It would be interesting to see whether there is something comparable for P700.

Porter: What enhancement does resonance Raman spectroscopy give?

Lutz: *In vitro* measurements on chlorophyll *b* showed a maximum enhancement of 5×10^4, so we are certainly rid of the major, non-resonating, potential contributions which should be from water and the proteins.

Robinson: You see the bands that are associated with the pigment and not just everything. You could separate the fluorescence from the resonance Raman in the red band by going to extremely high spectral resolution (say, about 0.1 wavenumbers) because the number of photons per unit wavenumber will be greater for the resonance Raman than for the fluorescence because of the usual width of the latter. There would be a better chance of collecting the Raman photons in the presence of a continuous background light at lower resolution. No Raman photon would be lost.

Lutz: We are working with a spectral slit width of 6 cm^{-1}. The intrinsic widths of our resonance Raman bands are not lower than 12 cm^{-1}. Decreasing the spectral band-width further would decrease the numbers of Raman and fluorescence photons by the same amount.

Porter: Are these spectra fully resolved?

Lutz: I think so. A potentially-efficient technique for rejecting fluorescence while detecting Raman scattering is to make use of the shorter lifetime of the Raman process, about 10^{-14} s, by using a laser emitting pulses much shorter than the fluorescence lifetime and by adequately gating the detection (see, e.g., Van Duyne *et al.* 1974). Using picosecond lasers one should thus reject much of the fluorescence of isolated and of antenna chlorophyll but not much of the very short-lived fluorescence from reaction centres. However, the time constant for thermal decay in the small volume being irradiated is about 10^{-3} s. If the necessary amount of energy is dispensed in this volume in pulses shorter than 10^{-3} s, a sharp increase in temperature will result, which is not the case working with pulses longer than 10^{-3} s (Salet *et al.* 1970).

Seely: Excitation of the Soret band presumably excites at least two electronic excited states. Would this tend to duplicate the Raman lines belonging to groups that are differently coupled to the excited states?

Lutz: No, those vibrational modes we observe clearly concern molecules in their ground electronic states. Simultaneous resonances with more than one electronic transition may merely result in enhancing more normal modes of the molecule in its ground state, depending on the symmetries of the transitions.

Clayton: Has anyone tried to resolve fluorescence spectra vibrationally in the same way as for monomeric chlorophylls at low temperature?

Lutz: Litvin *et al.* (1969) and Avarmaa & Rebane (1975) published vibrationally resolved absorption and fluorescence spectra of chlorophyll diluted in alkane matrixes, at low temperature (Shpol'skii technique). These spectra are difficult to interpret because of a multiplicity of lines for a given vibrational mode due to inequivalence of host sites in the matrix. More promising is site-selection fluorescence spectroscopy with narrow line excitation (Avarmaa & Rebane 1975; Fünfschilling & Williams 1975). Both these methods, however, put stringent conditions on the environment of chlorophyll and may not be applicable *in vivo*.

References

ANDERSON, J. M. (1975) The molecular organization of chloroplast thylakoids. *Biochim. Biophys. Acta 416*, 191–235

AVARMAA, R. & REBANE, K. (1975) Fine structured spectra of chlorophyll molecules in solid solutions. *Studia Biophys. 48*, 209–218

FENNA, R. E. & MATTHEWS, B. W. (1975) Chlorophyll arrangement in bacteriochlorophyll protein from *Chlorobium limicola*. *Nature (Lond.) 258*, 573–577

FÜNFSCHILLING, J. & WILLIAMS, D. F. (1975) Vibrationally resolved low temperature fluorescence spectra of porphin and chlorophylls *a* and *b* in organic glass matrices. *Photochem. Photobiol. 22*, 151–152

LITVIN, F. F., PERSONOV, R. I. & KOROTAEV, O. N. (1969) Fine structure of the electronic spectra of chlorophyll in crystalline solution at 4°K. *Dokl. Akad. Nauk. S.S.S.R. 188*, 1169–1171 (Engl. transl. 118–120)

LUTZ, M. (1974) Resonance Raman spectra of chlorophyll in solution. *J. Raman Spectrosc. 2*, 497–516

LUTZ, M. (1975) Resonance Raman spectroscopy of the chlorophylls in photosynthetic structures at low temperatures, in *Lasers in Physical Chemistry and Biophysics* (Joussot-Dubien, J., ed.), pp. 451–463, Elsevier, Amsterdam

LUTZ, M. (1977) Antenna chlorophyll in photosynthetic membranes: a study by resonance Raman spectroscopy. *Biochim. Biophys. Acta 460*, 408–430

SALET, C., LUTZ, M. & BARNES, F. S. (1970) Paramètres physiques caractérisant le domage thermique sélectif de mitochondries en microirradiation par laser. *Photochem. Photobiol. 11*, 193–205

SAUER, K., LINDSAY SMITH, J. R. & SCHULTZ, A. J. (1966) The dimerization of chlorophyll *a*, chlorophyll *b* and bacteriochlorophyll in solution. *J. Am. Chem. Soc. 88*, 2681–2688

SHIPMAN, L. L., COTTON, T. M., NORRIS, J. R. & KATZ, J. J. (1976) An analysis of the visible absorption spectrum of chlorophyll *a* monomer, dimer and oligomers in solution. *J. Am. Chem. Soc. 98*, 8222–8230

VAN DUYNE, R. P., JEANMAIRE, D. L. & SHRIVER, D. F. (1974) Mode-locked laser Raman spectroscopy—a new technique for the rejection of interfering background luminescence signals. *Anal. Chem 46*, 213–222

The field of possible structures for the chlorophyll *a* dimer in photosystem I of green plants delineated by polarized photochemistry

WOLFGANG JUNGE and HELMUT SCHAFFERNICHT

Max-Volmer-Institut für Physikalische Chemie und Molekularbiologie, Technische Universität Berlin, West Germany

Abstract Photoselection experiments with immobilized photosystem I particles have been done to determine the mutual orientation of pigments in the reaction centre. When these particles are excited and interrogated with linearly polarized light, the flash-induced transient absorption changes (mainly from the chlorophyll *a* dimer) reveal linear dichroism, which yields information on the mutual orientation between the excited and the interrogated transition moments. The interpretation of the data, however, is ambiguous, (1) for reasons of principles inherent in the photoselection technique when applied to complex systems and (2) because of incomplete knowledge about the relative contribution of *x*- and of *y*-polarized transitions of chlorophyll *a* to absorption or to absorption changes at a given wavelength. We find it impossible to attribute any particular structure to the photooxidizable dimer based on photoselection data alone. Instead we present a field of possible structures, imposing constraints on proposed models for the dimer structure.

Photochemical reaction centres from green plants and bacteria are distinguished by high efficiency of photochemical energy conversion and by stability towards excessive light (for general reviews see Govindjee 1975 and Olson & Hind 1977).

In contrast to the time domain where the events have been resolved down to picoseconds (e.g. Rockley *et al.* 1975) the spatial arrangement of pigments within reaction centres is still obscure. The major components of reaction centres are a chlorophyll dimer, which is the primary photooxidation product (Norris *et al.* 1971, 1974), the primary electron acceptors, and closely associated pigments with antennae function. Attempts to isolate biochemically photosystem I from green plants have resulted in polypeptides with some 40 chlorophylls *a* per dimer, a few carotene molecules but no pheophytins (Boardman & Anderson 1964; Vernon & Shaw 1971; Bengis & Nelson 1975; Ke *et al.* 1975; Malkin 1975; Thornber *et al.* 1977). Only one report described as few as between five and nine chlorophylls (Ikegami & Katoh 1975). In contrast to a light-harvesting chlorophyll–protein complex from a bacterium which was crystallized and analysed by X-ray diffraction (Fenna & Matthews 1975), no photochemically-active reaction centre from bacteria and green plants has been crystallized until now and so structural studies have had to rely on more indirect techniques. On the whole three different lines have been followed. (1) Chlorophyll aggregates have been modelled *in vitro* in order to simulate the chemical and the spectroscopic properties of the *in vivo* dimer (e.g. Katz & Norris 1973; Fong 1974; Shipman *et al.* 1976; Boxer & Closs 1976). (2) Linear dichroism of either oriented photosynthetic lamellae (e.g. Breton & Roux 1971; Geacintov *et al.* 1974; Breton 1976) or isotropic suspensions (by photoselection) (Junge & Eckhof 1973, 1974; Mar & Gingras 1976; Shuvalov *et al.* 1976; Vermeglio & Clayton 1977; Junge *et al.* 1977; Junge & Schaffernicht 1978) has been studied with the aim of discovering the mutual orientation of pigments. (3) The zero-field splitting parameters of reaction-centre triplets have been studied (Clarke *et al.* 1976, 1978) also with the purpose of investigating the orientation of pigments.

It is evident that neither information on the mutual angles between chlorophylls nor information on possible chlorophyll aggregates *in vitro* alone will lead to unequivocal pictures of the internal structure of the dimer, say, in photosystem I from green plants. However, fields of possible structures and of forbidden structures may be constructed with an iterative approximation to the true structure. In this paper we shall delineate the field of possible configurations of the photosystem I dimer which are compatible with data obtained by photoselection studies on immobilized photosystem I particles.

EXPERIMENTAL RESULTS

We have previously reported results of photoselection studies with immobilized photosystem I particles (Junge *et al.* 1977; Junge & Schaffernicht 1978). These particles, containing down to 40 chlorophylls per reaction centre, were prepared according to Bengis & Nelson (1975) and were investigated in a rapid kinetic spectrophotometer, as reviewed elsewhere (Junge 1975). The geometry of the light beams is illustrated in the upper part of Fig. 1. An isotropic ensemble of photosystem I particles, immobilized on DEAE-Sephadex granules, was excited with a linearly polarized flash of light at non-saturating energy. The absorption changes, mainly due to chlorophyll a_I resulting from the photoselected anisotropic subset, were monitored with light polarized either parallel or perpendicular to the exciting light. These absorption changes (see, e.g. the lower part of Fig. 1) reveal different magnitudes (dichroism) depending on the polarization of the measuring beam. The extent of the linear dichroism

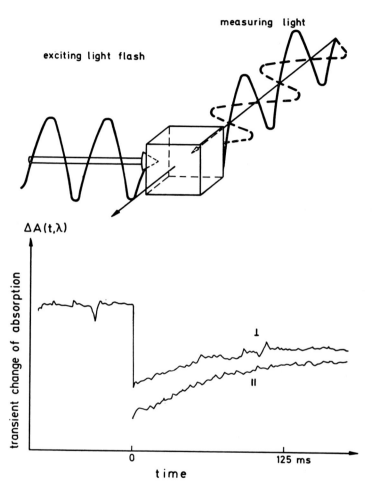

FIG. 1. Geometrical arrangement of the light paths in a photoselection experiment (above) and the resulting dichroism of the transient absorption changes of chlorophyll a_1 (below). The traces in the lower part were obtained with photosystem I particles prepared after Bengis & Nelson (1975) and immobilized on DEAE-Sephadex. The particles were excited at 487 nm with a short flash of light (at $t = 0$) and absorption changes were measured at 701 nm for two different polarizations of the measuring light as indicated (for details, see Junge *et al.* 1977).

gives information about the mutual orientation of the excited and the interrogated oscillators (for the theoretical basis, see Appendix).

Fig. 2 and Table 1 summarize the experimental results.

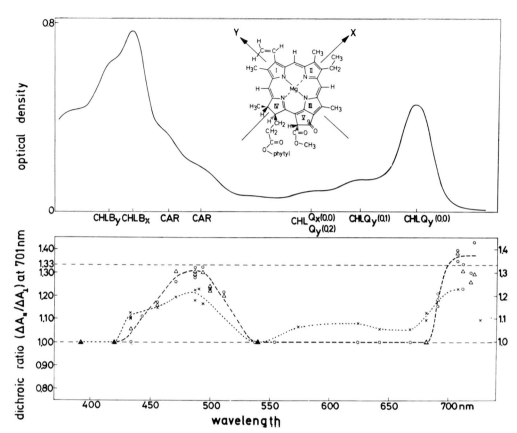

FIG. 2. Top: The conventional assignment of the axes of chlorophyll a; the curve shows the absorption spectrum of photosystem I particles. Bottom: The excitation spectrum of the photoinduced linear dichroism at 701 nm for (×) photosystem I particles. (○) depleted particles and (△) depleted particles fixed with glutaraldehyde. For experimental details, see Junge *et al.* (1977).

TABLE 1

The dichroic ratio in photoselection experiments with immobilized photosystem I particles as a function of the wavelength of excitation and observation. Flash light at 720 nm preferentially excites the dimer (plus possibly antennae peaking at 700 nm as the dimer) but flash light at 490 nm preferentially excites carotenoids (for details, see Junge & Schaffernicht 1978)

Excitation wavelength (nm)	Dichroic ratio			
	Wavelength of measuring beam (nm)			
	430	656	680	700
720	1.21 ± 0.02	1.00 ± 0.07	1.00 ± 0.07	1.28 ± 0.02
490	1.19 ± 0.07	1.18 ± 0.05	1.90 ± 0.06	1.31 ± 0.01

Fig. 2 shows the structure of chlorophyll *a* (with the conventional assignment of the *x*- and the *y*-axes) and the absorption spectrum of photosystem I particles with the attribution of bands to chlorophyll *a* and carotene. The lower part of Fig. 2 illustrates the excitation spectrum of the photoinduced dichroism of the absorption changes of chlorophyll a_I at 701 nm.

The major feature of this excitation spectrum is that excitation at the red end of the absorption spectrum produces a dichroic ratio of about 4/3 which does not increase at the tail of the spectrum (say, at excitation at 724 nm). As this figure of 4/3 characterizes a circularly-degenerate system we concluded that the oscillators contributing to the absorption at 724 nm are arranged in one of the following forms (Junge *et al.* 1977):

(1) If the dimer (or 'special pair' which is characterized by the delocalization of the unpaired electron [Norris *et al.* 1971]) is the major species contributing to absorption at 724 nm, the *y*-axes of its two chlorophyll *a* molecules have to be mutually perpendicular.

(2) If, in addition to the dimer, there are antennae peaking at 700 nm as there are in the dimer, then the *y*-axes of the dimer may be parallel but with the *y*-axes of the special antennae perpendicular to the former. (It is unlikely that there are more than two such special antennae chlorophyll *a*!)

The argument is independent of the state of transition dipole coupling among the Q_y transitions of the dimer (not shown here). In conclusion, the results from Fig. 2 define a plane of circular degeneracy within the reaction centre of photosystem I particles (referred to as 'the plane' in what follows).

Table 1 shows the dichroic ratio at the four major peaks of the absorption changes of chlorophyll a_I (which comprises the dimer plus the antennae it influences spectrally) on excitation into 'the plane' at 720 nm and into carotenoid bands at 490 nm. The exclusive role of excitation into carotenoids is already apparent from Fig. 2. We learn that excitation into the plane does not produce dichroic ratios greater than 4/3 at any wavelength, but excitation into carotenoid bands produces a dichroic ratio of 1.9 at 680 nm. This observation shows that the carotenoid molecules are parallel to each other rather than perpendicular (see Junge & Schaffernicht 1978). Moreover, Table 1 reveals a dichroic ratio near 4/3 for excitation into the 'linear' carotenoids and observation of the absorption changes of *y*-bands of the dimer at 700 nm, which is at least compatible with a perpendicular configuration of the two *y*-axes [see (1) above]. As there is no real proof for a perpendicular orientation of the *y*-axes of the dimer we shall only retain the existence of the plane and concentrate on the non-forbidden field of dimer configurations which are compatible with the dichroism observed for the predominantly *x*-polarized transitions of the dimer (at 430 nm) on excitation into the plane (at 720 nm). In so doing we shall treat

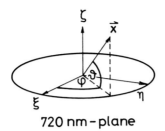

720 nm – plane

Fig. 3. A coordinate system based on the plane of circular degeneracy on excitation of photo-system I particles at wavelengths greater than 720 nm.

the inclination of the two y-axes of the dimer as a variable parameter.

THE INCLINATION OF THE x-AXES OF THE TWO CHLOROPHYLL a MOLECULES OF THE DIMER TO THE PLANE

We shall now discuss possible dimer configurations in terms of the polar coordinate system illustrated in Fig. 3. Any vector is characterized by its inclination to the plane (the angle ϑ is the complement to the polar angle) and by the angle between its projection to the plane and the y-axis of chlorophyll no. 1 of the dimer (ϕ, the azimuth angle).

The absorption changes of chlorophyll a_I around 430 nm are mixed with contributions from the dimer and from its radical cation. Moreover, there are transitions with x-polarization and with y-polarization, and exciton interaction has to be taken into account. We shall consider a total of eight oscillators, distinguished by their origin from the first or the second chlorophyll a in the dimer (subscript 1 or 2), by their polarization (subscript x or y) and by their origin from the neutral dimer (no superscript) and its radical cation (superscript $+$), respectively:

$$B_{x1}, \ B_{x2}, \ B_{x1}^{+}, \ B_{x2}^{+}, \ B_{y1}, \ B_{y2}, \ B_{y1}^{+}, \ B_{y2}^{+}.$$

For the moment we have neglected the possibility that a component resulting from the 'neutral' half of the radical cation exists. (This will be corrected for later.)

The complexity is somewhat relieved since it is probable that the photo-oxidation does not alter the structure so that all y-transitions are within the plane (and hence produce the same dichroic ratio). On the other hand the complexity is increased by the possibility that the x-transition of the neutral dimer may be under exciton interaction, giving rise to new transitions B_{XI}, B_{XII}, related to B_{x1} and B_{x2}, as shown in Fig. 4.

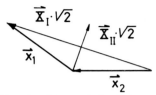

FIG. 4. Geometrical relationship between two monomer transition moments ($\vec{x_1}$ and $\vec{x_2}$), which are excitonically coupled, and two transition moments of the dimer ($\vec{x_I}$ and $\vec{x_{II}}$) resulting from this interaction.

At a given wavelength within the 430 nm band of chlorophyll a_1 the polarization anisotropy in photoselection experiments is then the weighted sum of the polarization anisotropies of the respective sub-bands (see equation (A4) in the Appendix.) If, for excitation into the plane, we do not discriminate between the *y*-polarized transitions, the observed polarization anisotropy r_{obs} is given by (1).

$$r_{obs} = (f_y + f_y^+)r_y + \tfrac{1}{2}f_x^+(r_{x1}^+ + r_{x2}^+) + f_{XI}r_{XI} + f_{XII}r_{XII} \tag{1}$$

Here we assume that the observed absorption of the radical cation of the dimer is the time average over the two *x*1 and *x*2 polarized monomer cation configurations. (This is because of the delocalization of the unpaired electron [Norris *et al.* 1971].) The properties of the weight factors f_i are discussed in the Appendix.

The simple exciton theory (Davidov 1962) as outlined for dimers by Tinoco (1963) shows that the transition moments of the dimer are the vectorial sum and the difference, respectively, of the monomer transition moments multiplied by $(2)^{-1/2}$ (see Fig. 4). The frequency separation between exciton bands is proportional to the dipolar interaction energy between the transition moments of the monomers. In terms of wavelength, the splitting in the region around 430 nm will be about 2.5-times less than that in the region around 680 nm (at equal interaction energy). This, plus the experimentally observed fact that the Soret region of chlorophyll *a* in the chloroplast membrane closely resembles the Soret region of monomeric chlorophyll *a* in organic solvents, makes us believe that, if there is exciton interaction, it does not cause too much splitting between the possible exciton bands of the dimer in the Soret region. If the two exciton bands practically overlap, then their contribution to the polarization anisotropy becomes (2):

$$r_{exc} = f_x(|X_I|^2 r_{XI} + |X_{II}|^2 r_{XII})/(|X_I|^2 + |X_{II}|^2) \tag{2}$$

Defining the monomer *x*-axes in the coordinate system of Fig. 3 by the azimuth angles χ_1 and χ_2 and the inclination angles ϑ_1 and ϑ_2 it follows from simple exciton theory (see Fig. 4) that:

$$|X_{I,II}|^2 = |x|^2[1 \pm \cos\vartheta_1 \cos\vartheta_2 (\cos\chi_1 \cos\chi_2 + \sin\chi_1 \sin\chi_2) \pm \sin\vartheta_1 \sin\vartheta_2] \quad (3)$$

and therefore:

$$|X_I|^2 + |X_{II}|^2 = 2|x|^2 \quad (4)$$

The inclination angles of the exciton components to the plane ϑ_I and ϑ_{II} are related to the inclination angles of the two monomers ϑ_1 and ϑ_2 as follows:

$$\sin\vartheta_{I,II} = (\sin\vartheta_1 \pm \sin\vartheta_2)/2^{1/2}|X_{I,II}| \quad (5)$$

The polarization anisotropy of an oscillator inclined at an angle ϑ_i to the plane (on excitation into the plane) is given by equation (A8) (see Appendix):

$$r_i = (1 - 3\sin^2\vartheta_i)/10 \quad (A8)$$

Inserting equations (2) and (4) into equation (1) we obtain (6) for the observed polarization anisotropy.

$$r_{obs} = (f_y + f_y^+)r_y + \tfrac{1}{2}f_x^+(r_{x1} + r_{x2}) + \tfrac{1}{2}f_x(|X_I|^2 r_{XI} + |X_{II}|^2 r_{XII}) \quad (6)$$

Inserting equations (5) and (A8) therein with $r_y = 1/10$ and taking into account that $f_y + f_y^+ + f_x + f_x^+ = 1$ (see Appendix), we obtain equations (7) and (8):

$$\sin^2\vartheta_1 + \sin^2\vartheta_2 = R^2 \quad (7)$$
$$R^2 = (20/3)(1/10 - r_{obs})/(f_x + f_x^+) \quad (8)$$

Equations (7) and (8) describe the field of possible inclinations of the x-axes of the two chlorophyll a molecules in the dimer to the plane that is compatible with the observed dichroic ratio (according to Table 1 it is 1.21, corresponding to $r_{obs} = 0.0654$, see Appendix).

Equation (7) describes a circle with coordinates $\sin\vartheta_1$ and $\sin\vartheta_2$. It is evident that the circle is wider (and hence the inclination of x-axes larger) the smaller the contribution of x-polarized transitions to the observed absorption changes at 430 nm is (i.e. the smaller $f_x + f_x^+$).

A quantitative evaluation of equations (7) and (8) is given in Fig. 5. The inclination angle of x-axes is plotted against the relative contribution of x-transitions for two extreme cases: (1) one x-axis in plane ($\vartheta_2 = 0$) and the other one maximally inclined ($\vartheta_1 = \vartheta_{max}$) and (2) both equally inclined ($\vartheta_1 = \pm \vartheta_2$).

The question arises as to *the field of possible band compositions around* 430 nm. Monomeric chlorophyll in organic solvents shows an absorption peak at 430 nm (mainly B_x) with a shoulder (B_y) at 410 nm at about half height. Hence it is reasonable to assume that the ratio of the weight factors for the neutral species at 430 nm is $f_x/f_y \geqslant 2/1$. The absorption spectrum of the radical cation

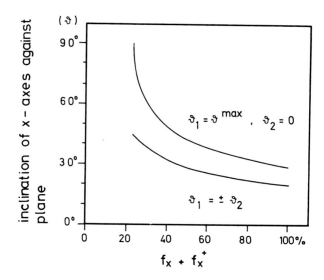

FIG. 5. The dependence of the inclination of the *x*-axes of the two chlorophyll *a* molecules in the dimer of photosystem I on the relative contributions $(f_x + f_x^+)$ of *x*-polarized transitions to the absorption changes at 430 nm, calculated according to equations (7) and (8) with $r_{obs} = 0.0654$.

of monomeric chlorophyll *a* in solution (Borg *et al.* 1970) shows the Soret blue band shifted to about 400 nm with about half the absorption of the neutral form at 430 nm. π-Electron calculations on bacteriochlorophyll cation (Otten 1971) revealed that the Soret has mixed polarization with the *y* component at shorter wavelength. Based on the foregoing information the field of possible values for $f_x + f_x^+$ can be constructed. One example

$$f_x/f_y/f_x^+/f_y^+ = 2/1/(-1)/(-1/2) \quad \text{yields} \quad (f_x + f_x^+) = 2/3$$

(The sign convention for the weight factors is given in the Appendix.) It is apparent that any reasonable combination of *f*-values (even those taking into account the fact that the radical cation of the dimer produces an absorption analogous to a monomeric neutral chlorophyll *a* in addition to the one analogous to the monomer cation) ranges between 0.5 and 1.

Consequently we conclude that the highest inclination of one *x*-axis is 43° to the plane (with the other one in plane) or, if both are equally inclined, it is 29° for both axes. On the other hand, the minimal inclination for one axis is 29° (with the other *x*-axis in plane) and it is 20° if both *x*-axes are equally in-

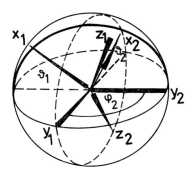

FIG. 6. Illustration of the three angles defining the mutual orientation of the two chlorophyll a molecules in the dimer of photosystem I: ϕ_2, ϑ_1 and ϑ_2.

clined. It is evident that the uncertainty about the band composition does not dramatically act on the inclination of the x-axes.

ON THE INCLINATION OF THE z-AXES OF THE TWO CHLOROPHYLL a MOLECULES IN THE DIMER

The angle between the two z-axes defines the inclination of the two porphyrin ring planes against each other. Since we assume—in analogy to the models by Fong (1974) and by Shipman $et\ al.$ (1976)—that the two rings are 'face-to-face', the complement of the angle between the z-axes and 180° is the inclination between the ring planes.

If we represent the two porphyrin rings by their major axes $x_{1,2}$, $y_{1,2}$ and $z_{1,2}$ in the framework of the polar coordinate system of Fig. 3 shown in Fig. 6, then vector calculus shows that the angle between the z-axes is given by (9),

$$\measuredangle\,(z_1, z_2) = \text{arc cos}(\cos\phi_2\,\sin\vartheta_1\,\sin\vartheta_2 - \cos\vartheta_1\,\cos\vartheta_2) \qquad (9)$$

where ϕ_2 is the azimuth of the y_2-axis (against the y_1-axis) and ϑ_1 and ϑ_2 are the inclination angles of the respective x-axes against the plane. If one x-axis is in the plane (e.g. $\vartheta_2 = 0$) and the other one is maximally inclined ($\vartheta_1 = \vartheta_{\text{max}}$) then the z,z-angle is simply the complement of ϑ_{max}:

$$\measuredangle\,(z_1, z_2) = \text{arc cos}(-\cos\vartheta_{\text{max}}) \qquad (9a)$$

According to the foregoing section this implies that the angle between the porphyrin ring planes then varies between 29° and 43°.

If, on the other hand, both x-axes are out of the plane and equally inclined, then:

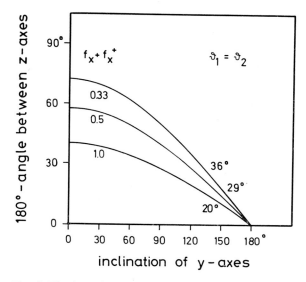

FIG. 7. The dependence of the angle between the two porphyrin rings in the dimer of photosystem I on the angle between the (in-plane oriented) y-axes (ϕ_2) with the inclination of the x-axes against the plane (ϑ_1 and ϑ_2) as parameter (calculated according to equation 9b).

$$\sphericalangle\, (z_1, z_2) = \text{arc } \cos(\pm \cos\phi_2 \sin^2 \vartheta - \cos^2\vartheta) \tag{9b}$$

Fig. 7 shows a numerical evaluation of equation (9b) (for the positive sign) for three different inclinations of the x-axes: $\vartheta = 36°$ (corresponding to $f_x + f_x^+ = 0.333$), 29° (0.5) and 20° (1). The angle between the porphyrin rings is represented as a function of the inclination between the two y-axes ϕ_2. An evaluation of equation (9b) for the negative sign produces a mirror image of the curves in Fig. 7 reflected at the line $\phi_2 = 90°$.

SUMMARY AND COMPARISON WITH STRUCTURAL INFORMATION BY OTHER METHODS

Table 2 summarizes the possible configurations of the two chlorophyll *a* molecules in the dimer of photosystem I that are compatible with our data. How are these configurations related to the *in vitro* dimer structures described by Fong (1974) and by Shipman *et al.* (1976)? And how do they compare with the angle between porphyrin rings of 46–50° which Clarke *et al.* (1976) reported for the dimer in various bacterial reaction centres? (We are not overestimating the latter's data in their relevance for photosystem I for two reasons: first, they bear on bacterial centres and, secondly, it has still to be proven that the triplet

TABLE 2

Summary of the possible configurations of the two porphyrin rings in the dimer of photosystem I that are compatible with our data. (For an illustration of the three angles, see Fig. 6.)
The ambiguity of the configurations is partly inherent in the photoselection technique and
partly due to lack of information on the spectral composition of the absorption changes of
chlorophyll a_1 at 430 nm (see text).

Inclination of x-axes against 'the plane'		Angle between y-axes	Angle between porphyrin rings
(ϑ_1)	(ϑ_2)	(ϕ_2)	
29–43°	0°	Arbitrary	29–43°
20–29°	20–29°	0° or 180°	0° or 40–58°
20–29°	20–29°	60° or 120°	20–30° or 35–50°
20–29°	20–29°	90°	28–40°

state is confined to the dimer. Nevertheless a comparison seems worth-while.)

Common to both *in-vitro* models is their C_2-symmetry and a plane parallel
orientation of the two porphyrin rings. The angle between the y-axes, however,
it at variance: although 120° in Fong's model, it is 180° in the one by Shipman
et al.

We come to the following conclusions:

If one x-axis is strongly inclined and the other one parallel to the plane (first
row in Table 2), then the angle between the y-axes is free, but the porphyrin
planes are inclined at 29–43°. This is not compatible with either model;

If the x-axes are both inclined at the same angle there are several possibilities:
if the y-axes are inclined at 180° (or 0°), as in the model by Shipman *et al.*,
then the porphyrin rings could be inclined at 0° or at 40–58° to each other.
The first possibility is essentially the structure by Shipman *et al.* (1976); if it is
real then we have to conclude from our data that photosystem I reaction
centres contain, in addition to the dimer peaking at 700 nm, one pair of antennae chlorophyll *a*, peaking at 700 nm as the dimer, with its y-axes perpendicular to the ones of the dimer;

if the y-axes are inclined by 60° or by 120° as in Fong's model (1974), then the
porphyrin ring planes have to be inclined at 20–30° or at 35–50°, which is at
variance with the structure he proposed (Fong 1974);

if the two y-axes are perpendicular, as we proposed (Junge *et al.* 1977), then the
porphyrin-ring planes have to be inclined at 28–40°.

We learn that there are several ways to account for a high inclination (some
50°) between the porphyrin rings as postulated by Clarke *et al.* (1976). The
model by Shipman *et al.* (1976) is not incompatible with our data. Fong's model

(1974) has to be modified for non-parallel porphyrin-ring planes to become compatible.

APPENDIX: SOME BASIC RELATIONS ON LINEAR DICHROISM IN PHOTOSELECTION

Photoselection experiments are done by exciting an isotropic ensemble of oscillators with linearly polarized light and observing the absorption changes (or fluorescence transients) with light polarized parallel or perpendicular to the exciting light. If we designate the respective absorption changes $\Delta A_\parallel$ and $\Delta A_\perp$, the dichroism is measured by either the dichroic ratio DR (equation A1)

$$DR = \Delta A_\parallel / \Delta A_\perp \tag{A1}$$

or the polarization anisotropy r (equation A2).

$$r = (\Delta A_\parallel - \Delta A_\perp)/(\Delta A_\parallel + 2\Delta A_\perp) = (DR - 1)/(DR + 2) \tag{A2}$$

If the absorption changes observed at a given wavelength are composite with contributions from various pairs of excited and interrogated oscillators, which differ in their pairwise orientation, then the apparent absorption changes for each polarization are just the sum of the contributions from those pairs. Since any pair contributes absorption changes into three orthogonal polarization directions, two of which are perpendicular to the exciting light, the contribution of a given pair (indexed i) to the total absorption changes is given by (A3).

$$f_i = (\Delta A_{\parallel,i} + 2\Delta A_{\perp,i})/ \sum_i (\Delta A_{\parallel,i} + 2\Delta A_{\perp,i}) \tag{A3}$$

It is easy to see that the sum of f_i over all pairs is unity, although each f_i may exceed unity. This is because the sign of this weight factor may be positive or negative. It is positive if a given pair contributes absorption changes of the same sign (increase or decrease of absorption, respectively), as the majority of pairs does.

Straightforward calculus shows that in these conditions the observed polarization anisotropy is the weighted sum of the polarization anisotropies for each pair (A4) (Weber's summation law (1961)), with the observed dichroic ratio related to the former by equation (A5).

$$r_{\text{obs}} = \sum_i f_i r_i \tag{A4}$$

$$DR_{\text{obs}} = (2r_{\text{obs}} + 1)/(1 - r_{\text{obs}}) \tag{A5}$$

If a photoselection experiment is done in perfect experimental conditions (i.e. 100% polarizers, parallel light beams, clear samples and non-saturating excitation energy), then the polarization anisotropy of two linear oscillators inclined at an angle β to each other is given by (A6) and consequently the dichroic ratio by (A7).

$$r_{\text{lin-lin}} = (2 - 3\sin^2\beta)/5 \tag{A6}$$

$$DR_{\text{lin-lin}} = (3 + \tan^2\beta)/(1 + 2\tan^2\beta) \tag{A7}$$

Details of the derivation may be obtained from Albrecht (1961).

Equations (A3–A6) may be merged to construct the observed polarization anisotropy for all cases where several pairs contribute to absorption changes at a given wavelength. One special case is worth mentioning. This is if either two mutually perpendicular (excited) oscillators interact with one linear (interrogated) oscillator or if one excited oscillator interacts with two perpendicular interrogated oscillators. Then the system shows circular degeneracy and the polarization anisotropy becomes (A8) and (A9), where β now denotes

$$r_{\text{circ}} = (1 - 3\sin^2\beta)/10 \tag{A8}$$

$$DR_{\text{circ}} = 2(1 + \cos^2\beta)/(4 - \cos^2\beta) \tag{A9}$$

the inclination angle of the linear oscillator to the plane of the two mutually perpendicular ones, i.e. the plane of circular degeneracy.

Photoselection experiments are invariant against translational displacement of the oscillators within each pair.

ACKNOWLEDGEMENT

This work was supported by the European Commission and by the Deutsche Forschungsgemeinschaft. We thank Mrs A. Schulze for the graphs.

References

ALBRECHT, A. C. (1961) Polarizations and assignments of transitions: the method of photoselection. *J. Mol. Spectrosc. 6*, 84–108

BENGIS, C. & NELSON, N. (1975) Purification and properties of the photosystem I reaction centre from chloroplasts. *J. Biol. Chem. 250*, 2783–2788

BOARDMAN, N. K. & ANDERSON, J. M. (1964) Isolation from spinach chloroplasts of particles containing different proportions of chlorophyll *a* and chlorophyll *b* and their possible role in the light reactions of photosynthesis. *Nature (Lond.) 203*, 166–167

BORG, D. C., FAJER, J., FETTON, R. H. & DOLPHIN, D. (1970) The π-cation radical of chlorophyll *a*. *Proc. Natl. Acad. Sci. U.S.A. 67*, 813–820

BOXER, S. G. & CLOSS, G. L. (1976) A covalently bound dimeric derivative of pyrochlorophyllide *a*. A possible model for reaction center chlorophyll. *J. Am. Chem. Soc. 98*, 5406–5408

BRETON, J. (1976) Dichroism of transient absorption changes in the red spectral region using oriented chloroplasts. II. *P*-700 absorbances changes. *Biochim. Biophys. Acta 459*, 66–75

BRETON, J. & ROUX, E. (1971) Chlorophyll and carotenoid states *in vivo*. I. A linear dichroism study of pigment orientation in spinach chloroplasts. *Biochem. Biophys. Res. Commun.* 45, 557–563

CLARKE, R. H., CONNORS, R. E. & FRANK, H. A. (1976) Investigation of the structure of the reaction centre in photosynthetic bacteria by optical detection of triplet state magnetic resonance. *Biochem. Biophys. Res. Commun.* 71, 671–675

CLARKE, R. H., CONNORS, R. E. & FRANK, H. A. (1978) *Proc. R. Irish Acad. Sci.*, in press

DAVIDOV, A. S. (1962) *Theory of Molecular Excitons*, McGraw-Hill, New York

FENNA, R. E. & MATTHEWS, B. W. (1975) Chlorophyll arrangement in a bacteriochlorophyll protein from *Chlorobium limicola*. *Nature (Lond.)* 258, 573–577

FENNA, R. E., MATTHEWS, B. W., OLSON, J. M. & SHAW, E. K. (1974) Structure of a bacterio-chlorophyll–protein from the green photosynthetic bacterium *Chlorobium limicola*: crystallographic evidence for a trimer. *J. Mol. Biol.* 84, 231–240

FONG, F. K. (1974) Molecular basis for the photosynthetic primary process. *Proc. Natl. Acad. Sci. U.S.A. 71*, 3692–3699

GEACINTOV, N. E., VANNOSTRAND, F. V. & BECKER, J. F. (1974) Polarized light spectroscopy of photosynthetic membranes in magneto-oriented whole cells and chloroplasts. *Biochim. Biophys. Acta 347*, 443–463

GOVINDJEE (1975) *Bioenergetics of Photosynthesis*, Academic Press, New York

IKEGAMI, I. & KATOH, S. (1975) Enrichment of photosystem I reaction center chlorophyll from spinach chloroplasts. *Biochim. Biophys. Acta 376*, 588–592

JUNGE, W. (1975) Flash spectrophotometry in the study of plant pigments, in *Chemistry and Biochemistry of Plant Pigments* (Goodwin, T. W., ed.), 2nd edn., vol. 2, pp. 233–333, Academic Press, New York

JUNGE, W. & ECKHOF, A. (1973) On the orientation of chlorophyll *a* in the functional membrane of photosynthesis. *FEBS (Fed. Eur. Biochem. Soc.) Lett. 36*, 207–211

JUNGE, W. & ECKHOF, A. (1974) Photoselection studies on the orientation of chlorophyll *a*, in the functional membrane of photosynthesis. *Biochim. Biophys. Acta 357*, 103–117

JUNGE, W. & SCHAFFERNICHT, H. (1978) in *Photosynthesis '77* (Hall, D. O., Coombs, J. & Goodwin, T. W., eds.), pp. 21–32, The Biochemical Society, London

JUNGE, W., SCHAFFERNICHT, H. & NELSON, N. (1977) On the mutual orientation of pigment in Photosystem I particles from green plants. *Biochim. Biophys. Acta 462*, 73–85

KATZ, J. J. & NORRIS, J. R. (1973) Chlorophyll and light-energy transduction in photosynthesis. *Curr. Top. Bioenerg. 5*, 41–72

KE, B., SUGUHARA, K. & SHAW, E. R. (1975) Further purification of 'Triton subchloroplast fraction 1'. Isolation of a cytochrome-free high P700 particle and a complex containing cytochromes *f* and *b6*, plastocyanin and iron-sulfur protein(s). *Biochim. Biophys. Acta 408*, 12–25

MALKIN, R. (1975) Photochemical properties of a photosystem I subchloroplast fragment. *Arch. Biochem. Biophys. 169*, 77–83

MAR, T. & GINGRAS, G. (1976) Photodichroic studies of the photoreaction center from *Rhodospirillum rubrum*. I. Attribution of P870 to two non parallel dipoles. *Biochim. Biophys. Acta 440*, 609–621

NORRIS, J. R., UPHAUS, R. A., CRESPI, H. L. & KATZ, J. J. (1971) Electron spin resonance of chlorophyll and the origin of signal I in photosynthesis. *Proc. Natl. Acad. Sci. U.S.A. 68*, 625–628

NORRIS, J. R., SCHEER, H., DRUYAN, M. E. & KATZ, J. J. (1974) An electron-nuclear double resonance (Endor) study of the special pair model for photo-reactive chlorophyll in photosynthesis. *Proc. Natl. Acad. Sci. U.S.A. 71*, 4897–4900

OLSON, J. M. & HIND, G. (1977) *Chlorophyll–Proteins, Reaction Centres and Photosynthetic Membranes*, Brookhaven National Laboratory, Upton

OTTEN, H. A. (1971) Absorption changes in the reaction centre of photosynthetic bacteria and electron calculations on bacteriochlorophyll, its monocation and anion. *Photochem. Photobiol. 14*, 589–596

ROCKLEY, M. G., WINDSOR, M. W., COGDELL, R. J. & PARSON, W. W. (1975) Picosecond detection of an intermediate in the photochemical reaction of bacterial photosynthesis. *Proc. Natl. Acad. Sci. U.S.A. 72*, 2251–2255

SHIPMAN, L. L., COTTON, T. M., NORRIS, J. R. & KATZ, J. J. (1976) New proposal for the structure of special-pair chlorophyll. *Proc. Natl. Acad. Sci. U.S.A. 73*, 1791–1799

SHUVALOV, V. A., ASADOV, A. A. & KRAKHMALEVA, I. N. (1976) Linear dichroism of light-induced absorbance changes of reaction centres of *Rhodospirillum rubrum*. *FEBS (Fed. Eur. Biochem. Soc.) Lett. 76*, 240–245

THORNBER, J. P., ALBERTE, R. S., HUNTER, F. A., SHIOZAWA, J. A. & KAN, K.-S. (1977) in *Chlorophyll–Proteins, Reaction Centers and Photosynthetic Membranes* (Olson, J. M. & Hind, G., eds.), pp. 132–148, Brookhaven National Laboratory, Upton

TINOCO, M. (1963) The exciton contribution to the optical rotation of polymers *Radiat. Res. 20*, 133–139

VERMEGLIO, A. & CLAYTON, R. K. (1977) Kinetics of electron-transfer between primary and secondary-electron acceptor in reaction centers from *Rhodopseudomonas sphaeroides*. *Biochim. Biophys. Acta 449*, 500–515

VERNON, L. P. & SHAW, E. R. (1971) Subchloroplast fragments: Triton X–100. *Methods Enzymol. 23*, 277–289

WEBER, G. (1961) in *Fluorescence and Phosphorescence Analysis* (Hercules, D. M., ed.), pp. 217–240, Interscience, New York

Discussion

Paillotin: The circular degeneracy implies two perpendicular transitions absorbing in the same way and it may be an intrinsic property of both the dimer and the antenna—that is one possibility. But one can also imagine that to each transition of the dimer there corresponds a similar but perpendicular transition of the antenna.

Junge: As I mentioned, the possibility that photosystem I particles contain a few antennae chlorophyll *a* which peak around 700 nm (as does the dimer) cannot be ruled out. And there is the possibility that the species observed at low temperature by Dr Butler, which is distinct from the photooxidizable dimer, is his far-red antennae chlorophyll. But I do not believe that there are more than two such chlorophylls because of the absorption spectrum of the particles.

Butler: Did you attempt any of the photoselection experiments at low temperature?

Junge: Not at temperatures as low as in your experiments.

Katz: What is the structure of the antenna that enables it to absorb at 700 nm but makes it different in optical properties from synthetic special-pair chlorophyll that absorbs at 695 nm? The longest wavelength for absorption by a dimer [Chl a]$_2$ is about 680 nm, but the species Chl·H_2O·Chl absorbs at about 695 nm. (I am not concerned with photoactivity *per se* but with optical properties of your presumed antenna.) According to our exciton calculations (Shipman & Katz 1977) the absorption maximum in the red of a stack of the

Strouse variety is a function of the length of the stack. For two chlorophyll molecules, as in a special pair, the absorption maximum is calculated to be at 695 nm; for three, it will be at 705 nm (see p. 347). This suggests the possibility that, on cooling, water is forced into chlorophyll dimers or oligomers. I doubt that the P705 species is a dimer in which two chlorophyll molecules are linked by a keto carbonyl–magnesium interaction. Such a species does not have the required absorption spectrum.

Barber: Did you observe a 730 nm emission band?

Junge: Not yet.

Katz: The 680 nm peak is so broad that many chlorophyll species could contribute to it. To restrict the number as you have done begs an important question.

Porter: The number should be less than 20.

Katz: There are reasons to believe that oligomers with an absorption maximum near 680 nm may be present; at least such a possibility should be entertained. There is also reason to suppose that sufficient water may be present that can be intercalated in a highly specific way. Why did you not do these experiments on a system that could be prepared in the laboratory and that is better characterized?

Junge: That is a popular argument. Why aren't you working with chloroplasts? Both approaches together (and only together) will reveal the architecture of the reaction centre to us.

Katz: One cannot, in principle, use visible absorption spectroscopy for structural purposes without additional information. Studies on reasonably well characterized *in vitro* systems should be useful in the interpretation of experimental results obtained on plant material.

Junge: I agree; that is why we should try to do similar experiments with your dimers embedded in glasses.

Cogdell: Why is there such a high dichroic ratio with the carotenoids at 680 nm?

Junge: There may be two possible origins for this. The one we favour at present (based on a scan of the dichroic ratio over the narrow 680 nm band) is that these absorption changes are due to chlorophyll shifting its absorption maximum in response to the photooxidation of the dimer. Then the high dichroic ratio on excitation of the carotenoids tells us (1) that the carotenoids are highly ordered—their long axes fall within a rather narrow cone—and (2) that those antennae chlorophylls responding by a small band-shift are highly oriented with respect to the carotenoids.

The other possibility is that the absorption changes around 680 nm are due to several species with different dichroic ratios which contribute absorption changes with positive and negative sign. In the neighbourhood of the isosbestic point of a difference spectrum very high or very low apparent dichroic ratios may appear that are not easily related to the dichroic ratios of the underlying components of the composite absorption band. Kunze & I (1977) reported such effects in flash photolysis experiments with the CO-complexed cytochrome c oxidase. However, it is obvious that this is accompanied by strong variations of the dichroic ratio scanned over the complex band in a difference spectrum.

Robinson: Is it certain that the singlet–singlet energy transfer is 100% dipole? In more-weakly-absorbing molecules (i.e. with oscillator strengths of 0.1 and 0.2 as opposed to 1.0) other kinds of interactions with different angular dependences become important. Has this been ruled out?

Knox: There is a large monopole-type correction—'monopole' in the London (1942) sense at short distances (i.e. 1.0–1.2 nm).

Robinson: By monopole do you mean an angularly-independent transfer?

Knox: In this context 'monopole' means that one substitutes the transition moments with a set of point *transition charges* at each carbon atom—so it really means *all multipoles*. The correction to the transfer rate is about 50–100% at short distances (Chang 1977).

Robinson: Any argument about the angles may have to be modified.

Knox: The more accurate treatment makes the angular transfer pattern appear much more anisotropic—it pushes the large lobes out further and pulls in the small ones (Chang 1977). Also at these short distances some exchange correction to the dipole–dipole interaction would be expected. That has not been calculated. I doubt if there is quadrupole or octupole interaction in the usual sense.

Robinson: The exchange corrections can be angularly peculiar.

Knox: Yes; and then there is the cross-term between exchange and dipole–dipole (but *nobody* worries about that!).

Junge: What are the implications for the construction of the resulting transition moments of a dimer from, say, x-polarized transition moments of two monomers?

Knox: I cannot say without making a detailed analysis. The overall x- and y-symmetry of these transitions is not changed by the fact that one has to make a monopole approximation.

Breton: How is the P700 dimer oriented in the membrane?

Junge: We did only few experiments on oriented chloroplasts, mainly because you seem to have exhausted the information derivable therefrom. The major difficulty in relating our results to yours on the orientation of the dimer to the membrane resides in the ambiguity in assigning with sufficient precision the inclination of the plane of circular degeneracy to the thylakoid membrane.

This ambiguity is hard to resolve: for instance, your studies admit the possibility that the transition moments of the chlorophyll *a* peaking at and above 690 nm are inclined at any angle between 0° and some 20° to the membrane. With this uncertainty it may be that a transition moment which we assign as inclined at less than the magic angle (35°) to the plane of degeneracy is inclined at more than the magic angle to the membrane. Concerning the plane of degeneracy we find that the *y*-axes of the two chlorophyll *a* molecules in the dimer are inclined at less than 13° and the *x*-axes, grossly, at less than 30°.

Breton: We find a dichroic ratio smaller than unity.

Junge: In photoselection with excitation at 720 nm we find a dichroic ratio of unity for this band.

Robinson: Such experiments as yours (especially when time-dependent polarized fluorescence is perfected) and the resonance Raman studies are elegant but I have an uneasy feeling that the interpretations are complicated. Have Dr Katz's dimers, which have been well characterized structurally, been studied in this way?

Katz: Other systems can be generated from ordinary chlorophyll *a*, which is much more readily available, that have the optical and redox properties of special-pair chlorophyll. These have been well characterized. For example, a P700 model can be made by cooling chlorophyll *a* and ethanol in toluene to −100 °C. The species formed at low temperature absorbs maximally in the red at 700 nm. It has been characterized by infrared spectroscopy and its spin-sharing capabilities have been examined by e.p.r. It is more convenient to prepare this system than the linked dimer. However, some physical measurements (e.g. sedimentation) are more easily made on linked dimer systems than in an organic glass. Other measurements (e.g., dichroism or Raman spectroscopy) could be made on a glass at low temperatures without too much difficulty. In the interpretation of *in vivo* spectroscopic data it is a comfort to have some data on defined systems containing chlorophyll species presumed to be present *in vivo* (Cotton *et al.* 1978).

Junge: I agree; that is why I discussed a 'field of possible structures...' I accept that we need more information.

Katz: À propos resonance Raman experiments, a chlorophyll *a* oligomer in n-octane shows hardly any fluorescence. It appears likely that resonance-enhanced Raman spectroscopy of this species could be readily studied by excitation into the red band. A properly prepared chlorophyll *a* oligomer is much less fluorescent than many chlorophyll–protein complexes.

Lutz: Even minimal fluorescence often proves to be too much for the Raman spectroscopy.

Knox: Vacek *et al.* (1977) recently reported high fluorescence polarization of CP I (about 14%). R. L. Van Metter (unpublished results, 1977) has also seen a high value. Both groups used, I believe, excitation at 640 nm.

Katz: With regard to the 435 nm band, the *ab initio* calculations on ethyl chlorophyllide (Shipman *et al.* 1976) indicated that the Soret band is far more complex than the red band. Whereas the four-orbital model is adequate for the red band and gives roughly the same results as *ab initio* calculations, this is by no means true for the Soret band—there are many energy levels.

Junge: The least that is known is their polarization behaviour from fluorescence experiments (see Bär *et al.* 1961). The Soret band was found to be mixed with a predominance of *x*-polarized components towards shorter wavelength.

Katz: Weber (1961) proposed that any time fluorescence is excited at the far-red end of an absorption maximum the fluorescence maximum is anomalous.

Porter: That has been much questioned since then.

References

BÄR, F., GANG, H., SCHNABEL, E. & KUHN, H. (1961). *Z. Elektrochem. 65*, 346–352

CHANG, J. C. (1977) Monopole effects on electronic excitation interactions between large molecules I. Application to energy transfer in chlorophylls. *J. Chem. Phys. 67*, 3901–3909

COTTON, T. M., LOACH, P. A., KATZ, J. J. & BALLSCHMITER, K. (1978) Studies of chlorophyll-chlorophyll and chlorophyll-ligand interactions by visible absorption and infrared spectroscopy at low temperatures. *Photochem. Photobiol. 27*, 735–750

KUNZE, U. & JUNGE, W. (1977) Ellipticity of cytochrome α-3 and rotational mobility of cytochrome *c*-oxidase in cristae membrane of mitochondria. *FEBS (Fed. Bur. Biochem. Soc.) Lett. 80*, 429, 434

LONDON, F. (1942) On centres of van der Waals attraction. *J. Phys. Chem. 46*, 305–316

SHIPMAN, L. L. & KATZ, J. J. (1977) Calculation of the electronic spectra for chlorophyll *a*– and bacteriochlorophyll *a*-water adducts. *J. Chem. Phys. 81*, 577–581

SHIPMAN, L. L., COTTON, T. M., NORRIS, J. R. & KATZ, J. J. (1976) New proposal for the structure of special-pair chlorophyll. *Proc. Natl. Acad. Sci. U.S.A. 73*, 1791–1799

VACEK, K., WONG, D. & GOVINDJEE (1977) Absorption and fluorescence properties of highly enriched reaction center particles of photosystem I and of artificial systems. *Photochem. Photobiol. 26*, 269–276

WEBER, G. (1961) in *Fluorescence and Phosphorescence Analysis* (Hercules, D. M., ed.), pp. 217–240, Interscience, New York

Effects of ions and gravity forces on the supramolecular organization and excitation energy distribution in chloroplast membranes

L. ANDREW STAEHELIN and *CHARLES J. ARNTZEN

*Department of Molecular, Cellular and Developmental Biology, University of Colorado, Boulder and *USDA/SEA, Department of Botany, University of Illinois, Urbana*

Abstract This study was designed to explore the possible relationship between chloroplast membrane stacking or particle aggregation in stacked membrane regions (or both) and excitation energy distribution between photosystems I and II. To this end we have quantitatively examined the effects of different concentrations of univalent ions on the above-mentioned parameters, using a combination of freeze-fracture and thin-section electron microscopy for structural analysis and chlorophyll fluorescence measurements to assay energy-transfer processes. Membrane stacking was found to saturate at about 150mM-NaCl. Maximal EF_s-face particle density and chlorophyll fluorescence occurred at about 100mM-NaCl, although only 50% of the potential EF_s-face particles were located in stacked membrane regions at this salt concentration. Centrifugation (30000 **g**, 1 h) could significantly increase the amount of stacked membranes at salt concentrations between 20- and 60-mM-NaCl; in contrast, centrifugation had little effect on cation-regulation of chlorophyll fluorescence properties. These and other findings suggest that neither chloroplast membrane stacking nor the aggregation of EF-face particles into stacked regions is directly related to the mechanism of excitation energy distribution between the two photosystems (as measured by chlorophyll fluorescence changes) although both structural and functional changes may be mediated by the same membrane component. It is proposed that the salt-induced stacking of chloroplast membranes and the concomitant aggregation of EF-face particles is mediated by the screening of negative surface charges on the membrane pigment–protein subunits, by the establishment of specific interactions between light-harvesting pigment–protein complexes and by 'entropic ordering' forces.

---

*Mention of a trademark name or a proprietary product does not constitute a guarantee or warranty of the product by the US Department of Agriculture and does not imply approval of the product to the exclusion of others that may also be suitable.

147

One of the most striking morphological features of higher plant chloroplasts is the differentiation of their thylakoids into grana and stroma membrane regions. Within grana adjacent thylakoids adhere to each other to form stacked membrane regions; these are interconnected by unstacked (stroma) membranes. Although grana stacks were already recognized by light microscopists nearly 100 years ago, their functional significance has remained an enigma.

During the past decade, numerous chloroplast membrane fractionation studies have demonstrated that the lighter stroma membranes are enriched in photosystem I (PS I) activity and have low photosystem II (PS II) content, but the grana membranes possess both PS I and PS II activities (Armond & Arntzen 1977; Arntzen 1978). In freeze-fracture replicas it can be seen that this functional differentiation of thylakoids is matched by a non-random distribution of intra-membrane particles between stacked and unstacked membrane regions (Staehelin et al. 1977). Complementing these studies are analyses showing a dependence of the functional properties of chloroplasts on their ionic environment (Barber 1976). It has been demonstrated that appropriate uni- and bi-valent cations can increase the quantum efficiency of the Hill reaction and stimulate the rates of photophosphorylation and the light-induced uptake of protons. These data are consistent with the idea that cations regulate the coupling of redox reactions to the energy-conserving mechanism of the membrane.

Izawa & Good (1966) first showed that the morphological organization of thylakoid membranes is also controlled by cations. They demonstrated that isolated chloroplasts are converted into a grana-free membrane system in low-salt conditions. One consequence of this unstacking and unfolding of thylakoid membranes is the intermixing and randomization of the intramembrane particles of stacked and unstacked membrane regions (Staehelin 1976). Addition of appropriate cations can partially reverse these effects (Izawa & Good 1966; Murakami & Packer 1971; Staehelin 1976).

Measurement of chlorophyll fluorescence has been widely used to monitor cation effects on energy distribution in isolated chloroplast membranes (Papageorgiou 1975). At room temperature, chlorophyll fluorescence arises primarily from the pigment bed of PS II; the amplitude of fluorescence is determined, in part, by the oxidation/reduction state of the photosystem II primary electron acceptor, Q. A time-dependent change in the amplitude of chlorophyll fluorescence occurs when dark-adapted chloroplasts are illuminated. This period of variable fluorescence corresponds to the time required to reduce Q. In 1969 Homann observed that cations increase the room-temperature intensities of chlorophyll fluorescence from isolated chloroplasts independently from the redox state of Q. Murata (1969) and others have extended these studies and concluded that cations can alter the distribution of absorbed excitation energy between PS I and PS II, with maximum exciton arrival at PS II centres occurring at > 3mM-bivalent cations and > 100mM-univalent cations. Since these concentrations are essentially the same as those needed to restack experimentally unstacked thylakoids and concomitantly resegregate the intramembrane particles (Staehelin 1976), it seemed possible that the regulation of excitation energy distribution may be directly related to membrane stacking. In an attempt to test this hypothesis, we have examined how these two parameters change when experimentally unstacked thylakoids are restacked in the presence of different concentrations of univalent ions. Our findings indicate that thylakoid membrane stacking does not correlate directly to cation-induced changes in chlorophyll fluorescence, even though both phenomena are mediated by identical components of the chloroplast membrane. The main functional importance of stacking may be related to the fact that stacking provides a mechanism for segregating elements of the electron-transport chain from other membrane components, thereby bringing them closer together to increase their overall functional efficiency of electron transport in light-limited growth conditions.

MATERIALS AND METHODS

Seeds of dwarf peas (*Pisum sativum*, var. Laxton Progress #9) were grown under fluorescent lamps on vermiculite moistened with half-strength Hoagland's solution. Chloroplasts were isolated, experimentally unstacked, processed for thin-section and freeze-fracture electron microscopy, and the micrographs quantitatively analysed as described elsewhere (Staehelin 1976; Armond *et al.* 1977). Experimental details of specific experiments have also been included in the text and in the figure legends.

For the measurement of chlorophyll fluorescence induction transients, isolated chloroplasts were suspended to a concentration of 5 μg chl/ml in 5mM-Na-Tricine (pH 7.8) plus glycerol and salts as indicated. Illumination was with a Unitron microscope illuminator powered by a 6-A stable output power supply (Model C5-6 Power-One, Camarillo, California). The actinic beam was passed through a Corning 4-96 (broad band blue) optical filter; onset of illumination was controlled by a Uniblitz electronic shutter (Vincent Associated, Rochester, New York) with iris opening adjusted to provide a complete shutter opening time of 1.3 ms. Chlorophyll fluorescence was detected through two Corning 2-58 optical filters with a Photops UDT-500 photodiode (United Detector Technology, Santa Monica, California). The photodiode voltage output signal was stored on a Nicolet 535 digital signal averager

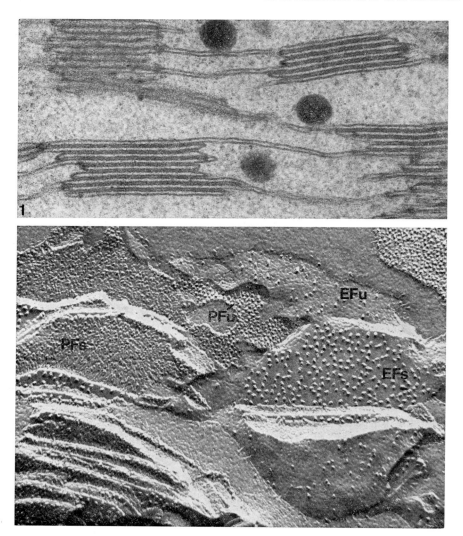

FIG. 1 (Top). Thin section through a portion of a spinach chloroplast, showing interconnected stacked (grana) and unstacked (stroma) thylakoid membranes. × 90 000

FIG. 2 (Bottom). Freeze-factured isolated thylakoids of a spinach chloroplast showing similar features as seen in Fig. 1. The flat, partly circular, membranes of two grana stacks (left and right) appear interconnected by more tubular membranes of a stroma lamella. The complementary type fracture faces marked PFs and EFs are characteristic of stacked membrane regions and the faces PFu and EFu belong to unstacked regions. Note the distinct aggregate of large 16 nm particles on the EFs face. × 75 000

(Nicolet Instruments, Madison, Wisconsin). The data points measured within 1 ms after full shutter opening were used to determine F_0 (original or initial fluorescence). The steady-state fluorescence-intensity signal observed at 1–2 s after onset of illumination was measured as maximal fluorescence (F_M). Stored transient signals from the Nicolet averager were plotted on an x–y recorder for a permanent record.

Fluorescence emission spectra were recorded with an Aminco-Bowman spectrophotofluorometer (Silver Springs, Maryland). Samples were frozen in liquid nitrogen in capillaries (1.0 mm inside diameter). The exciting wavelength for measurement was at 440 ± 20 nm. The observation (emission) wavelengths were scanned with an exit slit adjusted to give a 6 nm half-band width.

RESULTS

Fig. 1 illustrates the typical differentiation of thylakoid membranes of higher-plant chloroplasts into grana and stroma membrane regions. Within grana the membranes of adjacent thylakoids adhere to each other, giving rise to grana stacks that are interconnected by stroma membranes (thylakoids with free surfaces). This structural organization can be preserved in isolated chloroplasts if they are isolated in a medium containing >100mM-univalent or >2mM-bivalent cations. Fig. 2 shows a typical freeze-fracture image of an isolated spinach chloroplast, which demonstrates that grana and stroma membranes possess a substantially different organization at the supramolecular level. In this micrograph four distinct types of membrane fracture faces can be recognized: PF_s (fracture face of a protoplasmic or stroma membrane leaflet from a stacked, grana region), PF_u (face of a stroma leaflet of an unstacked stroma lamella), EF_s (face of an exoplasmic or lumenal membrane leaflet from a stacked grana region), and EF_u (face of a lumenal leaflet from an unstacked stroma lamella). The EF_s face is the most distinctive thylakoid fracture face owing to its dense population of large, clearly spaced particles on a relatively smooth background. Particle size histograms (Staehelin 1976) of the EF_s particles reveal that they fall into two major and broad size categories (10–14 nm and 14–18 nm), and the more-uniform EF_u particles average about 11 nm. In control spinach and pea chloroplasts about 80% of the EF face particles are found in the stacked membrane regions; adpressed membranes comprise about 60% of the total thylakoid membrane area. In contrast to the particles on the EF faces, those on the PF_s and PF_u faces are both smaller and more numerous (Fig. 2). Owing to their limited importance for the present study, their structural and functional features will only be discussed where needed.

Fig. 3 draws attention to a feature of higher-plant chloroplasts which has received little attention in the literature but which provides the structural basis

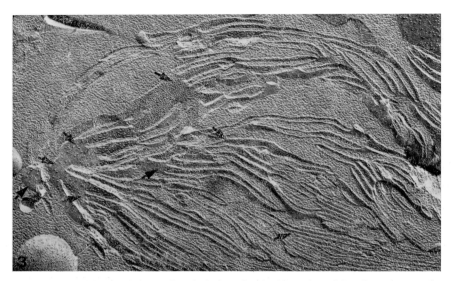

FIG. 3. Thylakoid membranes of an isolated spinach chloroplast. Note how the membrane region on the left becomes more and more branched (arrows) to give rise to virtually all membranes on its right. This ordered sequence of membrane branching events allows experimentally unstacked membranes to unfold without becoming entangled (see Fig. 4). × 25 000

for the unstacking and restacking experiments described in this paper. In particular, it demonstrates how the complex architecture of grana and stroma lamellae is formed by multiple branching of thylakoids that seem to have a single point of origin. Thus, when the thylakoids of an isolated chloroplast are suspended in a low-salt unstacking medium, the lamellae can easily unstack and unfold without becoming entangled (Fig. 4; Izawa & Good 1966; Staehelin 1976). This also explains why the unstacked and unfolded lamellae of a chloroplast do not fall apart. One of the most notable membrane changes accompanying this unfolding and unstacking is the disappearance of the structural differentiation of the thylakoids into stacked and unstacked regions, and the concomitant randomization of the intramembrane particles (Fig. 4).

As previously demonstrated (Staehelin 1976) the addition of >2mM-MgCl$_2$ or of >100mM-NaCl to a suspension of experimentally unstacked thylakoids leads to the reformation of stacked membrane areas and to a corresponding resegregation of the different categories of intramembrane particles. In particular, the resegregation and aggregation of the large EF face particles into the reformed EF$_s$ regions has attracted interest since the large particles are believed to consist of PS II cores surrounded by light-harvesting pigment–

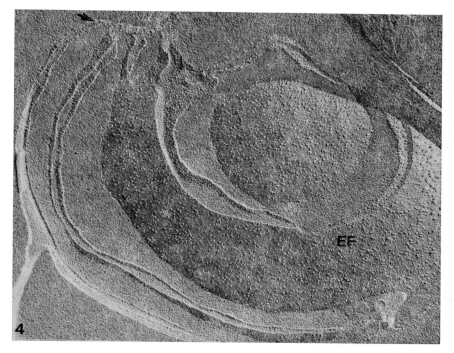

FIG. 4. Thylakoids from an isolated spinach chloroplast freeze-fractured after experimental unstacking and unfolding in a 50mM-Na-Tricine buffer, pH 7.6. All EF face particles appear randomly distributed, and no distinction between formerly stacked and unstacked membrane regions can be made. The arrow points to the membrane area connecting all lamellae of this chloroplast. × 40 000

protein complex (LHC) aggregates, and have been directly implicated in the stacking process (Arntzen 1978; Staehelin *et al.* 1977). The present study has been designed to explore in greater detail the role of the EF_s face particles in membrane stacking, and to obtain information on the functional significance of thylakoid membrane adhesion and the aggregation of EF_s particles into stacked regions.

Fig. 5 reveals that the addition of 20mM-NaCl to a suspension of experimentally unstacked thylakoid membranes can lead to the formation of the first distinct regions of membrane contact. These adhesion spots usually had the form of small, round patches; the contact regions constituted about 2% of the total membrane area (Figs. 5 and 11). Probably the most intriguing feature of these contact regions was that they seemed to *exclude* the large EF face particles (Fig. 5). This was intriguing since EF particles normally aggregate

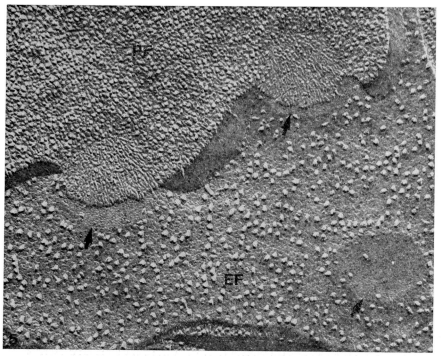

Fig. 5. Freeze-fractured pea thylakoids after experimental unstacking as for Fig. 4, and subsequent restacking in a 20mM-NaCl + 5mM-Tricine solution (pH 7.6). Three small, round contact regions (arrows) between the adjacent membranes can be seen. Note the nearly complete exclusion of large EF face particles from such regions (see also Fig. 10). × 85 000

in stacked membrane regions. Because of the anomaly, we refer to the stacked regions with excluded EF particles as pseudograna, even though they exhibit a normal complement of small PF face particles (Fig. 5).

When the NaCl concentration in suspensions of isolated chloroplasts was increased to 40 mmol/l, the size of the membrane adhesion areas also increased (Figs. 6 and 11), and a small, but significant, number of large particles (about 350 particles/μm^2) could be detected on the EF$_s$ faces of the contact regions (Fig. 10). At 60mM-NaCl (Fig. 7) the density of particles on the EF$_s$ faces was about the same as on the surrounding EF$_u$ faces (about 600 particles/μm^2; Fig. 10). At this salt concentration the percentage of stacked membrane regions rose to about 15% (Fig. 11). Between 60- and 150-mM-NaCl (Figs. 8–11) both the density of EF$_s$ face particles and the size of the stacked thylakoid membrane regions increased. It should be emphasized, however, that although

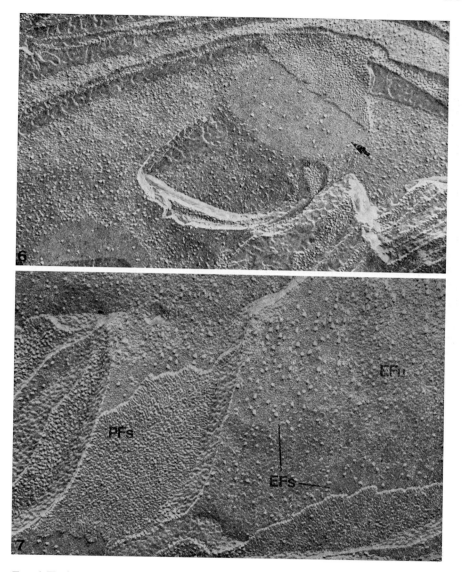

FIG. 6 (Top). Pea thylakoids experimentally unstacked as for Fig. 4, and then restacked in a medium containing 40mM-NaCl + 5mM-Tricine. Although the membrane contact areas (arrow) are considerably larger than in samples restacked in 20mM-NaCl (Fig. 5), the number of EF particles in such areas is still rather low (see also Fig. 10). × 50 000

FIG. 7 (Bottom). Pea thylakoids experimentally unstacked as for Fig. 4, and then restacked in 60mM-NaCl + 5mM-Tricine. The density of particles on the EF_s and EF_u faces is very similar (see also Fig. 10). × 68 000

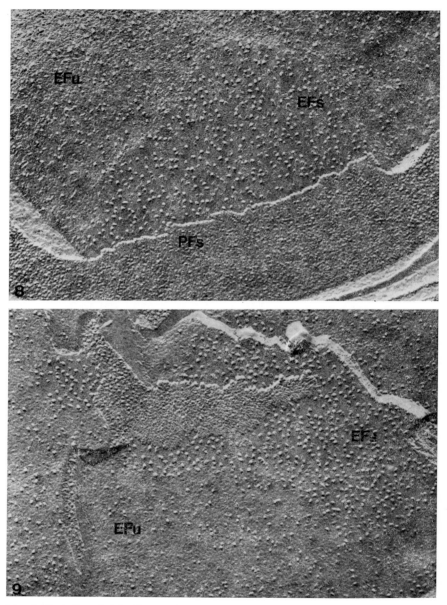

FIG. 8 (Top). Pea thylakoids experimentally unstacked as for Fig. 4, and then restacked in 80mM-NaCl + 5mM-Tricine. The density of particles on the EF_s and EF_u faces is near normal even though in these conditions the percentage of restacked membranes is only about half maximal (see also Figs. 10 and 11). × 64 000

FIG. 9 (Bottom). Pea thylakoids experimentally unstacked as for Fig. 4 and then restacked in 100mM-NaCl + 5mM-Tricine. The EF_s and EF_u faces are virtually indistinguishable from those of control membranes (see also Figs. 10 and 11). × 68 000

the increase in EF_s particle density ceased at about 100mM-NaCl, the univalent ion requirement for membrane stacking was much higher and levelled off only at 150 mmol/l.

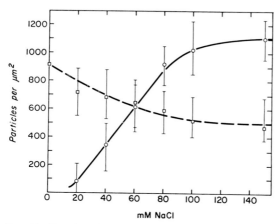

FIG. 10. Density of EF_s (o——o) and EF_u (□---□) particles of experimentally unstacked and then restacked pea thylakoid membranes as a function of NaCl concentration. Note that the particle density levels off just over 100mM-NaCl, and that the half maximal EF_s particle density is reached at about 50mM-NaCl (see also Figs. 5–9).

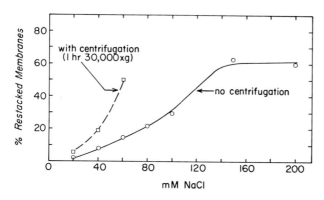

FIG. 11. Percentage of restacked membrane regions measured on thin-section micrographs of experimentally unstacked and then restacked pea thylakoids as a function of NaCl concentration. In the 'no centrifugation' (o——o) samples the maximal amount of restacked areas (60–65%, same as control thylakoids) is found at NaCl concentrations >150mM-NaCl; the half-maximal value is reached between 90- and 100-mM-NaCl. When the membranes are also pushed together by gravity forces (30 000 **g** for 1 h) during the restacking period, the amount of restacked membranes (□---□) is significantly increased. Technical difficulties have prevented us from obtaining accurate measurements of centrifuged samples containing >60mM-NaCl, but values of 70–80% restacked membrane areas may be assumed.

As proposed by Murakami & Packer (1971) and confirmed by Staehelin *et al.* (1977), thylakoid membrane stacking appears to be mediated by hydrophobic associations of molecules in adjacent membranes. However, Fig. 11 demonstrates that the amount of stacked membrane regions depends on the concentration of ions in the medium as well. If one assumes that the number of molecules carrying surface-exposed hydrophobic groups that can mediate membrane stacking remains constant in a given membrane preparation, then the univalent ion dependence of the stacking process could be explained as resulting from the reduction of membrane repulsion by Na^+ ion shielding of fixed negative charges. To test this hypothesis we have centrifuged isolated chloroplast membranes during restacking in NaCl solutions to force the membranes physically together and thereby increase the amount of stacked regions. The results shown in Fig. 11 illustrate that increased stacking at a given NaCl concentration can be brought about by restacking the membranes in a gravity field of 30 000 *g* for 1 h. This enhanced membrane stacking may last for up to 30 min and possibly longer. Although glutaraldehyde fixation of the samples immediately after centrifugation can be used to stabilize the membranes for thin-section electron microscopy, this treatment is unsuitable for freeze-fracture preparations, since glutaraldehyde changes the partitioning of the particles between the EF and PF faces in restacked thylakoids. For this reason no systematic freeze-fracture analysis of the centrifuged membranes could be done.

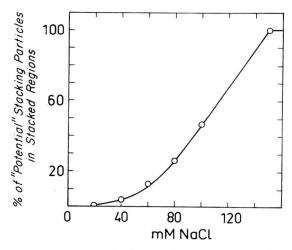

FIG. 12. Percentage of 'potential' large EF_s particles in stacked regions of experimentally unstacked and then restacked membranes as a function of NaCl concentration (calculated from data in Figs. 11 and 12). Note that the increase becomes linear above 80mM-NaCl and that a maximum value is reached at 150mM-NaCl.

The data of Figs. 10 and 11 have been further analysed by calculating how many of the potential EF$_s$ particles are in a stacked membrane region at any given NaCl concentration. For calculating this percentage it is assumed that all the potential stacking particles are in the stacked regions at 150mM-NaCl. The data (Fig. 12) demonstrate that few of the potential EF$_s$ particles are in adpressed lamellae over NaCl concentration ranges of 0–60 mmol/l. Above 80mM-NaCl, the number of potential EF$_s$ particles increased linearly until the maximum value was reached at 150 mM-NaCl.

Analyses of chlorophyll fluorescence characteristics were used to determine cation effects on excitation energy distribution in isolated chloroplast samples treated identically to those used for structural studies. Since relatively high glycerol concentrations must be used to obtain maximal sample clarity in the freeze-fracture preparations, our initial fluorescence studies were an evaluation

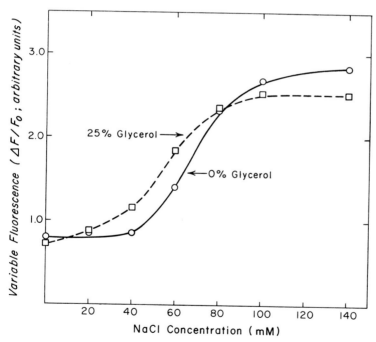

FIG. 13. The effect of a range of NaCl concentrations on variable fluorescence ($\triangle F$, normalized to initial fluorescence, F_0) in isolated pea chloroplasts suspended in 5mM-Na$^+$-Tricine (pH 7.8), NaCl as indicated and either 0 or 25% glycerol.

of cation effects on chloroplasts in the presence and absence of glycerol. The data of Fig. 13 show the effects of a range of salt concentrations on variable fluorescence yield (ΔF; normalized in each experiment by dividing by F_0). As has previously been reported (Papageorgiou 1975; Barber 1976), Na^+ ions increased fluorescence yield when added to low-salt chloroplasts. It should be emphasized that the experimental protocol for these experiments was the same as that used for structural studies: chloroplasts were osmotically disrupted in 20mM-NaCl and then were incubated on ice in 5mM-Na-Tricine (pH 7.8) containing the salt concentration indicated. At the time when the membrane samples were to be centrifuged to obtain a pellet for freeze-fracture analysis, aliquot portions were removed and diluted to 5 μg chl/ml with the buffer-salt solution ($\pm$ glycerol) appropriate for each sample. These aliquot portions were used for fluorescence measurements.

When glycerol was included in the samples used for analysis, there were small variations in the pattern of salt-induced increase in chlorophyll fluorescence intensity compared with the control sample (Fig. 13). These variations were

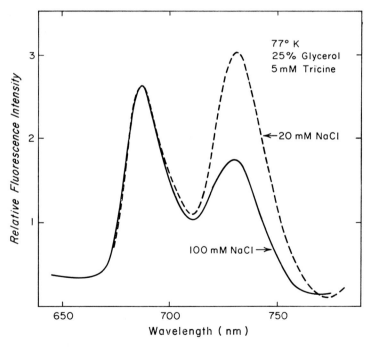

FIG. 14. The 77 K chlorophyll fluorescence emission spectrum of isolated chloroplasts suspended in 5mM-Na^+-Tricine (pH 7.8), 25% glycerol plus either 20- or 100-mM-NaCl.

primarily a shift to slightly lower salt concentrations required to give the cation induced fluorescence increase.

At 77 K, the chloroplast chlorophyll fluorescence emission spectrum is characterized by two major peaks. Fluorescence at 685 nm arises from the photosystem II pigment bed whereas that at 730 nm arises from photosystem I (Barber 1976; Papageorgiou 1975). In high (25%) glycerol, high Na^+ concentrations decreased the 730/685 nm intensity ratio (Fig. 14). Similar changes were observed at 15 or 35% glycerol. The amplitude of the salt-induced change in peak emission ratio was slightly less in samples containing no glycerol: 730/685 ratio $=$ 1.02 for 20mM-Na^+, 0 glycerol; 0.82 for 100mM-Na^+, 0 glycerol, as compared to 1.25 for 20mM-Na^+, 25% glycerol; 0.68 for 100mM-NaCl, 25% glycerol.

Since centrifugation of salt-treated samples caused large changes in the proportion of stacked membranes observed by electron microscopy, the effect of centrifugation on fluorescence characteristics of these samples was determined. In these experiments the centrifuged pellets were held on ice in the dark until immediately before assay; pelleted samples were then dispersed in sufficient solution (containing salts at the same concentrations as had been present before centrifugation) to give 5 μg chl/ml and the fluorescent transients were then quickly recorded. Representative recordings of the fluorescent transients for centrifuged (C) and non-centrifuged (NC) samples at salt concentrations ranging from 0 to 100mM-NaCl are shown in Fig. 15. Qualitatively, centrifugation had little effect on the transients. The fluorescence rise was sigmoidal at 60–100mM-NaCl in both samples and exponential in rise at 30mM-NaCl. Only at 40mM-NaCl (near the half-maximal Na^+ concentration required for the fluorescence increase) and to a lesser extent at 60–100mM-NaCl did centrifu-

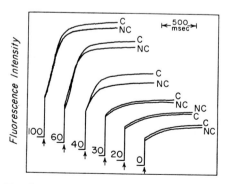

FIG. 15. Tracings of fluorescence induction transients obtained for chloroplasts incubated in 0, 20, 30, 40, 60 or 100 mM-NaCl and either centrifuged or not centrifuged at high speed before assay. Details of the sample preparations are described in Table 1.

TABLE 1

The effect of varying NaCl concentrations in the suspending media of isolated chloroplasts on the chlorophyll fluorescence characteristics of these plastids. All chloroplasts were osmotically shocked in 20mM-NaCl and then resuspended in 5mM-Na-Tricine (pH 7.8) containing the salt as indicated. Control samples were incubated on ice for 1 h before dilution (at the same salt concentrations) for fluorescence assay. Centrifuged samples were resuspended in Tricine–salt solutions and were then centrifuged at 20 000 g for 1 h. The pellets obtained were resuspended and assayed immediately. The chlorophyll concentration of the samples used for fluorescence assay was 5 ± 1 μg chl/ml; exact chlorophyll concentration of each of the samples was determined. The data were normalized to 5 μg chl/ml for each sample assuming direct proportionality of the fluorescence intensities to chlorophyll concentrations.

[NaCl] in chloroplast suspension (mmol/l)	Fluorescence intensity (in arbitrary units)					
	Control sample			Centrifuged sample		
	F_0	F_M	$\triangle F$	F_0	F_M	$\triangle F$
0	12	17	5	13	19	6
20	16	23	7	14	20	6
30	17	24	7	18	26	8
40	17	30	13	18	35	17
60	16	39	23	17	44	27
100	16	44	28	16	46	30

gation exhibit a clearly measurable, though limited, effect over the control by increasing the apparent effectiveness of the salt in stimulating variable fluorescence.

A quantitative evaluation of the components of the fluorescence transients of Fig. 15 is presented in Table 1. At low salt concentrations (20–30mM-NaCl) there was a significant increase in F_0 over the low-salt control in both centrifuged and non-centrifuged preparations. This increase is similar to the increase of F_0 observed in low MgCl$_2$ solutions by Henkin & Sauer (1977). Increasing the salt concentration from 30mM- to 100mM-NaCl greatly stimulated variable fluorescence (ΔF).

DISCUSSION OF RESULTS

The observations reported in this paper both confirm and extend previous reports on the effects of cations on chloroplast membrane architecture (Izawa & Good 1966; Staehelin 1976), and on excitation energy distribution between PS II and PS I (Murata 1969; Papageorgiou 1975). We have confirmed that the addition of cations to thylakoids suspended in a low-salt medium can lead to restacking of the membranes, to the aggregation of large EF face particles into stacked membrane regions, and to an increase in exciton arrival at PS II. We have presented a quantitation of these effects for different univalent ion concentrations and a correlative analysis of the structural and the chlorophyll fluorescence data. In the following discussion we shall first describe the evidence suggesting that some of the components of the light harvesting pigment–protein complex are closely associated with all the cation-mediated phenomena. Then we shall present a hypothesis to explain the observed sequence of salt-induced changes in membrane structure. We shall then compare our fluorescence data with comparable reports in the literature, and, finally, examine the possible relationship between the structural and fluorescence results.

Involvement of LHC in the cation effects

In previous studies (Armond *et al.* 1976, 1977; Davis *et al.* 1976) the appearance of the light-harvesting pigment–protein complex (the LHC) in developing chloroplast membranes has been analysed. It was concluded that there was a parallel appearance of the LHC, cation regulation of energy distribution between the two photosystems (Davis *et al.* 1976), grana stacking (Armond *et al.* 1976, 1977), and cation regulation of the mobility of the large EF face particles (Armond *et al.* 1977). The large EF face particles were also shown to consist of a central PS II core surrounded by one, two or four LHC aggregates (Armond *et al.* 1977; see also Fig. 18). In studies with the isolated, purified LHC, Burke *et al.* (1978) found that cations cause the aggregation of separated LHC subunits. The resulting aggregates, when examined by thin sectioning, appeared in lamellar structures that resembled grana stacks. The cation concentrations required to elicit this aggregation were similar to those required to induce stacking in native chloroplasts. Considered together, the above-mentioned studies lead to the conclusion that the LHC contains cation-binding sites, that cation binding to these sites causes LHC interaction, and that these interactions control the establishment of grana stacks in intact membranes. This conclusion is in agreement with other data in the literature (see reviews of Anderson 1975; Arntzen 1978).

Salt-induced changes in membrane structure

Fig. 16 summarizes diagrammatically the changes in thylakoid configuration and in energy transfer between the photosystems that can be elicited by changes in salt concentrations. All the structural changes seem to reflect changes in the balance of adhesive and repulsive forces between membrane components within the plane of each membrane and between adjacent membranes. Thus, when

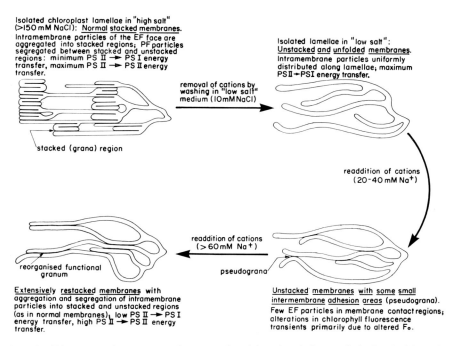

FIG. 16. Diagrammatic summary of structural and functional changes in isolated chloroplast membranes subjected to experimental unstacking and restacking conditions.

the membranes are transferred from a solution containing 150mM-NaCl to one containing only 10mM-NaCl, the resulting reduced screening of the negative charges associated with membrane surface components (particularly LHC) leads to increased membrane repulsion and thus to unstacking of the thylakoids. With the structural constraints of the stacked membrane regions gone, the electrostatic repulsion between the negatively charged components also leads to their dispersal within the plane of the membrane, and to the randomization of all membrane particles (Fig. 4).

The addition of cations to the low-salt thylakoid system increases the shielding of the negative charges on the membrane surfaces, enabling the membranes to come closer together and to establish hydrophobic interactions or van der Waals bonds, or both, between the LHC subunits (Staehelin *et al.* 1977, Fig. 18). The following reaggregation of the EF face particles into the restacked membrane regions that occurs when the univalent salt concentration is increased to >60 mmol/l can best be explained by the principle of 'entropic ordering' illustrated in Fig. 17.

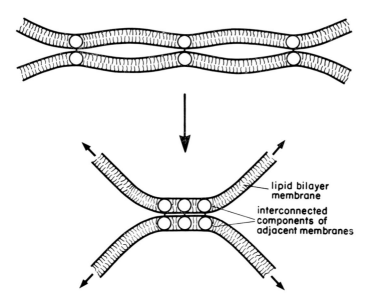

lipid bilayer
membrane

interconnected
components of
adjacent membranes

FIG. 17. Schematic diagram illustrating how intermembrane adhesion between particles of adjacent fluid membranes and entropic forces (small arrows) can lead to the aggregation and ordering of the particles in the contact region.

At this point we may ask why the aggregation of EF$_s$ face particles saturates at about 100mM-NaCl (Fig. 10) although membrane stacking only saturates at about 150 mmol/l (Fig. 11). Thus, although the EF$_s$ face particle density reaches a maximum at 100mM-NaCl, the percentage of potential EF$_s$ face particles in stacked regions at 100mM-NaCl is actually less than 50% (Fig. 12). The simplest explanation seems to be that below 100mM-NaCl the EF$_s$ particles (LHC aggregates) still carry residual, uncompensated negative charges that lead to particle repulsion. Above 100mM-NaCl and below 150mM-NaCl repulsion between the EF$_s$ particles is gone, but some residual negative charges associated with components other than the EF$_s$ particles continue to repulse adjacent membranes and thereby supplement the separative, entropic forces. In support of the notion that charge repulsion prevents thylakoid membranes from forming extensive stacked membrane regions at intermediate salt concentrations, we have observed that increased stacking can be achieved at these salt concentrations by overcoming the charge repulsion with gravitational forces (Fig. 11).

This leaves us with the question of how the small stacked membrane regions devoid of EF face particles (pseudograna) are formed at 20–40mM-NaCl (Figs.

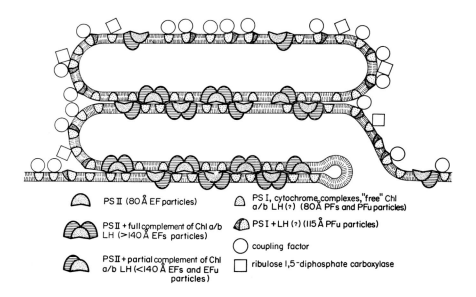

PS II (80 Å EF particles)

PS I, cytochrome complexes, "free" Chl a/b LH (?) (80 Å PFs and PFu particles)

PS II + full complement of Chl a/b LH (>140 Å EFs particles)

PS I + LH (?) (115 Å PFu particles)

PS II + partial complement of Chl a/b LH (<140 Å EFs and EFu particles)

coupling factor

ribulose 1,5-diphosphate carboxylase

FIG. 18. Schematic illustration of the supramolecular organization of thylakoid membranes of higher plants and green algae. Note the different composition of the stacked (grana) and unstacked (stroma) membrane regions. This differentiation appears to result from the adhesion between LHC–PS II complexes in adjacent membranes, and the concomitant physical exclusion of components not directly associated with the electron-transport chain. Our results support the hypothesis that it is this functionally more efficient packaging of PS II, PS I and cytochrome complexes in stacked regions that is the main function of membrane stacking.

5, 6 and 10). We believe that at these relatively low salt concentrations some LHC aggregates become dissociated from the larger PS II–LHC units and that these free LHC aggregates (which now appear as 7 nm PF particles) associate to form very tight regions of membrane adhesion that physically exclude most of the larger PS II–LHC units (the EFs particles).

Cation effects on chlorophyll fluorescence

Since the freeze-fracture samples contained glycerol for optimal freezing and image quality, the salt effects on chlorophyll fluorescence were examined both in the presence and absence of glycerol. Cations increased the intensity of chlorophyll fluorescence in isolated chloroplasts suspended over a range of glycerol concentrations. In general, NaCl seemed to be more effective in stimulating the fluorescence intensity at high (in contrast to low) glycerol concentrations. Both the change in fluorescence emission spectra at 77 K and the increase in variable fluorescence (ΔF), in the presence or absence of glycerol, are in agreement with the widely held view that cations regulate excitation energy distribution between photosystems I and II (Murata 1969; Papageorgiou 1975; Barber 1976). We conclude that the inclusion of glycerol in our reaction media does not qualitatively alter the physiological effects of cations on the membranes. The structural data obtained in the same conditions by freeze-fracture analysis can, therefore, be correlated directly to cation-induced functional changes in native membranes.

The observation that low salt levels (0–30mM-Na$^+$) stimulated the immediate (F_0) level of fluorescence (Fig. 16, Table 1) is in agreement with earlier studies of Mg^{2+} effects on chloroplasts by Henkin & Sauer (1977). In their analysis, stimulation of both F_0 and ΔF was observed, but the ΔF effect required higher MgCl$_2$ concentrations. They attributed these observations to cation effects on two different fluorescing sites. We can conclude that the same phenomena hold true for univalent ions. It is most striking that the low salt concentrations that give rise to pseudograna also produce a stimulation of the F_0 level of fluorescence. Whether this reflects a common mechanism, and what type of mechanism this may be, are not known at present.

Correlation between structural and fluorescence data

Although a direct relationship between the formation of pseudograna and F_0 fluorescence levels at low salt concentrations can possibly be envisaged, we have been unable to establish any positive correlation between the development of stacked membrane regions and cation-induced fluorescence levels at intermediate to high salt concentrations (40 to 150mM-NaCl). As seen in Fig. 13, the Na$^+$-induced increase in variable fluorescence reached a half maximal value at about 50mM-NaCl, and saturated at 100mM-NaCl. The cation requirement for grana stacking, on the other hand, was much higher (half maximal increase at about 100 mmol/l, saturated response at 150 mmol/l; Fig. 11). Only in centrifuged samples did the percentage of membrane stacking increase dramatically at low Na$^+$ concentrations, but the centrifugation effects on fluorescence were slight.

A somewhat closer relationship appears to exist between the salt-induced increase in EF$_s$ face particle density (Fig. 10) and the increase in variable fluorescence (Fig. 13). Both these phenomena have a half maximal Na$^+$ requirement of about 50 mmol/l and saturate at about 100 mmol/l. However, the fact that at 100mM-NaCl only 50% of the potential EF$_s$ face stacking particles are found in the stacked membrane regions indicates that particle aggregation is not responsible for the observed fluorescence changes. However, we believe that the pattern of EF face particle migration along the membrane is a sensitive and direct probe of the degree of interaction (and repulsion) of LHC subunits as controlled by cations.

CONCLUSIONS

Neither thylakoid membrane stacking nor the aggregation of EF face particles into stacked membrane regions is a primary factor regulating the distribution of excitation energy between the photosystems I and II as measured by chlorophyll fluorescence changes. The fluorescence parameters probably reflect molecular changes that occur within the EF face particles (and are too small to be visualized by freeze-fracture techniques) and affect their relationship with immediate neighbours (PF face and EF face particles). Thylakoid membrane stacking, in turn, may provide a mechanism for concentrating the components of the electron-transport chain and thereby increase the overall efficiency of the system in light-limiting conditions.

ACKNOWLEDGMENTS

Thanks are due to Marcia DeWit and to Cathy Ditto for their excellent technical assistance. This work was supported in part by the National Institute of General Medical Sciences under Grant GM 22912 to L.A.S. and by the Department of Energy under Contract EE-77-S-02-4475.A000 to C.J.A.

References

ANDERSON, J. M. (1975) The molecular organization of chloroplast thylakoids. *Biochim. Biophys. Acta 416*, 191–235

ARMOND, P. A. & ARNTZEN, C. J. (1977) Localization and characterization of photosystem II in grana and stroma lamellae. *Plant Physiol. 59*, 398–404

ARMOND, P. A., ARNTZEN, C. J., BRIANTAIS, J.-M. & VERNOTTE, C. (1976) Differentiation of chloroplast lamellae: light harvesting efficiency and grana development. *Arch. Biochem. Biophys. 175*, 54–63

ARMOND, P. A., STAEHELIN, L. A. & ARNTZEN, C. J. (1977) Spatial relationship of photosystem I, photosystem II, and the light-harvesting complex in chloroplast membranes. *J. Cell Biol. 73*, 400–418

ARNTZEN, C. J. (1978) Dynamic structural features of chloroplast lamellae, in *Current Topics in Bioenergetics* (Sanadi, D. R. & Vernon, L. P., eds.), pp. 111–160, Academic Press, New York

BARBER, J. (1976) Ionic regulation in intact chloroplasts and its effect on primary photosynthetic processes, in *The Intact Chloroplast* (Barber, J., ed.), pp. 89–134, Elsevier/North Holland, New York

BURKE, J. J., DITTO, C. L. & ARNTZEN, C. J. (1978) Involvement of the light harvesting complex in cation regulation of excitation energy distribution in chloroplasts. *Arch. Biochem. Biophys. 187*, 252–263

DAVIS, C. J., ARMOND, P. A., GROSS, E. L. & ARNTZEN, C. J. (1976) Differentiation of chloroplast lamellae: onset of cation regulation of excitation energy distribution. *Arch. Biochem. Biophys. 175*, 64–70

HENKIN, B. M. & SAUER, K. (1977) Magnesium ion effects on chloroplast photosystem. II. Fluorescence and photochemistry. *Photochem. Photobiol. 26*, 277–286

HOMANN, P. H. (1969) Cation effects on the fluorescence of isolated chloroplasts. *Plant Physiol. 44*, 932–936

IZAWA, S. & GOOD, N. E. (1966) Effects of salts and electron transport on the conformation of isolated chloroplasts. II. Electron microscopy. *Plant Physiol. 41*, 544–553

MILLER, K. R. & STAEHELIN, L. A. (1976) Analysis of the thylakoid outer suface: coupling factor is limited to unstacked membrane regions. *J. Cell Biol. 68*, 30–47

MURAKAMI, S. & PACKER, L. (1971) The role of cations in the organization of chloroplast membranes. *Arch. Biochem. Biophys. 146*, 337–347

MURATA, N. (1969) Control of excitation transfer in photosynthesis. III. Magnesium ion-dependent distribution of excitation energy between two pigment systems in spinach chloroplasts. *Biochim. Biophys. Acta 189*, 171–181

PAPAGEORGIOU, G. (1975) Chlorophyll fluorescence: an intrinsic probe of photosynthesis, in *Bioenergetics of Photosynthesis* (Govindjee, ed.), pp. 319–371, Academic Press, New York

STAEHELIN, L. A. (1976) Reversible particle movements associated with unstacking and restacking of chloroplast membranes *in vitro*. *J. Cell Biol. 71*, 136–158

STAEHELIN, L. A., ARMOND, P. A. & MILLER, K. R. (1977) Chloroplast membrane organization at the supramolecular level and its functional implications. *Brookhaven Symp. Biol. 28*, 278–315

Discussion

Barber: Changes in chlorophyll fluorescence similar to those you described resulting from additions of salt to isolated thylakoids are seen in intact organisms when they change from State I to State II. In these conditions the efficiency of photosynthesis also changes. Is there any evidence for particle movements that can be related to changes between State I and State II?

Staehelin: I view the transition from State I to State II as resulting from the displacement of magnesium from the luminal side of the PS II–LHC complex. Thinking in gross morphological terms, if the hydrophobic interactions between the PS II cores and the LHC aggregates remain constant but a salt linkage between the two is broken or weakened, the light-harvesting complex might become partly separated from the photosystem II core. If the photosystem I elements lie adjacent, interaction between them and the light-harvesting complex could increase while that between the light-harvesting complex and photosystem II decreases. The whole system would be stabilized by hydrophobic interactions both in the plane of the membrane and between the membranes oj a stack; these temporary changes in ionic concentrations might not affect the overall structure much.

Tredwell: In your model the fluorescence quantum yield might be expected to increase as the magnesium ion concentration is reduced, since the light-harvesting complexes are being separated from one of the quenching species, namely the photosystem II reaction centres.

Joliot: It is easier to imagine energy transfer between units inside the photosystem II particle than between these particles which are too far apart from each other. Experiments with *Cyanidium* (Diner & Wollman 1978) suggest that energy transfer between photosystem II units occurs in a small core of chlorophyll molecules. Could there be more than one photocentre per particle?

Staehelin: From other studies on re-greening systems (Armond *et al.* 1977) we conclude that the core element is between 7.0 and 8.0 nm in size. I don't know how many photosystem II elements can be packed into that size particle.

Butler: I have the impression that several orders of magnitude separate the gross structural changes that you observe, Dr Staehelin, from the rather subtle structural changes which one infers from studies of fluorescence and energy distribution. Energy transfer between adjacent photosystem II units or between photosystem II and photosystem I units will occur only if the units are in close juxtaposition and the changes of energy transfer which can be observed in chloroplasts (e.g. in the addition of bivalent cations) probably result from slight changes of physical structure which are not detected by electron microscopy.

Staehelin: The aggregate of light-harvesting complexes probably consists of two elements: a pigment component and an adhesive component. At present we cannot distinguish which component is responsible for which salt effects. This might explain the fluorescence results.

Thornber: How many chlorophyll molecules does one light-harvesting complex contain?

Staehelin: We cannot answer that yet. We know that when there are no light-harvesting complexes the size of these particles drops to 7.0–8.0 nm. As the light-harvesting complex is incorporated into the membrane the particles systematically increase in size up to 13–16 nm.

Porter: And these aggregates must be composed of about 20–30 particles?

Staehelin: Yes.

Joliot: Wollman has shown that with *Cyanidium*, which contains phycocyanin, the size of the particles attributed to system II is 10 nm, that is, close to the size of the particles of green plants which lack CP II.

Staehelin: We have found that photosystem II particles of *Anabaena cylindrica* also measure 10.0 nm in diameter. Furthermore, during heterocyst development in this alga, which involves the loss of photosystem II activity, the 10.0 nm particles disappear (Giddings & Staehelin 1979).

Junge: How often would the two-dimensional quasi-crystalline arrays that you showed appear in 100 electron micrographs?

Staehelin: In winter using Mexican spinach we see such chloroplasts in about one in 10 experiments—we might have to look at between 100 and 1000 chloroplasts before we find one. Unfortunately, we cannot produce those arrays experimentally.

Porter: I am still doing my calculations: this aggregate contains about 30 particles, each containing about seven chlorophyll molecules and this makes up something about the size of a photosynthetic unit?

Staehelin: Possibly. Could there be four centres in each large EF_s particle? From the luminal side we observe up to four subunits. The four centres could form the core with one, two, or four antennae aggregates forming the outer shell.

Anderson: We do not know whether the size of the particle represents the true *in vivo* size because of the shadowing.

Staehelin: No; that is the source of much confusion in the literature. To measure the particles we measure the width of the shadow, not the width of the heavy metal that we put on—that would give a false value. More significant than the shadowing problem is the possibility of plastic deformation when the membrane is torn apart; we have no measure of that. But the size of the particles is certainly related to what is present.

Anderson: Negative staining gives a smaller size because the diameter of the chloroplast coupling factor is 15.0 nm by freeze-*etching* but only 10.0 nm by negative staining.

Staehelin: That will always be so because in negative staining the stain will penetrate between the subunits and around the structures.

Wessels: You suggest that the light-harvesting complex is responsible for stacking?

Staehelin: Yes—or at least a component of the complex, as I suggested earlier.

Wessels: Dr Thornber (1975) has reported, however, that some stacking occurs in mutants which lack the light-harvesting complex. Could the 'adhesion component' still be present in these mutants?

Staehelin: We tried to test that. In those systems which exhibit stacking but which lack chlorophyll *b* the particles are no longer 16 nm but 10–12 nm across. Since they still possess four smaller lobes (Miller *et al.* 1976), I believe that the large particles normally contain light-harvesting complexes with at least two components.

Malkin: You said that 20% of the large particles are in the stroma but there have been several reports that the stroma does not contain significant amounts of the photosystem II primary reactants. For example, Erixon & Butler (1971) could find little C550 in the stromal preparation and we could not detect any significant amounts of P680 (Malkin & Bearden 1975). If this is so, what are all those particles doing in the stroma?

Staehelin: Separation of the stromal and granal fractions indicates 20% and 80% photosystem II, respectively, in each. Experiments that I prefer consist of the application to isolated chloroplasts of an external agent (e.g. antibodies or iodinating agents etc.) which does not permeate the membrane but which can kill exposed photosystem II units. Results of such experiments consistently indicate that about 20% of the photosystem II activity is knocked out by these externally applied agents—i.e. 20% is exposed and 80% is protected in the stacked regions.

Joliot: Why are the membranes in a mutant which lacks CP I completely stacked, as has been shown by Goodenough & Levine (1969)?

Staehelin: If the system lacks CP I particles the relative proportion of stacking components will increase, thus leading to a higher percentage of stacked membranes.

Barber: Membranes will stack at their most hydrophobic regions, that is where the charge density is low or well shielded. If CP I is a major charge-carrying component of the membrane, then its absence will result in a more hydrophobic membrane and lead to stacking. Furthermore, in general terms, it is not necessary to propose a special 'stacking component'; membranes will stack when conditions render some regions more hydrophobic than others. Changing the conditions of the medium may unstack the membranes in one region but stack them in another.

Staehelin: Isn't this semantics? You are saying that *regions* on the membrane are hydrophobic: I am saying that *components* of the membrane are hydrophobic.

Barber: I am proposing a physical mechanism for stacking as opposed to a special chemical linkage effect.

Staehelin: The physical fact is that, when the membranes begin to adhere to each other and restack, we see a re-segregation of particles and an exclusion of certain components.

Barber: This is not surprising since the organization and movement of particles are controlled by the same basic laws that control stacking, as discussed in detail in my paper (pp. 283–298).

Staehelin: What is the difference between those 'particles' and the components I discussed? The exposed hydrophobic regions most likely do not consist of lipids, and so they must be proteins. If you accept that, why not assume that certain proteins have this feature?

Barber: When we did electrophoretic measurements on chloroplast membranes (Nakatani *et al.* 1978) we found that the light-harvesting chlorophyll *a/b*–protein complex apparently does not contribute significantly to the surface charge; that is, we found that thylakoids isolated from the wild-type barley and from a mutant lacking chlorophyll *b* had similar electrophoretic mobility.

Staehelin: So how do you explain the fact that isolated light-harvesting complexes mixed with magnesium ions in solution aggregate in huge clumps?

Barber: Experiments of this type with isolated detergent-coated complexes may not be applicable to salt effects observed with intact membrane systems.

Thornber: For example if SDS is present then magnesium links the SDS molecules.

Staehelin: But this is in Triton. Without magnesium the complexes do not aggregate; only when one adds 10mM-magnesium do they clump.

Barber: Nevertheless our particle electrophoretic studies indicate that although the light-harvesting chlorophyll *a/b*–protein complex is a major membrane constituent it does not seem to contribute significantly to the external surface charges on the thylakoid membrane.

Butler: Is it possible that the large mobile particles are hydrophobic proteins which are involved in the fusion of adjacent thylakoids but are only indirectly related to chlorophyll proteins? It is natural that we should try to find entities in electron micrographs that we can label photosystem I and II but I worry about the 'Siren's call' of your micrographs. They show well that the large particles tend to congregate in the stacked regions of the thylakoids but the energy-transfer properties could be rationalized more readily by assuming that the photochemical apparatus is in the membrane regions between the large particles.

Joliot: Wollman (1978) observed in *Cyanidium* a correlation between the number of the 10 nm particles and the amount of photosystem II centres.

Staehelin: The large particle contains both the photosystem II core and the light-harvesting protein. The question seems to be whether the latter also contains the adhesive component. Membranes stack when the light-harvesting complex is inserted. The isolated complex aggregates in clumps at the same concentration of ions as that needed for membrane adhesion. Also, gel electrophoresis of the light-harvesting complex in the cold in mild conditions gives a component of molecular weight 70 000 which contains a pigment protein and a glycoprotein. Although we have not proven this yet, we believe that the glycoprotein is the part of the molecule responsible for adhesion. We are at present purifying the complex and putting it into reconstituted vesicles to see whether we can induce the membranes to adhere together. After that we want to purify the glycoprotein and the pigment protein, insert them into vesicles and see whether the membranes adhere.

Junge: Between the stacked and the unstacked configuration there is much structural difference but little difference on the functional side. What is the teleological reason for stacked membranes? As the transition moments of the chlorophylls absorbing at the red end of the spectrum are oriented in plane it would be most favourable for maximum use of light if membranes were oriented perpendicular to the propagation vector of the light. As chloroplasts orient themselves within the cell according to the light it is conceivable that the parallel alignment of thylakoid membranes in the form of a stack serves to bring them into the same orientation to the incoming light.

Staehelin: Possibly; when a chloroplast adjusts to light, the whole chloroplast appears to rotate.

Barber: The advantage of stacking may be to reduce the strikingly large surface area/volume ratio of the thylakoid compartment. This would have the effect of reducing the leak of ions and enabling light-induced electrical and chemical gradients to be established across the membrane. Even if the thylakoids have a very low permeability to ions the effect of a large surface area and small internal volume would combine to form a leaky system and interfere with the efficiency of the energy-conserving processes which occur at this membrane.

Junge: But we find that the leak conductivity for protons, which determines the efficiency of photophosphorylation, is only little changed on going from the unstacked to the stacked configuration.

The relaxation time of the electrochromic absorption changes at 520 nm on flashing light does not represent the relaxation of a proton gradient, provided the repetition rate of flashes is low. With flashes at 0.1 s^{-1} the proton gradient relaxes at 5–10 s while the electric field relaxes at some 100 ms (e.g. Junge & Ausländer 1974). Only at higher frequencies and as a consequence of acidification of the internal phase by more than one pH unit, protons start to dominate the electric conductance of the thylakoid membrane (Gräber & Witt 1976).

Duysens: The grana may act as a pump—diffusion may not be able to move protons or NADP(H) sufficiently.

Anderson: I agree; I cannot believe that there is no hydrophilic region between the stacked membranes. How does the NADP get out if it is all hydrophobic?

Staehelin: The NADP is not produced there; it is formed in the unstacked membrane region.

Anderson: I disagree completely. The coupling factor and the mechanism for phosphorylation may be in the unstacked regions only, but NADP is formed in the stacked regions. What are photosystems I and II doing in such quantities in the stacked region if not making NADP?

Staehelin: Antibodies against the NADP-producing enzymes destroy the whole activity when the coupling factor is first removed. I have always assumed that the antibody-labelling experiments were done in conditions that prevented unstacking; but maybe they were not, and Dr Anderson's theory would be correct.

I agree with Dr Barber in as much as stacking would provide for more efficient membrane packing. But that is only one reason for explaining the functional significance of thylakoid stacking; the second is that components do segregate and aggregate in the stacked regions. It is surely not coincidence that the elements of the electron-transport chain concentrate in the stacked regions and components that are not as directly associated with these reactions are excluded.

References

ARMOND, P. A., STAEHELIN, L. A. & ARNTZEN, C. J. (1977) Spatial relationship of photosystem I, photosystem II, and the light harvesting complex in chloroplast membranes. *J. Cell Biol. 73*, 400–418

BARBER, J. (1978) Energy transfer and its dependence on membrane properties, in *This Volume*, pp. 283–298

DINER, B. & WOLLMAN, F. A. (1978) A functional comparison of the photosystem II center-antenna complex of a phycocyanin-less mutant of *Cyanidium caldarium* with that of *Chlorella pyrenoidosa. Plant Physiol.*, in press

ERIXON, K. & BUTLER, W. L. (1971) Light-induced absorbance changes in chloroplasts at —196 °C. *Photochem. Photobiol. 14*, 427–433

GIDDINGS, T. H. & STAEHELIN, L. A. (1979). *Biochim. Biophys. Acta*, in press

GOODENOUGH, U. W. & LEVINE, R. P. (1969) Chloroplast ultrastructure in mutant strains of *Chlamydomonas reinhardi* lacking components of the photosynthetic apparatus. *Plant Physiol. 44*, 990–1000

GRÄBER, P. & WITT, H. T. (1976) Relations between the electrical potential, pH gradient, proton flux and phosphorylation in the photosynthetic membrane. *Biochim. Biophys. Acta 423*, 141–163

JUNGE, W. & AUSLÄNDER, W. (1974) Electric generator in photosynthesis of green plants. I. Vectorial and protolytic properties of electron-transport chain. *Biochim. Biophys. Acta 333*, 59–70

MALKIN, R. & BEARDEN, A. J. (1975) Laser-flash-activated electron-paramagnetic resonance studies of primary photochemical reactions in chloroplasts. *Biochim. Biophys. Acta 396*, 250–259

MILLER, K. R., MILLER, G. J. & McINTYRE, K. R. (1976) The light-harvesting chlorophyll–protein complex of photosystem II. Its location in the photosynthetic membrane. *J. Cell Biol. 71*, 624–638

NAKATANI, H. Y., BARBER, J. & FORESTER, J. A. (1978) *Biochim. Biophys. Acta*, in press

THORNBER, J. P. (1975) Chlorophyll–proteins: light-harvesting and reaction center components of plants. *Annu. Rev. Plant Physiol. 26*, 127–158

WOLLMAN, F. A. (1978) Correlation of biophysical measurements with ultrastructural observations of *Cyanidium caldarium* wild type and IIIC mutant lacking phycobilisomes. *Plant Physiol.*, in press

Fluorescence of light-harvesting chlorophyll a/b–protein complexes: implications for the photosynthetic unit

ROBERT S. KNOX and RICHARD L. VAN METTER*

Department of Physics and Astronomy, University of Rochester, Rochester, New York

Abstract   To the extent that extracted light-harvesting chlorophyll proteins (LHCPs) retain the chlorophyll configuration which they had *in vivo*, information on the optical properties of LHCPs is useful for an assessment of the transfer process of the primary excitation energy in photosynthesis. Within this context we report and discuss the implication of three kinds of data on spinach chloroplast LHCP. First, an analysis of the spectroscopic dependence of absorption, polarization and circular dichroism (reported recently by R.L.V.) suggests a model affording the possibility of easy chlorophyll *a* intercomplex transfer with chlorophyll *b* groups acting as local antitraps.   Second, the ratio of LHCP emission and absorption probabilities obeys the Stepanov relation over a relatively wide range, an observation which suggests rapid Chl *b*–Chl *a* excitation equilibration. Finally, an LHCP absolute fluorescence yield as great as 10% has been measured, which provides a possible upper limit for the yield of the antenna fluorescence.

The chlorophylls of green plants and alga are found in two or possibly three varieties of protein complexes (Thornber *et al.* 1977). Here we shall be concerned with the light-harvesting chlorophyll a/b–protein (LHCP), known also as CP II, containing three chlorophyll *a* (Chl *a*) and three chlorophyll *b* (Chl *b*) molecules per 29 500 molecular weight of protein. We report on an attempt to establish aspects of the arrangement of the chromophores in this complex by analysing optical data, particularly fluorescence emission. We assume that the LHCP has a unique structure at least within the one species, spinach, on which we made our measurements. This seems to be a reasonable assumption in view of the existence of a complete structure determination of a corresponding

---

Present address: Physics Division, Research Laboratories, Eastman Kodak Co., Rochester, New York 14650

complex in the green bacterium *Prosthecocloris aestuarii* (Fenna & Matthews 1975, 1977; Fenna *et al.* 1977).

In discussing the possible significance of our results we shall generally make the implicit assumption that the complexes studied have retained their *in vivo* structure. This must remain an assumption whose validity can be assessed at present only by the invariance of the results to isolation techniques and by spectral similarities between the isolated material and the original chloroplasts.

MATERIAL AND METHODS

LHCP were isolated from spinach chloroplasts by gel electrophoresis and hydroxyapatite column chromatography. Absorption was measured with a Perkin-Elmer Coleman 124 double-beam spectrophotometer and fluorescence polarization by photon-counting techniques (for details see Van Metter 1977 *a,b*). The absolute fluorescence yields reported later were calculated from spectra taken with the automated research fluorimeter maintained in the Eastman Kodak spectrophotometry laboratory (Rochester, N.Y.) under the supervision of L. Costa.

MODEL BASED ON OPTICAL PROPERTIES

We shall summarize the model developed by one of us (Van Metter 1977 *a, b*), who was able to resolve the absorption spectrum of spinach LHCP into six Gaussian components (631, 651, 662, 670, 677 and 684 nm), in general agreement with the results of Brown *et al.* (1974) on tobacco-leaf LHCP. With this as a guide, we selected four major components (one Chl *b* at 651; two Chl *a* at 670; one Chl *a* at λ_1) to fit simultaneously the absorption and 730 nm fluorescence polarization excitation spectrum in the region 650–700 nm. In addition to the parameter λ_1, three relative angles of the emitting dipoles were allowed to vary. The results were $\lambda_1 = 676.5$ nm (consistent with the more accurate absorption deconvolution) and a set of relative angles which indicate that of the three Chl *a* transition dipoles no two are even nearly parallel, the smallest angle between dipoles being about 50°. The resulting polarization excitation spectrum and the curve that best fits the data are shown in Fig. 1. In this analysis we assumed that absorption into any of the excited levels in the region 650–700 nm leads to emission from Boltzmann-distributed excited states of Chl *a*. This assumption was supported by the fact that the relative quantum yield in that region was constant within experimental error.

From circular dichroism measurements Van Metter showed that the Chl *b* transition moments exhibited exciton coupling with no such coupling dis-

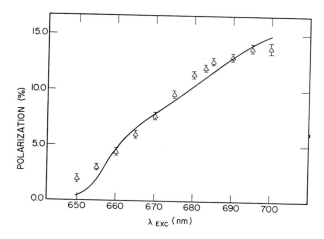

FIG. 1. The LHCP 730 nm fluorescence polarization excitation spectrum ($\triangle$) and a best fit to the data between 650 and 700 nm (———) based on a four-component model (see text) with relative Chl *a* angles and lowest-energy component position taken as adjustable parameters (Van Metter 1977*a*).

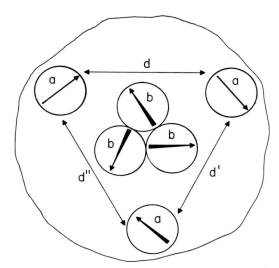

FIG. 2. Hypothetical arrangement of chromophores in LHCP: b, Chl *b*; a, Chl *a*. The distances *d*, *d'*, *d''* are unknown but relative angles within the Chl *a*s and Chl *b*s may be specified. The diameter of the figure is about 4.4 nm (Van Metter 1977*a*).

cernable among the Chl *a* moments. This led him to propose the model depicted in Fig. 2: the Chl *a*s are situated on the periphery, separated by unknown distances which are small enough for efficient Förster transfer but large enough to produce negligible exciton splitting, and the Chl *b*s are close together, also at unknown separations but close enough to produce an exciton splitting of about 10 nm at 650 nm, which corresponds to roughly an 0.8 nm separation in a dimer with transition moments of typical size (0.1 nm times the electronic charge). A unique determination of the Chl *b* separations was not possible because the spectrum could not be separated from the Chl *a* background. However, there are strong indications that the Chl *b* is associated in a strongly coupled trimer: for instance, simultaneous deconvolution of the absorption with the Chl *b* width as a free parameter leads to a width which is smaller than the monomer width by a factor of about $1/\sqrt{3}$. Such an exciton narrowing is analogous to spin resonance line narrowing and has recently been shown by Hemenger (1977) to be expected in the case of strong exciton coupling.

ABSORPTION–EMISSION RELATIONSHIP

In certain general conditions, the absorption and emission spectra of a homogeneous solution of molecules are simply related to each other. This relationship, first explored carefully by Kennard (1926) and given a modern interpretation by Stepanov (1957), is most conveniently expressed in terms of the function $\mathscr{F}(\nu)$ (equation 1) where $W(\nu)$ is the luminescent power at frequency

$$\mathscr{F}(\nu) \equiv \ln[W(\nu)/\nu^3 \varkappa(\nu)] = -(h\nu/k_B T) + D(T) \tag{1}$$

ν ($h\nu$ times the emission probability per second), $\varkappa(\nu)$ is the absorption coefficient at frequency ν, T is the absolute temperature, k_B is Boltzmann's constant and $D(T)$ is a function independent of frequency. The function $\mathscr{F}(\nu)$ is indeed found to be linear in ν for many molecules but it occasionally displays a slope characteristic of an effective temperature T^* other than the ambient. The two most-commonly proposed explanations of this anomaly are inhomogeneity in the sample and 'warm fluorescence', the latter meaning emission before complete thermal equilibration occurs in the excited electronic state. Further details and references are given elsewhere (Van Metter & Knox 1976), where the exceptionally-high effective temperatures ($T^* \gtrsim 400$ K) measured in dilute chlorophyll solutions (Szalay *et al.* 1974) are discussed.

The constant fluorescence yield, apparent quasi-equilibration, and well-overlapping absorption and fluorescence bands of LHCP prompted us to calculate $\mathscr{F}(\nu)$ for the *a/b* complexes. The results were surprisingly good (Fig. 3). No significant deviation in $\mathscr{F}(\nu)$ from a straight line occurs over the entire range in which absorption and emission are jointly measurable (650–700 nm) and the effective temperature deduced is 296.8 ± 4.5 K (340 K in certain anomalous samples; see below). This supports the idea that the LHCP sample is homoge-

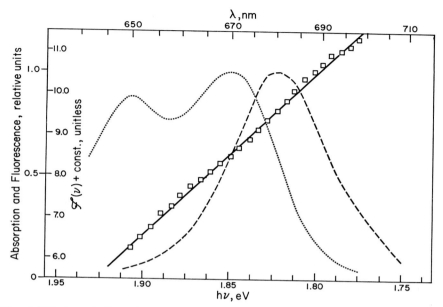

FIG. 3. Relation of absorption and emission data in terms of the Stepanov function $\mathscr{F}(\nu)$; □, calculated from data; —, a fit to equation (1) with $T = 297.7$ K; absorption (----) and fluorescence (— — —) are shown for reference.

neous and that each behaves as a single large molecule with equilibration not only among the excited vibrational states of the electronic excited states but also among the chromophores by excitation transfer. In particular, the short-wavelength portion of the region analysed contains the peak of the Chl b absorption. Further evidence supporting the whole picture is the fact that excitation into the individual Soret bands of the Chl a and Chl b in the complex produce experimentally-indistinguishable fluorescence spectra.

In vivo studies of the Stepanov relation in the same wavelength region that we have been considering (Szalay *et al.* 1967) indicated reasonable values of the effective Stepanov temperature in *Chlorella*, *Anacystis*, and *Porphyridium* after corrections for quantum yield were included. Evidently the T^* anomalies are reserved for the chlorophyll molecule itself.

FLUORESCENCE YIELD

We can report only preliminary results on the absolute value of the fluorescence yield ϕ_F of the LHCP. In a first sample chromatographically prepared with two passes through the column, ϕ_F was 10% as measured on the automated system at Eastman Kodak (fully corrected for instrument response). Seven later preparations, passed only once through the column, had yields varying from 2 to 6%. The spectra of these later preparations were almost identical to that of the first except for a slight overall red shift in three samples which became apparent through a slightly-anomalous Stepanov temperature of about 340 K. We are continuing these measurements and attempting to determine whether a simple quencher was at work on the later samples. Any LHCP yield smaller than the usual *in vivo* saturation yield of about 8% is suspect. For the samples with normal Stepanov temperatures, we believe that quenchers, rather than structural changes, are causing this variation in yield because the absorption and fluorescence spectra are independent of the yield.

IMPLICATIONS FOR EXCITONS

Depolarization studies

Exciton diffusion is sufficiently rapid that it does not limit the trapping of energy in the photosynthetic unit (see, e.g., the review by Knox 1977). This conclusion is based on the fact that numerous reasonable models of the antenna chlorophylls channelling energy to an efficient trap predict an even lower fluorescence yield than that which is observed. In addition to the yield, another quantity which reflects the history of the excitation on its way to the trap is the degree of polarization (p) of the antenna fluorescence. Any attempt to use p to evaluate the order of a system or the extent of energy transfer faces a fundamental difficulty: order and poor transfer mimic each other's effects. Extensive energy transfer in a well-ordered system will not necessarily result in a decrease of p, as it does in a random system. On the other hand, a random system with poor energy transfer will also show no decrease in p. When p is small, as it is *in vivo* (typically 0.03 to 0.08; see Arnold & Meek 1956; Mar & Govindjee 1972; Becker 1975) one may argue only that *some* energy transfer is taking place and that the transition dipoles involved are not purely parallel. Arguments which associate p with a 'number of hops' are particularly dangerous, because revisiting of the initially excited molecule is important (Knox 1968). The subtle interplay of membrane-orientation effects and transition-dipole order has been studied extensively (Geacintov *et al.* 1974; Becker 1975).

Our results (Van Metter 1977a) indicate that in any array of LHCP at normal temperatures most information derivable from polarization is lost as soon as energy transfer has taken place in *one* complex. Fig. 4 compares the excitation wavelength dependence of p in intact spinach chloroplasts (Becker 1975) with that in spinach LHCP. In the region of Chl b absorption (645–655 nm) p has dropped nearly to zero, indicating that the Chl b trimer has essentially isotropic

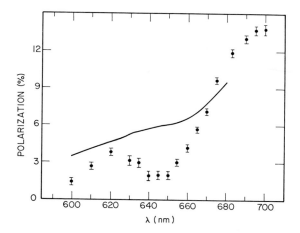

FIG. 4. Comparison of the fluorescence polarization excitation spectrum of LHCP (●) with that of spinach chloroplasts (——) as measured by Becker (1975). For LHCP, emission was viewed at 730 nm with a spectral width of ± 14 nm and the excitation spectral width was ± 5 nm (Van Metter 1977a).

absorption and transfer probabilities. The larger value of p in this region in the chloroplasts is probably due to emission from regions of the photosynthetic unit built from different complexes, such as CP I, the P700-containing complex, in which the large value of $p = 0.17$ is obtained under 640 nm excitation (Vacek *et al.* 1977).

Polarization measurements at low temperatures may be useful in determining the ordering of complexes, at least with respect to the mutual orientations of of their transition moments at 677 nm.

Structural model

The model described above lends itself well to an overall organizational model of the LHCP. Since the Chl *a* has the lower excitation energy and is located near the periphery, this energy-transferring species makes the better contact between complexes. The Chl *b*s form local antitraps which contribute absorbed energy directly to the Chl *a*. The energy then flows among the Chl *a*s along naturally-occurring 'energy channels to the trap', in the terminology of Borisov & Fetisova (1971). The logic of this arrangement is best appreciated by considering its converse. If Chl *a* formed trimers and the Chl *b* were at the periphery, the trimers would act as traps and long-distance transfer among them would be forced to occur over distances as large as the diameter of the complex.

Although the proposed LHCP structure is too tentative and unspecific for serious long-range energy-transfer modelling, it is perhaps worth noting that the most regular two-dimensional arrangement of objects such as those of Fig. 2 results in a lattice of Chl a molecules having hexagonal symmetry with a coordination number of four, a lattice known as the Kagomé.

Exciton coherence

The results presented above support the contention that exciton coherence is short-lived in the photosynthetic unit. Although the subject of coherence is much too extensive for review here, it is important to make two distinctions which are not always clear in the literature. First, coherence and strong coupling (exciton splitting) are different things. Coherence is a time-varying attribute of a state of motion, but exciton splitting is a fixed property of energy eigenstates. Second, coherence times (τ_c) and vibrational relaxation times (τ_v) are different things; an exciton may lose phase correlation without losing vibrational energy. Since vibrational relaxation is generally an incoherent process, however, we may expect that τ_v is an upper limit to τ_c.

Although the circular dichroic data for Chl b indicate some exciton splitting within the LHCP, one may not conclude that exciton coherence plays any role in energy transfer at room temperature. The widths of all the transitions involved are such that coherence in the sense of 'R^{-3} transfer' will disappear in times of the order of 10^{-14} s (Paillotin 1972; Kenkre & Knox 1974, 1976). Because of the complexity of the vibronic states of the assumed Chl b trimer, it is unlikely that appropriate simple initial conditions could be produced which would even make it possible to check this point. The trimer is, therefore, best regarded as a unit which absorbs and transfers energy to other molecules, in particular the Chl a of the same LHCP. Here again the widths of the transitions indicate a small coherence time. If the b–a transition rates are in the typical range of 10^{12}–10^{13} s^{-1}, whatever coherence exists will have been lost in the first 1–10 % of the transfer time. Complicated experiments in the sub-picosecond range must be done to sort out the coherence, with results likely to be of much greater interest to exciton physics than to photosynthesis.

The observed shift of one Chl a transition is probably due to a different local environment seen by one molecule, rather than exciton splitting, since there is negligible circular dichroism associated with the three Chl a. All the remarks about coherence among the Chl b states hold even more strongly with respect to Chl a.

The relevance of the 'good' LHCP Stepanov relation to the exciton coherence question is indirect. Essentially the validity of this relation supports the assumptions about rapid distribution of excitation which were used in arriving at the structural model. It also indicates that excited-state thermal equilibrium is reached in the complex in times short compared with the fluorescence lifetime (0.3 to 1.5 ns), a fact which provides an upper limit on individual vibrational relaxation times and a weak upper limit on the coherence time.

ACKNOWLEDGEMENTS

This research was supported by the National Science Foundation under Grant PCM-75-19638. Thanks are due to Professor G. E. Hoch for his advice and laboratory facilities and to Mr John Shepanski for the preparation of certain of the chlorophyll–protein complexes.

References

ARNOLD, W. & MEEK, E. (1956) The polarization of fluorescence and energy transfer in grana. *Arch. Biochem. Biophys. 60*, 82–90

BECKER, J. F. (1975) *Study of Biological Membranes oriented in a Homogeneous Magnetic Field*, Ph. D. thesis, New York University

BORISOV, A. YU. & FETISOVA, Z. G. (1971) Studies of resonance energy migration in a heterogeneous pigment complex I. Heterogeneity as a factor accelerating the localization of electron excitation in traps. *Mol. Biol. (Mosc.) 5*, 509–517 [Engl. transl.: *Mol. Biol. 5*, 405–411 (1971)]

BROWN, J. S., ALBERTE, R. S. & THORNBER, J. P. (1974) Comparative studies on the occurrence and spectral composition of chlorophyll–protein complexes in a wide variety of plant material, in *Proceedings of the Third International Congress on Photosynthesis*, vol 3 (Avron, M., ed.), pp. 1951–1962, Elsevier, Amsterdam

FENNA, R. E. & MATTHEWS, B. W. (1975) Chlorophyll arrangement in a bacteriochlorophyll protein from *Chlorobium limicola. Nature (Lond.) 258*, 573–577

FENNA, R. E. & MATTHEWS, B. W. (1977) Structure of a bacteriochlorophyll *a*–protein from *Prosthecocloris aestuarii. Brookhaven Symp. Biol. 28*, 170–182

FENNA, R. E., TEN EYCK, L. F. & MATTHEWS, B. W. (1977) Atomic coordinates for the chlorophyll core of a bacteriochlorophyll *a*–protein from green photosynthetic bacteria. *Biochem. Biophys. Res. Commun. 75*, 751–756

GEACINTOV, N. E., VAN NOSTRAND, F. & BECKER, J. (1974) Polarized light spectroscopy of photosynthetic membranes in magneto-oriented whole cells and chloroplasts: fluorescence and dichroism. *Biochim. Biophys. Acta 347*, 443–463

HEMENGER, R. P. (1977) Optical spectra of molecular aggregates near the strong coupling limit. *J. Chem. Phys. 67*, 262–264

KENKRE, V. M. & KNOX, R. S. (1974) Theory of fast and slow excitation transfer rates. *Phys. Rev. Lett. 33*, 803–806

KENKRE, V. M. & KNOX, R. S. (1976) Optical spectra and exciton coherence. *J. Luminescence 12/13*, 187–193

KENNARD, E. H. (1926) On the interaction of radiation with matter and on fluorescence exciting power. *Phys. Rev. 28*, 672–283

KNOX, R. S. (1968) Theory of polarization quenching by excitation transfer. *Physica 39*, 361–386

KNOX, R. S. (1977) Photosynthetic efficiency and exciton transfer and trapping, in *Primary Processes of Photosynthesis* (Barber, J., ed.), pp. 55–97, Elsevier/North-Holland Biomedical Press, Amsterdam

MAR, T. & GOVINDJEE (1972) Decrease in the degree of polarization of chlorophyll fluorescence upon the addition of DCMU to algae, in *Proceeding of the Second International Conference on Photosynthesis* (Stresa, Italy, 1971), vol. 1 (Forti, G., Avron, M. & Melandri, B. A., eds.), pp. 271–281, Junk, The Hague

PAILLOTIN, G. (1972) Motion of excitons in photosynthetic units, in Mar & Govindjee (1972), 331–336

STEPANOV, B. I. (1957) A universal relation between the absorption and luminescence spectra of complex molecules. *Dokl. Akad. Nauk S.S.S.R. 112*, 839–843 [*Dokl. Akad. Nauk S.S.S.R. Phys. Sect. (Engl. Transl.) 2*, 81–84]

SZALAY, L., RABINOWITCH, E., MURTY, N. R. & GOVINDJEE (1967) Relationship between the absorption and emission spectra and the 'red drop' in the action spectra of fluorescence *in vivo. Biophys. J. 7*, 137–149

SZALAY, L., TOMBACZ, E. & SINGHAL, G. S. (1974) Effect of solvent on the absorption spectra and Stokes shift of absorption and fluorescence of chlorophylls. *Acta Phys. Acad. Sci. Hung. 35*, 29–36

THORNBER, J. P., ALBERTE, R. S., HUNTER, F. A., SHIOZAWA, J. A. & KAN, K.-S. (1977) The organization of chlorophyll in the plant photosynthetic unit. *Brookhaven Symp. Biol. 28*, 132–148

VACEK, K., WONG, D. & GOVINDJEE (1977) Absorption and fluorescence properties of highly enriched reaction center particles of photosystem I and of artificial systems. *Photochem. Photobiol. 26*, 269–276

VAN METTER, R. L. (1977a) *A Study of Optical Properties of Chlorophyll in Solution and in a Protein Complex.* Ph. D. thesis, University of Rochester, New York

VAN METTER, R. L. (1977b) Excitation energy transfer in the light-harvesting chlorophyll a/b protein. *Biochim. Biophys. Acta 462*, 642–658

VAN METTER, R. L. & KNOX, R. S. (1976) On the relation between absorption and emission spectra of molecules in solution. *Chem. Phys. 12*, 333–340

Discussion

Porter: Rather than 'exciton' would it be better to talk about 'excitation transfer'?

Knox: Frenkel (1936) coined the word exciton; he described both localized excitons and wave-like excitons. Exciton is the term for either type of excitation.

Porter: By 'exciton' do you imply coherence?

Knox: No.

Paillotin: Two types of coherence time can be distinguished. One is related to the memory function that describes a non-Markoffian diffusion process (and in photosynthesis this time is very small). But a second time may be associated with the following process: the first excited state to be created is probably delocalized. Some time is needed for this excitation to be localized. This time may be longer than the former one and so may affect picosecond experiments (i.e. during the first 5–10 ps).

Knox: Consider a one-dimensional exciton in a one-dimensional crystal with a trap. What happens if a linear combination of localized states is created which has a long wavelength and which would correspond to Frenkel's original

(1931) exciton? In such an exciton the probability that the excitation is on one molecule is $1/N$. In addition, there is the correlation between excitons on two molecules: that is, the off-diagonal density matrix element ϱ_{ij} for the ith and jth molecules which reflects a phase correlation between these immediately after absorption. You are saying, Dr Paillotin, that there is an equation of motion for this off-diagonal matrix element ϱ_{ij} as well as for the diagonal one. ϱ_{ii} is the probability that the excitation is on the ith molecule, and one usually writes diffusion equations just for the diagonal matrix elements. These plus the equation for the off-diagonal elements are the coupled stochastic Liouville equations. In particular, the off-diagonal element has a decay rate, γ (see equation 1). The question is whether γ is the same as α, the decay rate of the

$$\frac{\partial \varrho_{ij}}{\partial t} = -\gamma \varrho_{ij} + \dots \tag{1}$$

memory function—that is an extremely complicated issue. (The relationships among various coherence parameters have been studied in detail by Kenkre [1975].) This decay rate, γ, may be included in the memory function but may be dynamically distinct and its associated decay time may be longer than $1/\alpha$. If this was the thrust of your comment, then I agree with you. Now, can we observe γ in the photosynthetic system? To try to answer that we Fourier transform the problem to $\vec{k}$ space whereupon γ becomes the rate of scattering between the exciton states. So, if we can create wave-like excitons and determine their scattering rates, we can in principle discover what their coherence time is. At low temperatures in certain molecular crystals (e.g. tetrachlorobenzene) these are the only scattering rates that are seen. To measure quantitatively one needs very low temperature and sharp lines; one almost has to use optically detected magnetic resonance techniques, generally suitable only for triplets.

Beddard: In the few sub-picosecond absorption experiments that have been done no coherence effect other than an optical coherence artifact has been observed.

Knox: In what system was this?

Beddard: Shank *et al.* (1976) observed this when measuring the induced absorption of carboxyhaemoglobin.

Knox: How was coherence defined?

Beddard: They saw coherence only between the light pulses as a spike at the growing edge of the absorption which they attributed to collision of the two light beams.

Katz: Your model, Dr Knox, seems to me to be inconsistent with the visible absorption spectrum of the sample. You have deconvoluted the absorption spectrum of your chlorophyll–protein complex into 677 nm and 670 nm components. Taken at face value that means to me a mixture of monomeric chlorophyll with some self-aggregated chlorophyll (to account for the absorption near 680 nm). In your model the chlorophyll *a* molecules are remote from other chlorophyll *a* molecules and, therefore, should absorb only at short wavelength. The model includes no features that I can see that correspond to the 'long-wavelength' form of chlorophyll *a* in your spectrum.

Knox: Could there not be a diagonal perturbation on the energy sufficient for the 7 nm shift?

Katz: Not unless there was a physical reason for it.

Knox: But if the shift were due to exciton interactions would there not be more prominent circular dichroism in the chlorophyll *a*?

Katz: That implies that one can correlate intensities in a c.d. spectrum with the concentration of the components, but we do not know what the extinction coefficients are.

Knox: I agree that I have not given any independent reason for you to believe that there is a single isolated chlorophyll having an environmental shift of 7 nm.

Duysens: I like your model, but one point worries me, however: the low observed fluorescence yield from your samples. In some you found a 10% yield. Since in other samples you found 2%, artifacts may occur that quench the fluorescence. Even in the '10%' sample part of the particles may quench strongly. Has the lifetime of the fluorescence been measured? If the fluorescence yield of the non-quenched particles were 30% (as *in vitro*), this lifetime would be 5 ns.

Knox: The sample with 10% fluorescence yield was passed through the hydroxyapatite column twice. Those samples with 2–6% yield were passed through only once: some of them showed an anomalous Stepanov temperature also. So some kind of quencher is likely.

Barber: One imagines that the yield would be greater than 10%.

Searle: The chlorophyll fluorescence lifetime in the F_{III} particle (the light-harvesting particle prepared with digitonin) was about 4 ns (Searle & Tredwell, this symposium), which corresponds to a yield of 20–25%.

Knox: We consider the 10% figure as a minimum.

Thornber: Dr Lutz and I have been trying to work out from his resonance Raman spectroscopic data how there might be six types of chlorophyll *a* molecules in the chlorophyll *a/b*–protein complex. Have you any information about the changes in the spectrum of the complex during degradation that might help us?

Knox: No. At practical level, let me add that we had to maintain samples at 2 °C whenever we made any long measurements. We determined the Stepanov temperature rapidly at room temperature (which is why I considered the slopes corresponding to 298 K 'normal') but for the long photon-counting runs we had to ensure that the complex did not degrade by keeping the system at 2 °C. An increase in fluorescence polarization signals the first stage of degradation; that, to us, means less energy transfer. Simultaneously, the absorption spectrum stays the same. Thereafter the chlorophyll *a* molecules begin to move away, as indicated by a relatively larger chlorophyll *b* absorption—that is another reason why we think that the chlorophyll *b* molecules are on the 'inside'.

Seely: Have you considered or eliminated polarized models with, say, three chlorophyll *b* molecules at one end and two 670 nm chlorophyll *a* molecules and a 677 nm chlorophyll *a* at the other end?

Knox: No. We merely *believe* that they are grouped together closely on the basis of the apparently very rapid excitation energy distribution within the complex. We do not exclude other relative *b–a* distributions.

Seely: Many combinations are possible. Also, there could be a good deal of compensation in the c.d. spectrum: the absence of an obvious band does not necessarily mean the absence of chirality.

Knox: I agree. In the case of the C_3 grouping for chlorophyll *b*, we were guided by the results of Kriebel & Albrecht (1976) and Ebrey *et al.* (1977) on *Halobacterium*.

Thornber: The deconvoluted spectrum showed a band at about 680 nm. Could it be that the complex has six chlorophyll *a*s and six chlorophyll *b*s? Do your data restrict the number to three?

Knox: Inclusion of a band at 684 nm was suggested to us by one of your deconvolutions (Brown *et al.* 1974). As to the number of chlorophylls, it *could* be six. However, if the complex becomes too big we may then have to start worrying about exciton dynamics within the complex in calculating the polarization. With only three molecules, the time of Förster energy transfer at our postulated chlorophyll *a* separation of 2.5 nm is 1.5 ps (which is sufficiently fast for equilibration during the lifetime).

References

BROWN, J. S., ALBERTE, R. S. & THORNBER, J. P. (1974) Comparative studies on the occurrence and spectral composition of chlorophyll–protein complexes in a wide variety of plant material, in *Proceedings of the Third International Congress on Photosynthesis*, vol. 3 (Avron, M., ed.), pp. 1951–1962, Elsevier, Amsterdam

EBREY, T. G., BECHER, B., MAO, B. & KILBRIDGE, P. (1977) Exciton interactions and chromophore orientation in the purple membrane. *J. Mol. Biol. 112*, 377–397

FRENKEL, J. (1931) On the transformation of light into heat in solids I, II. *Phys. Rev. 37*, 17–44 and 1294

FRENKEL, J. (1936) On the absorption of light and the trapping of electrons and positive holes in crystalline dielectrics. *Physik. Z. Sowjetunion 9*, 158–186

KENKRE, V. M. (1975) Relations among theories of excitation transfer II. Influence of spectral features on exciton motion. *Phys. Rev. B12*, 2150–2160

KRIEBEL, A. N. & ALBRECHT, A. C. (1976) Excitonic interaction among three chromophores: an application to the purple membrane of *Halobacterium halobium*. *J. Chem. Phys. 65*, 4575–4783

SEARLE, G. F. N. & TREDWELL, C. J. (1978) Picosecond fluorescence from photosynthetic systems *in vivo*, in *This Volume*, pp. 257–277

SHANK, C. V., IPPEN, E. P. & BERSOHN, R. (1976) Time-resolved spectroscopy of hemoglobin and its complexes with subpicosecond optical pulses. *Science (Wash. D. C.) 193*, 50–51

Energy transfer in a model of the photosynthetic unit

JULIANNA A. ALTMANN, GODFREY S. BEDDARD and GEORGE PORTER

Davy Faraday Research Laboratory, The Royal Institution, London

Abstract A simple model of the photosynthetic unit has been constructed and used for simulated Förster-type energy migration, fluorescence and intersystem crossing, in order to gain insight into the conditions that influence both the form and the lifetime of the fluorescence decay *in vivo*. The model consists of a two-dimensional random lattice with one central trap. The simulation was done by means of repetitive Monte Carlo-type computations. The results obtained show that the form of the decay curve changes from exponential to non-exponential, as the chlorophyll concentration (molecules/nm^2) is increased. The fluorescence lifetimes ($\tau_{1/e}$) were also found to decrease substantially with only slight increases in concentration. At a concentration comparable to that of chlorophyll in the chloroplast, both the form of the fluorescence decay and the lifetime are in fair agreement with experiment *in vivo*. The reasons for non-exponentiality of the decay as well as the properties of energy migration are discussed.

Preliminary work involving the dependence of trapping rate on donor concentration is also presented.

Chlorophyll molecules in the photosynthetic unit act collectively in delivering the absorbed light energy to the photochemical reaction centre. Previous work in modelling the photosynthetic unit has concentrated on the use of regular lattices. However, in view of recent data on the organization of chlorophyll *in vivo* (Govindjee 1975) the use of randomly distributed molecules on a lattice appears to be more realistic.

The simplest approach is to consider a two-dimensional lattice with equivalent sites and one central trap. This model can be realized as a thin film or bilayer membrane in which identical (chlorophyll) molecules are randomly positioned and the dipoles are randomly oriented.

We used this model to explore whether the energy transfer can be described by the Förster mechanism and, further, to investigate the effect of both the chlorophyll concentration and the ratio of the number of chlorophyll molecules to traps on the rate of migration. We found that the fluorescence decay curves, obtained by simulation using the present simple model, were in fair agreement to those obtained experimentally.

COMPUTATIONAL DETAILS

We calculated both the probability of trapping and the fluorescence-decay times by a Monte Carlo method by simulating Förster energy migration between similar randomly placed sites on a two-dimensional lattice.

A square lattice was used with area of 1.2×10^3 nm^2 and the number of sites per lattice was chosen to be 300, in accordance with *in vivo* results (Govindjee 1975). This gives a concentration of one site/4 nm^2. The sites were given a real size of 1.0 nm^2, the closest approach of two molecules being 1.0 nm. In the calculations of the fluorescence decays each molecule was allowed to fluoresce, intersystem cross, or transfer energy to one of several neighbours at a distance R_i. The rate of transfer, k_i, was calculated with Förster kinetics: $k_i = \tau_0^{-1} R_0^6 R_i^{-6}$ for which we used the natural lifetime of chlorophyll (τ_0) of 19.5 ns (Beddard *et al.* 1975), intersystem crossing rate of 1×10^7 s^{-1}, and an R_0 of 6.5 nm (Govindjee 1975). Each site on the lattice was considered equivalent except one trap molecule near the centre of the lattice which accepts the excitation irreversibly. Each random walk began with excitation of any one of the molecules chosen at random and was terminated by trapping, fluorescence or intersystem crossing. At any point in the walk the probability of jumping to any of the neighbours, fluorescence, intersystem crossing or trapping was calculated and one process was chosen at random from a uniform random-number generator (Hammersley & Hanscomb 1975). The farthest neighbour chosen had a rate of transfer of 5×10^{-4} that of the nearest-possible neighbour. The random-number generator was carefully checked for uniformity and lack of sequences. No restriction was placed on the number of times each site could be visited (except the trap). We varied the area of the lattice to achieve different concentrations in the fluorescence-decay calculations.

In the second part of the study we calculated the probability of trapping by allowing the sites either to transfer energy or to trap the energy irreversibly. In all the calculations the data were accumulated by performing many walks.

RESULTS AND DISCUSSION

Fluorescence decays

The results for the fluorescence-decay curves are shown in Fig. 1 and Table 1. We obtained the points shown in Fig. 1 by repeating the calculations many

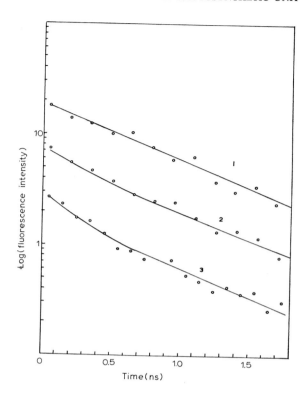

FIG. 1. Computer-simulated semilogarithmic fluorescence-decay curves obtained from random lattices, for packing densities of (1) 2.0×10^{-5}, (2) 2.5×10^{-5}, and (3) 4.0×10^{-5} molecules/nm^2, respectively. The lattice contains 300 molecules in 1.2×10^3 nm^2 and one trap. Each curve was constructed from about 2000 data points.

TABLE 1

Variation of concentration of donor molecules (nm^{-2}) and $1/e$ times of calculated fluorescence decays

Concentration (molecules/nm^2)	Lifetime (1/e) (ps)[a]	Form of decay
2.0×10^{-5}	850	Exponential
2.5×10^{-5}	540	Non-exponential
4.0×10^{-5}	430	Non-exponential

[a]Mean lifetime in vivo (Porter et al. 1977), $1/e = 450$ ps; values given are $\pm$ 50 ps.

times and averaging the data points at various time intervals to obtain a good signal-to-noise ratio. These curves contain about 2000 points each, collected from 3500 walks. The points were averaged over 50, 100 and 150 ps time intervals.

On calculating the fluorescence decays, about 10% of the walks yield fluorescence when we used a $1/\tau_0$ value of 2×10^{-7} s^{-1}. We found that *in vivo* the decay time is 1.5 ns when the traps are closed. As the present investigation was concerned with a simple model of the photosynthetic unit, the fluorescence rate was arbitrarily increased to yield a decay time of 1 ns when the trap is absent. This has the effect of increasing the data collection rate and making the calculation feasible.

As the Table illustrates, the lifetimes becomes progressively shorter as the concentration of donors (chlorophyll) is increased. The decay changes from a $1/e$ time of 850 ps to 450 ps with only a two-fold increase in concentration. It follows that, by increasing the donor concentration, the trap concentration has increased accordingly, since the size of the lattice and not the number of sites used was changed. This implies that the decrease in lifetime is due to the combined effect of the increase in the trap concentration and the increase in the rate of energy migration.

As is apparent from Fig. 1 (in which log [fluorescence intensity] is plotted against time) only one curve indicates an exponential decay, this being at the lower concentration. At higher concentrations the curves can be analysed as an exponential whose exponent has the form $-(A + Bt^n)$, where A and B are constants and $n = 0.3 \pm 0.2$. At long times, in all curves, A has the value of 1.1 ns^{-1}, which is the expected rate when the trap is absent or when no migration is possible.

In a regular two-dimensional lattice, one would expect to observe an exponential fluorescence decay at all concentrations because of the identical topology of each point. Since we are concerned with random lattices having an array of non-equivalent sites, the time to migrate from any given site to the trap differs considerably. Therefore, the non-exponential part of the decay curves is due to the migration pathway and not to the fluorescence or inter-system-crossing processes. It follows that the trapping process may have a non-first-order dependence on time.

Similar transient effects may be observed in diffusion-controlled quenching of excited states in fluid solutions. Transient effects can also be observed in single-step Förster transfer-type quenching by one of many quenchers placed at random around an excited molecule. In both cases in the exponent $n = 0.5$.

The experimental data obtained by picosecond laser excitation of *Chlorella* and chloroplasts also exhibit non-exponential decay characteristics at short times (Porter *et al.* 1977). We suggest, therefore, that the form of the decay is due to random-migration pathways towards the trapping centre. This also implies that the trapping is irreversible at the first arrival to the trapping centre, as we assumed for the present calculations.

When the migration originates at any site distant from the trap, the various pathways to the trap become equivalent and the probability of trapping varies exponentially with time. Consequently, in the present calculations (see Fig. 1), one observes an exponential decay of fluorescence only at long times (> 1.5 ns).

Trapping rates

We have made preliminary calculations to investigate the rate of trapping as a function of both time and concentration of the donors. In the previous calculations the combined effects of variable trap concentration, energy migration, and both radiative and non-radiative processes were observed. In this part of the study, to investigate the exact form of the rate equation for the energy migration processes, we kept the trap concentration at 10^{-4} molecules/nm^2 and removed the other competing processes.

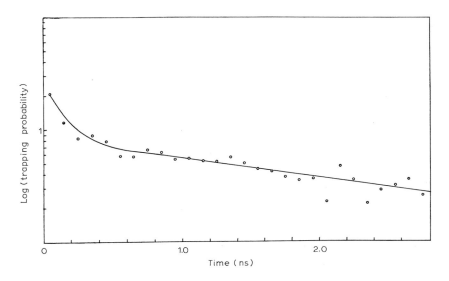

FIG. 2. Semilogarithmic plot of the probability of trapping (in arbitrary units) as a function of time. The lattice contains 300 molecules, giving a packing density of 2.5×10^{-5} molecules/nm^2. Curve contains 1980 data points.

Fig. 2 shows the results obtained for the probability of trapping at a donor concentration of 2.5×10^{-5} molecules/nm², plotted as the logarithm of the probability at time t against time. Up to 550 ps there is a distinct curvature in the semi-log plot. This non-exponential behaviour is again due to the topology of the sites in the random lattice, and is precisely the effect described earlier for the calculations of the fluorescence decay. The curve in Fig. 2 is composed of only about 2000 data points which is the reason for the slight scatter in the data. Previous experience has shown that this scatter is reduced substantially as the number of points collected is increased.

In Fig. 3 the effect of donor concentration on the trapping is illustrated for two extreme concentrations. Since the trap concentration was kept constant, the change in slope observed in the two curves is due solely to the difference in concentration of the donor, which is directly related to the energy migration.

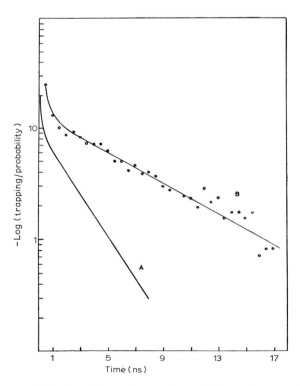

FIG. 3. Semilogarithmic plots of the probability of trapping (in arbitrary units) against time. Curve A is identical to the curve in Fig. 2. Curve B was obtained for a packing density of 1.25×10^{-5} molecules/nm², corresponding to 150 molecules and one trap on the lattice.

In Curve B at a concentration of 1.25×10^{-5} molecules/nm², the calculated trapping time (1/e of initial intensity) is about 2 ns. This is to be compared with about 0.4 ns at a concentration of 2.5×10^{-5} molecules/nm², corresponding to Curve A. At long times the calculated rates of trapping at the two concentrations shown are: 1.58×10^8 s^{-1} for the concentration of 1.25×10^{-5} molecules/nm², and 3.53×10^8 s^{-1} at 2.5×10^{-5} molecules/nm². It appears from this preliminary work that the trapping rate, at long times, varies approximately linearly with the donor concentration.

CONCLUSIONS

From these results two major conclusions may be drawn. First, transient terms can be important in the rate of trapping and can be observed in the fluorescence decays, both in the model and *in vivo*. This effect illustrates that quenching is more efficient at earlier times than one would expect at a given concentration of donors, and that trapping occurs at first arrival to the trapping centre. Secondly, the concentration of chlorophyll *in vivo* is critical for efficient energy migration in the antennae system. This implies that the most effective concentration must fall within a narrow range, to enable maximum energy transfer, while preventing self-quenching of the donors outside the trap.

References

BEDDARD, G., PORTER, G. & WEESE, M. (1975) Model systems for photosynthesis. 5. Electron-transfer between chlorophyll and quinones in a lecithin matrix. *Proc. R. Soc. Lond. A 342,* 317–325

GOVINDJEE (1975) *Bionergetics of Photosynthesis,* Academic Press, London

HAMMERSLEY, J. & HANSCOMB, D. (1975) *Monte Carlo Methods,* Methuen, London

PORTER, G., SYNOWIEC, J. & TREDWELL, C. (1977) Intensity effects on the fluorescence of *in vivo* chlorophyll. *Biochim. Biophys. Acta 459,* 329–336

Discussion

Paillotin: At low intensity the time decay of fluorescence is exponential. This has been shown in picosecond experiments by Campillo *et al.* (1976). Furthermore, by single-photon counting one can analyse the decay curves obtained at different temperatures into either one or two exponentials. But these two exponentials seem to be more related to the existence of photosystems than to a diffusion process.

Beddard: With single-photon counting detection with our dye laser of 6 ps duration as excitation we observed a low-intensity lifetime of 490 ps which is independent of excitation wavelength between 580 and 640 nm (see Fig. 1). We recorded the emission at room temperature from *Chlorella pyrenoidosa*. The data fit fairly well to a single exponential and slightly better to two exponentials. We cannot confirm or deny from this work the existence of a short transient component in the initial part of the decay, as Drs Searle & Tredwell describe in their paper.

Paillotin: Is the decrease in fluorescence mainly exponential with a short perturbation at short times or is the $t^{1/2}$ dependence predominant?

Tredwell: The fluorescence decay curves obtained from dark-adapted photosynthetic species, at excitation intensities below 5×10^{13} photon cm^{-2}, are predominantly exponential. In the time region between 0 and 100 ps after excitation we observed a faster component which cannot be ascribed to exciton fusion. Consequently, we use an $\exp(-At^{1/2})$ empirical decay law to describe the observed fluorescence decay at both high and low excitation intensities. This expression gives a good fit to the experimental curves over the first 1.5 ns, but decays too slowly at longer times. Fluorescence quantum yields and first $1/e$ lifetimes can, therefore, be calculated from this equation.

Robinson: Is it not surprising that on a two-dimensional random lattice at long times one obtains effectively a diffusion equation?

Beddard: Yes; we did not expect this.

Robinson: Nobody else seems to get a diffusion equation—for diffusion in two dimensions. Theoretically one would not expect that.

Beddard: I agree; I do not know why we apparently see diffusional behaviour in this system. We are at present calculating the 'diffusion coefficients'.

Robinson: With a three-dimensional random system Haan & Zwanzig (1978) are not convinced that one can use a diffusion equation to explain the results at any time.

Porter: In the diffusion equation do you include two time-dependent $\exp(-kt^{1/2})$ terms?

Robinson: If one explains the diffusion by a diffusion constant which is constant in time, the mean squared displacement is proportional to t.

Porter: If one uses a diffusion equation with this sort of system and very short times, one must use a time-dependent diffusion equation; one must include the $t^{1/2}$ term.

Robinson: At long times one can get either a real diffusion-type equation or not. In two dimensions one never gets a diffusion equation.

Beddard: At long times, although the probability of trapping appeared to decay exponentially there is some noise on the data and more analysis may prove it to be not exponential, as you suggest.

Porter: As you know this is an extremely complicated situation, involving $1/r^6$-dependent transfer in a random distribution. Zwanzig and others have tried to get an analytical solution but have not succeeded. At present, Dr Beddard's approach is the only one we have, even though it takes many hours of computer time.

Robinson: You are in a better position to try to fit these dynamic results to real experiments than the theorists are. Why can you not do 100 ps studies with photon-counting methods? K. G. Spears (personal communication), for example, claims he can easily do 200 ps.

Beddard: I can measure exponential decays shorter than 100 ps but I am not sure about non-exponential decays in this region. We have not tried that yet.

Porter: The intensity is then only 10^9 photons per pulse. There are other methods of detection that do not have any time-resolution problems but they are not so precise.

Seely: With regard to the liposomes and the vesicles, what was the wavelength of fluorescence and how large was the Stokes' shift? Was there a change with concentration of chlorophyll?

Beddard: Neither absorption nor emission changed with concentration up to about 0.1 mol/l (i.e. as much chlorophyll as could be dissolved in the lipid). The fluorescence wavelength was 680 nm; there was no unusually large shift. Absorption was at 672 nm.

Wessels: Could you detect a small absorption peak at about 730 nm in the chlorophyll-containing liposomes?

Beddard: No, not the way we made the preparations. In some conditions one can produce dimers and aggregates in the lipid bilayers.

Wessels: Did you separate the liposomes on a Sepharose column?

Beddard: We have done this but, for the experiments I described, the samples were used without separation on a column.

Clayton: To avoid concentration quenching did you try any additives other than lipids, such as hydrophobic proteins?

Beddard: We tried to dissolve some protein in chloroform but without success.

Cogdell: Some ATPases are soluble in chloroform.

Thornber: The light-harvesting proteins of higher plants are soluble in chloroform–methanol (Henriques & Park 1976).

Clayton: Chromatophores can be dissolved in 2-chloroethanol.

Beddard: We hope to try these proteins.

References

Campillo, A. J., Kollman, V. H. & Shapiro, S. L. (1976) Intensity dependence of the fluorescence lifetime of *in vivo* chlorophyll excited by a picosecond light pulse. *Science (Wash. D.C.) 193*, 227–229

Haan, S. W. & Zwanzig, R. W. (1978) 'Förster' migration of electronic excitation between randomly distributed molecules, in press

Henriques, F. & Park, R. B. (1976) Compositional characteristics of a chloroform-methanol soluble-protein-fraction from spinach chloroplast membranes. *Biochim. Biophys. Acta 430*, 312–320

Searle, G. F. W. & Tredwell, C. J. (1979) Picosecond fluorescence from photosynthetic systems *in vivo*, in *This Volume*, pp. 257–277

Dynamics of excitons created by a single picosecond pulse

G. PAILLOTIN and C. E. SWENBERG*

Service de Biophysique, Département de Biologie, Centre d'Études Nucléaires de Saclay, Gif-sur-Yvette, France and **Department of Chemistry and the Radiation and Solid State Laboratory, New York University, New York*

Abstract A theoretical analysis of bimolecular annihilation in finite domains is presented. A Pauli master equation is formulated for the case of varying incident delta function excitation sources. Expressions for the quantum fluorescence yield and its time dependence are derived. The relationship between the fluorescence yield and the number of hits per domain depends on two parameters: the rate constant of bimolecular exciton annihilation and the dimension of the domain in which this annihilation occurs. Recent experimental results imply that the exciton diffusion constant (D) is large ($D \gtrsim 10^{-3}$ cm^2 s^{-1}) and that the photosystem II domains may contain as many as five photosynthetic units.

An analysis of the time decay of the fluorescence indicates that, for a few hits per domain, the decay may be considered as exponential but for many hits it becomes non-exponential. Thus the fluorescence decay depends on the intensity of the excitation source and/or on the dimension of the domains. Conditions which change the effective size of the domain may change the shape of the fluorescence decay.

Some biological consequences and experimental applications of this theory are presented.

New information on energy transfer in the photosynthetic pigment arrays has become available through experiments with picosecond laser pulses. The fluorescence is strongly quenched when the intensity of the pulse is large (Mauzerall 1976a; Campillo *et al.* 1976a; Breton & Geacintov 1976) and it is now established that this new quenching process is the result of the fusion of singlet excitons (Swenberg *et al.* 1976), a process which has been studied intensively in molecular crystals (Swenberg & Geacintov 1973; Rahman & Knox 1973). At the same time it has been shown that the fluorescence lifetime decreases when the pulse intensity increases (Campillo *et al.* 1976b; Porter *et al.* 1977) and that the fluorescence decay curves obey an expression of the form $\exp(-At^{1/2})$ (Harris *et al.* 1976). Several models have been proposed which partly describe the experimental results: (i) the statistical model of Mauzerall

(1976*b*); (ii) the continuum model of Swenberg *et al.* (1976); and, for the time dependence, (iii) the phenomenological model of Harris *et al.* (1976).

We give here the main lines of a complete analysis of the singlet fusion process in the photosynthetic apparatus with an account of both the time dependence and the yield of fluorescence. The physical and mathematical details of this theory will be described elsewhere (Paillotin *et al.* 1978, in preparation) and we shall focus our attention on the biological applications of our model.

DEACTIVATION PROCESSES AND PHOTOSYNTHETIC DOMAINS

When one excitation only is created within a photosynthetic array, it may disappear by one of three different deactivation processes. Two of the processes —radiative and non-radiative deactivations (with respective rates k_F and k_D)— occur on each pigment molecule. The third occurs only at the level of specialized pigments associated with the reaction centres and corresponds to a photochemical reaction when the centre is open. The rate of excitation capture by the reaction centre, k_Q, generally depends on the state of the reaction centre (Paillotin 1976*a*). The set of pigments associated with a centre is called a photosynthetic unit. In the green plants the photosynthetic units contain from 200 to 300 chlorophyll molecules. Numerous lines of evidence indicate that these units are not isolated (Joliot & Joliot 1964; Paillotin 1976*a*) but are grouped in photosynthetic domains. A domain contains several units and in most cases the movement of excitations in such a domain is free (Paillotin 1976*b*).

When several excitations are created simultaneously within a domain a new process of deactivation arises which corresponds to singlet–singlet annihilation; when picosecond pulses are used, singlet–triplet annihilations do not play an important part (Geacintov & Breton 1977). Finally, when the exciton concentration is not too large (less than one excitation per 10 pigment molecules, for instance), the fusion of more than two excitons is negligible.

MASTER EQUATION

For a rigorous treatment of singlet–singlet annihilation in a photosynthetic domain it is necessary to solve a multidimensional diffusion equation. Such a problem is very complicated (see e.g. Teramoto & Shigesada 1967). Fortunately, in photosynthesis, the diffusion of excitons is not the limiting process and the singlet excitons are randomized before any deactivation occurs (Paillotin 1972; Knox 1977). Then, if the randomization time is neglected, at each instant of time the state of a domain is defined solely by the number i of excitons it contains. The time evolution of a set of excitons is governed by the rates of transition from a state with i excitons (i-state) to states containing fewer excitons.

Physiological deactivations (those which still exist at low intensities of excitation) remove only one exciton at a time—that is, they cause a transition from the i-state to the ($i-1$)-state. The rate of this transition is Ki where K is the sum of the rates of the deactivations (1).

$$K = k_F + k_D + k_Q \qquad (1)$$

Corresponding to the singlet–singlet annihilation is a transition from the i-state to the ($i-\alpha$)-state where $\alpha = 1$ or 2, according to whether one or two excitons are removed during the singlet–singlet collision (Geacintov et al. 1977). The rate of these transitions is given by (2), where $\gamma^{(\alpha)}$ denotes the pairwise rate of fusion.

$$\binom{i}{2}\gamma^{(\alpha)} = \frac{i(i-1)}{2}\gamma^{(\alpha)} \qquad (2)$$

Consider a domain containing d units and let us introduce the probability $P_i(n,t)$ that there are i singlet excitons at time t, given that there were n excitons at time $t = 0$. The number $F_n(t)$ of photons emitted by fluorescence at time t is then given by (3) [$F_n(t)$ is normalized to n] and the quantum yield of fluorescence ϕ_n by (4).

$$F_n(t) = \frac{k_F}{n} \sum_{i=1}^{n} iP_i(n, t) \qquad (3)$$

$$\phi_n = \int_0^\infty F_n(t)dt \qquad (4)$$

Now consider a set of identical domains; the number of photons $F(t)$ emitted by fluorescence at time t and the quantum yield ϕ may be derived (5 and 6, respectively) by obtaining mean values of $F_n(t)$ and ϕ_n over a Poisson distribution.

$$F(t) = \sum_{n=1}^{\infty} \frac{e^{-y}y^{n-1}}{(n-1)!} F_n(t) \qquad (5)$$

$$\phi = \sum_{n=1}^{\infty} \frac{e^{-y}y^{n-1}}{(n-1)!} \phi_n \qquad (6)$$

$F(t)$ is normalized in such a way that $F(0) = 1$ and y is the mean number of excitons created in a domain at time $t = 0$. $F(t)$ and ϕ are the two macroscopic quantities obtained from picosecond experiments and may be calculated if $P_i(n,t)$ is known.

According to equations (1) and (2) the time evolution of $P_i(n,t)$ is governed by the following master equation (7).

$$\frac{d}{dt} P_i(n, t) = \left[K(i + 1) + \gamma^{(1)} \frac{i(i + 1)}{2} \right] P_{i+1}(n, t) + \gamma^{(2)} \frac{(i + 2)(i + 1)}{2} P_{i+2}(n, t)$$

$$- \left[Ki + \gamma^{(1)} \frac{i(i - 1)}{2} + \gamma^{(2)} \frac{i(i - 1)}{2} \right] P_i(n, t) \qquad (7)$$

This equation can be solved exactly (Paillotin et al. 1978), and one obtains equations (8) and (9)

$$F(t) = \sum_{p=0}^{\infty} (- 1)^p \exp[- (p + 1)(p + r)\gamma t/2] A_p \qquad (8)$$

where

$$A_p = \sum_{k=p}^{\infty} (- 1)^k \binom{k}{p} \frac{(r + 1 + 2p)}{(r + p + 1 + k)(r + p + k) \dots (r + p + 1)} y^k (1 + \xi)^k$$

and

$$\phi = \phi_1 \sum_{k=0}^{\infty} (- 1)^k \frac{1}{(r + k)(r + k - 1) \dots (r + 1)} y^k \frac{(1 + \xi)^k}{(k + 1)} \qquad (9)$$

where
$$r = 2K/\gamma \qquad (10a)$$
$$\gamma = \gamma^{(1)} + \gamma^{(2)}$$
$$\xi = \gamma^{(2)}/\gamma \qquad (10b)$$
and
$$\phi_1 = k_F/K.$$

Besides the yield ϕ_1, $F(t)$ and ϕ (considered as functions of the mean number of hits per domain y) depend on three parameters: ξ equals the yield of the two-exciton annihilation when only the pairwise fusion processes are taken into account and introduces only a scale change, because it appears only in the product $y(1 + \xi)$. Then it is convenient to consider that $F(t)$ and ϕ are functions of z where

$$z = y(1 + \xi) \qquad (11)$$

The same thing occurs for γ and we may introduce instead of the time t the quantity τ:

$$\tau = \gamma t/2 \qquad (12)$$

The only non-trivial parameter is r (equation 10a) which is a measure of the efficiency of the one-exciton deactivation processes as opposed to the two-

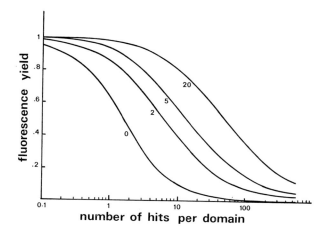

FIG. 1. Fluorescence yield as a function of the number of hits per domain for different values of r (0, 2, 5 and 20).

exciton annihilation. If a two-exciton state is considered, it may disappear by a one-exciton process (of rate $2K$) or by a two-exciton process; r is the ratio of these two rates.

YIELD OF FLUORESCENCE

Fig 1 shows the fluorescence quenching curves $\phi(y)$ calculated for $\xi = 0$ for different values of the parameter r.

If r is small ($r \ll 1$) one obtains from (9) and (11) the yield $\phi(z)$, an expression which was proposed by Mauzerall (1976b) to describe the fluorescence quenching.

$$\phi(z) = \phi_1 \frac{(1-e^{-z})}{z} \tag{13}$$

However, this law does not fit the experimental results (Geacintov *et al.* 1977). Furthermore, since γ depends on the dimension of the domain ($\gamma \approx 1/M$ where M is the number of pigments within a domain), the condition $r \ll 1$ is physically improbable. In fact, 10^9 s^{-1} is a typical value for K and, even if M is 200, this condition would imply too small a value for the time of fusion (less than 10^{-12} s).

Conversely, if r is large, equation (9) becomes (14), which is similar to the expression derived by Swenberg *et al.* (1976) and fits the experimental results of Geacintov *et al.* (1977).

$$\phi(z) = \phi_1 \frac{r}{z} \ln\left(1 + \frac{z}{r}\right)$$ (14)

In the most general case, the theoretical curves shown in Fig. 1 may be easily adjusted to the experimental curves. Experimentally one obtains ϕ as a function of x, the number of hits per photosynthetic unit. If d is the number of units per domain we have equations (15).

$$y = xd \quad \text{and} \quad z = (1 + \xi)xd$$ (15)

When ϕ is plotted as a function of $\ln x$ it is possible to adjust the theoretical and experimental curves by a mere translation along the abscissa. In this way one can obtain from the quenching curves of Geacintov et al. (1977) for photosystem II of green plants limiting values for $(1 + \xi)d$ and r (equations 16 and 17, respectively).

$$(1 + \xi)d \geqslant 10 \quad \text{or} \quad d \geqslant 5$$ (16)

$$r \geqslant 10$$ (17)

Since the r value is large one can calculate a lower limit for the diffusion coefficient D of the singlet excitations: $D \geqslant 10^{-3}$ cm^2 s^{-1}. Equation (16) indicates that each domain of photosystem II contains more than five units (see also Joliot et al. 1972).

Thus the picosecond experiments confirm that the motion of excitons in photosystem II is a free motion between a relatively large number of connected units. Other experiments are imaginable in conditions where the connection of units is modified (cation concentration, flashed leaves, mutants etc.) or by direct excitation of photosystem I.

TIME DEPENDENCE OF FLUORESCENCE

Equation (8) gives the time dependence of the fluorescence; it corresponds to a sum of exponentials. At low intensity (when y is small), the decay curve is exponential: $\exp(-r\tau) = \exp(-Kt)$. However, if y is large, one can write equation (8) as (18), which is in agreement with results obtained within the framework of the bimolecular kinetics (Swenberg et al. 1976).

$$F(t) = \frac{e^{-Kt}}{\left(1 + \frac{z}{r}\right) - \frac{z}{r}e^{-Kt}}$$ (18)

Fig. 2 shows the curve giving $\ln[F(t)]$ as a function of $t^{1/2}$ for $z/r = 5$. This curve was calculated from equation (14). Although it is coincidence, this

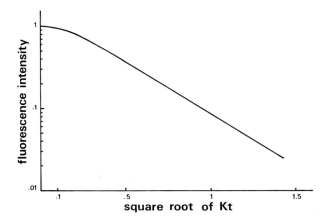

FIG. 2. Fluorescence intensity plotted on a logarithmic scale as a function of the square root of time $(Kt)^{1/2}$ $(z/r = 5$, see text).

curve appears to verify the law established by Harris *et al.* (1976) (for further discussion of this point see Knox 1977).

This empirical law applies if z/r is large. From (10) and (15) we derive (19).

$$\frac{z}{r} = \frac{(1 + \xi)}{2K} \gamma x d \tag{19}$$

(At time $t = 0$, the rate of deactivation by two-exciton processes is given by $^{1}/_{2} (1 + \xi) \gamma(xd)^2$, and the rate of deactivation by one-exciton processes is Kxd; z/r is the ratio of these two rates.)

If the number of hits per unit, the number of units per domain d, the rate of annihilation $\gamma(1 + \xi) = \gamma^{(1)} + 2\gamma^{(2)}$ are large and if K is small, then z/r is large. If, for instance, according to our previous discussion of the yield of fluorescence, the value of r for open centres is 20 and the value of $(1 + \xi)d$ is 10, then $z/r = 5$ if $x = 10$. For units with closed centres where K is about one-third as large, this condition is achieved with $x = 3$.

Note that a decrease in γ, which induces a decrease in the ratio z/r, may cause a transition from a non-exponential decay to an exponential one. For the photosystem II such a change of γ occurs, for instance, when the connection between units is weakened.

In the more general case (i.e. if z/r is neither too large nor too small) it is necessary to compare the experimental values of $F(t)$ to the theoretical ones given by (8). Such a fitting procedure is not, in practice, easy. In that respect, two relations (20 and 21) may be written, and thus the ratio of these two derivatives is equal to $(1 + z/r)$.

$$\frac{d}{d\tau} \ln F(\tau)_{\tau=0} = r + z \qquad (20)$$

$$\frac{d}{d\tau} \ln F(\tau)_{\tau\to\infty} = r \qquad (21)$$

CONCLUSION

In the previous section we have described the process of singlet fusion in finite photosynthetic domains. Our analysis is grounded on the basic mechanisms of singlet–singlet annihilation and gives theoretical expressions for the yield ϕ and the time decay curve $F(t)$ of the fluorescence which are in good agreement with the experiments. According to our model we can derive some important parameters which control energy transfer in photosynthesis:

r (equation 10a) gives either the value of the capture rate K or the value of the pairwise rate of fusion γ. From γ one can calculate a limiting value of the exciton diffusion coefficient D;

d, the dimension of the photosynthetic domains.

When applied to the results which were at our disposal (Geacintov et al. 1977) our theory leads us to conclude that the domains of photosystem II contain at least five units and that for this photosystem the exciton diffusion coefficient is large ($D > 10^{-3} \, \text{cm}^2 \, \text{s}^{-1}$). These data confirm, by direct physical evidence, the proposed model of free motion of excitons (Paillotin 1976b).

Barber et al. (1978) have recently shown that the fluorescence decay curves become exponential when cations ($MgCl_2$) are added. According to the foregoing discussion, this transition comes from a decrease of the value of z/r (19). Since in the same conditions the connection between photosystem II units increases (Bennoun 1974) (i.e. γ increases) it can be explained only if the dimension d of the domains decreases.

Then addition of cations seems to cause a transition from a state where the connections between units are weak but the domains large to a state where the connections are strong but the domains are small. This hypothesis needs more experiments to be established, but results such as these show how much information may be obtained from picosecond experiments when all the physical parameters that govern the singlet fusion are taken into account.

References

BARBER, A., SEARLE, G. F. W. & TREDWELL, C. J. (1978) Picosecond time-resolved study of MgCl₂-induced chlorophyll fluorescence yield changes from chloroplasts. *Biochim. Biophys. Acta 501*, 174–182

BENNOUN, P. (1974) Correlation between states I and II in algae and the effect of magnesium on chloroplasts. *Biochim. Biophys. Acta 368*, 141–147

BRETON, J. & GEACINTOV, N. E. (1976) Quenching of fluorescence of chlorophyll *in vivo* by long-lived excited states. *FEBS (Fed. Eur. Biochem. Soc.) Lett. 69*, 86–89

CAMPILLO, A. J., SHAPIRO, S. L., KOLLMAN, V. H., WINN, K. R. & HYER, R. C. (1976*a*) Picosecond exciton annihilation in photosynthetic systems. *Biophys. J. 16*, 93–97

CAMPILLO, A. J., KOLLMAN, V. H. & SHAPIRO, S. L. (1976*b*) Intensity dependence of the fluorescence lifetime *in vivo* chlorophyll excited by a picosecond light pulse. *Science (Wash. D.C.) 193*, 227–229

GEACINTOV, N. E. & BRETON, J. (1977) Exciton annihilation in the two photosystems in chloroplasts at 100°K. *Biophys. J. 17*, 1–15

GEACINTOV, N. E., BRETON, J., SWENBERG, C. E. & PAILLOTIN, G. (1977) A single pulse picosecond laser study of exciton dynamics in chloroplasts. *Photochem. Photobiol. 26*, 629–638

HARRIS, L., PORTER, G., SYNOWIEC, J. A., TREDWELL, C. J. & BARBER, J. (1976) Fluorescence lifetimes of *Chlorella pyrenoidosa. Biochim. Biophys. Acta 449*, 329–339

JOLIOT, A. & JOLIOT, P. (1964) Etude cinétique de la réaction photochimique libérant l'oxygène au cours de la photosynthèse. *C. R. Acad. Sci. Paris 258*, 4622–4625

JOLIOT, P., BENNOUN, P. & JOLIOT, A. (1972) New evidence supporting energy transfer between photosynthetic units. *Biochim. Biophys. Acta 305*, 317–328

KNOX, R. S. (1977) Photosynthetic efficiency and exciton transfer and trapping, in *Primary Processes of Photosynthesis*, vol. 2 (Barber, J., ed.), pp. 55–97, Elsevier, Amsterdam

MAUZERALL, D. (1976*a*) Multiple excitations in photosynthetic systems. *Biophys. J. 16*, 87–91

MAUZERALL, D. (1976*b*) Fluorescence and multiple excitation in photosynthetic systems. *J. Phys. Chem. 80*, 2306–2309

PAILLOTIN, G. (1972) Transport and capture of electronic excitation energy in the photosynthetic apparatus. *J. Theor. Biol. 36*, 223–235

PAILLOTIN, G. (1976*a*) Capture frequency of excitations and energy transfer between photosynthetic units in the photosystem II. *J. Theor. Biol. 58*, 219–235

PAILLOTIN, G. (1976*b*) Movement of excitations in the photosynthetic domains of photosystem II. *J. Theor. Biol. 58*, 237–252

PORTER, G., SYNOWIEC, J. A. & TREDWELL, C. J. (1977) Intensity effects on the fluorescence on *in vivo* chlorophyll. *Biochim. Biophys. Acta 459*, 329–336

RAHMAN, T. S. & KNOX, R. S. (1973) Theory of singlet-triplet exciton fusion. *Phys. Stat. Sol. (b)*, 715–720

SWENBERG, C. E. & GEACINTOV, N. E. (1973) Exciton interaction in organic solids, in *Organic Molecular Photophysics* (Birks, J. B., ed.), pp. 489–564, John Wiley, Chichester

SWENBERG, C. E., GEACINTOV, N. E. & POPE, M. (1976) Bimolecular quenching of excitons and fluorescence in the photosynthetic unit. *Biophys. J. 16*, 1447–1452

TERAMOTO, E. & SHIGESADA, N. (1967) Theory of bimolecular reaction processes in liquids. *Prog. Theor. Phys. 37*, 29–51

Discussion

Tredwell: At what excitation wavelength did you measure the decrease in the fluorescence quantum yield with excitation intensity?

Paillotin: 610 nm.

Tredwell: We used 530 nm excitation in our measurements; at this wavelength about two times less light is absorbed than at 610 nm. Considering the differences in excitation wavelength and in the amount of light absorbed, I wonder whether the experimental results are comparable.

We normally use excitation intensities of about 5×10^{13} photon cm^{-2} at 530 nm. According to the published fluorescence quantum yield curves (Geacintov & Breton 1977; Campillo *et al.* 1976; Mauzerall 1976), this intensity corresponds to less than one hit per photosynthetic unit. How can this intensity result in such a high degree of quenching?

Paillotin: Generally one uses the number of hits per unit, but suppose that 10 units are connected; then one has to multiply the number of excitons in the primary unit by 10. The rate of annihilation is proportional to the square of the concentration of excitons; thus it depends on the dimension of the domain.

Tredwell: But if there is significant exciton fusion why is the observed fluorescence quantum yield unaffected at this intensity (5×10^{13} photon cm^{-2})?

Barber: Addition of magnesium ions tends to increase interaction between photosystem II units—that is, the domain for photosystem II increases. We are looking at fluorescence of photosystem II. Therefore, addition of magnesium ions should increase the possibility of singlet annihilation and lead to a more non-exponential decay, which is the opposite of what is experimentally observed.

Paillotin: If there is more fusion then the decay curve is transformed from the exponential to the non-exponential case. Probably the interaction between units increases, but d decreases. That simple parameter, d, is the number of units per domain—i.e., the dimension of the domain in which the exciton migrates freely.

Joliot: Your estimate of the number of units per domain does not differ much from that obtained by variation of the cross-section of the centre as a function of the number of active centres. The difficulty that we face in our method is that we can demonstrate only that the number of units connected is greater than four because we observe that the cross-section increases four-fold when the fraction of active centres decreases from one to zero. How do you derive an upper limit when you say that photosystem II may contain as many as five units?

Paillotin: When r is large we are in the continuum case. In the continuum case (and that is why I use the ratio z/r) all the quenching curves are functions of z/r. One cannot tell if one has modified the dynamics of the system (r) or the number of units per domain (z). Judging from the reported quenching curves we are in the continuum case—with a rather large value for both r and z.

Beddard: Why do you ignore the effect of triplets which can be produced by annihilations when many photons are absorbed from the first laser pulse?

Porter *et al.* (1977) have shown this to be important.

Paillotin: We neglect the triplet state because its concentration is probably small. With a single picosecond pulse (10 ps in our experiments) only a small quantity of triplet state is created.

Porter: Is the triplet being created by intersystem crossing?

Paillotin: Yes.

Porter: The following processes are possible:

$$S_1 + S_1 \rightarrow T + T$$
$$S_1 + T \rightarrow T + T$$

Paillotin: The former is not neglected. The second is negligible on a single picosecond flash.

Porter: If every annihilation leads to triplet formation why is the yield of triplets low?

Breton: Some of the data are in my paper but are not discussed in detail along these lines. We have compared the effect of intense picosecond pulse trains and of microsecond laser flashes of similar duration on both the yield of carotenoid triplets and the fluorescence quenching. These two modes of excitation correspond to quite different conditions of creation of chlorophyll triplets. In the picosecond regime, the triplets can be created within a few tens of ps (corresponding to the lifetime of the singlets) by exciton fusion processes of the type Sir George described. In the microsecond case, triplets are created only by intersystem crossing during the lifetime of the singlet. Using these two types of excitation, we found no difference in the yield of carotenoid triplets and only minor differences on the fluorescence quenching efficiency. These observations suggest that the amount of chlorophyll triplets created in these two different conditions of excitation are similar.

Paillotin: We consider that there are two rates of deactivation: one is proportional to the number of excitons and the other to the number of combinations of two excitons. This comes from our assumption of randomization; but one can modify the theory to take some diffusion-limited processes into account.

Robinson: Are you saying that in the continuum limit you do not get non-exponential decay?

Paillotin: The formula of Swenberg *et al.* (1976) does not give an exponential decay; it has the form (1) where c is a constant. The Swenberg formula gives

$$1/F = c + \exp(Kt) \tag{1}$$

a good fit to the theoretical curve (see Fig. 2) of $\ln F$ against $(Kt)^{1/2}$ at small values of the latter, but at larger values there is a progressively bigger discrepancy.

Robinson: In some experiments on three-dimensional crystals (Fleming *et al.* 1977) the emission was extremely non-exponential and this non-exponentiality can be explained by singlet–singlet annihilation. We have calculated the singlet–singlet annihilation constant. It should be possible to do that for singlet–singlet annihilation in chloroplasts.

Knox: According to some calculations that we are doing with Dr Swenberg, the total amount of singlet–triplet quenching is extremely small—there is not enough intersystem crossing during one single lifetime even though we are using large singlet–triplet fusion rate constants.

Porter: I wonder why, because it is a common effect in many other molecules.

Knox: During the lifetime of the exciton, which is much shorter than that of the molecule in solution, there is much less crossing—if one assumes the same rate of intersystem crossing as in solution.

Porter: But it is not ordinary intersystem crossing.

Knox: I am talking about populating the triplets as a result of one pulse.

Tredwell: If the triplet population is formed by the interaction of singlet excitons, the process will be independent of the rate of intersystem crossing. Consequently, a large triplet population could be formed within the lifetime of the exciton population generated by the first excitation pulse (see Porter *et al.* 1977).

Paillotin: This has been taken into account in our model.

Robinson: But it is small and, therefore, annihilation is not important; that is what we seem to be saying.

Tredwell: Quenching by singlet–triplet exciton fusion is not insignificant; we estimate that at least 7% of the observed quenching can be ascribed to this process (see Porter *et al.* 1977).

Amesz: Are the calculations based on fluorescence measurements (Fig. 1) at low temperature?

Paillotin: Yes. But the same experiments were done at room temperature, at 200 K and 21 K, and the shape of the curve is the same (Geacintov *et al.* 1977).

Amesz: I ask this, because our experiments indicate that below 120 K the fluorescence kinetics in low-intensity continuous light can be analysed into two exponentials. This fact suggests that energy transfer between photosynthetic units does not occur below that temperature.

Joliot: Since you base your calculation on the fluorescence kinetics in continuous light, I do not understand how you can estimate the efficiency of energy transfer. The time course of the fluorescence emission is determined more by the competition between the rates of charge stabilization and back-reaction than by the efficiency of energy transfer. Fluorescence-induction curves can be used as an indicator of energy transfer only if all the hits on the active traps are efficient, which means that the positive and negative charges are stabilized after each hit; this is not so at low temperature.

Amesz: A low efficiency of charge stabilization would not necessarily affect the relation between quencher concentration and fluorescence intensity, unless the quenching efficiency by 'open traps' is lowered.

Paillotin: The fluorescence curve can rise exponentially even if there are connections.

Joliot: If your theory is correct, the number of units connected and the probability of energy transfer between units are independent of temperature.

Paillotin: Yes—on the sample we have studied. It is not a physical law; it may depend, for instance, on the way the sample is frozen.

Joliot: At both low (-60 °C) and high (0 °C) temperature, we observe about the same degree of connection between units.

Paillotin: It would be interesting to use leaves during greening, for instance. With this material the connections seem to appear after a short time.

Beddard: By using a master equation you are presumably assuming a regular array of chlorophyll molecules with nearest-neighbour interactions. How can you apply this to something that may not be regular?

Paillotin: I have made a totally different assumption to that which you have made, namely that the diffusion rate constant of an exciton is high and that randomization is efficient. To take into account the distribution of the molecules and the distribution of the exciton one would have to solve some extremely complicated equations. In my opinion the diffusion of an exciton is not a limiting process in photosynthesis.

Porter: This is an important assumption.

Knox: There seems to be a misunderstanding; the index i is not the label of any site—there is no nearest-neighbour approximation—i is the number of excitons in the unit at a given time.

Paillotin: But I can use Dr Beddard's results by introducing a non-linear dependence of the rate of deactivation (i.e. replacing Ki by a more complex function of i). That would probably be a good approximation of the very complicated diffusion-limited case.

Porter: Dr Knox, what is your opinion of Dr Paillotin's preference for randomization being complete before trapping and the rate of diffusion being unimportant?

Knox: Qualitatively I agree because all random-walk calculations give too rapid an approach to the trap.

Porter: What sort of random-walk calculations?

Knox: In regular lattices.

Porter: The calculations are just right for a random lattice. Nothing said at this meeting so far has led us to believe that it is a regular lattice.

Paillotin: No, but there are biological arguments in favour of efficient randomization. In a random walk, if the time needed for an exciton to reach the centre is the limiting factor, the rate of trapping is inversely proportional to the number of jumps made by the exciton. But if the time is not a limiting factor, the rate of trapping is proportional to the rate of the charge separation in the reaction centre. Experimentally the yield of fluorescence seems to depend on the state of the reaction centre and so favours this second hypothesis. (See also pp. 357–363.)

Staehelin: From these calculations can one obtain a value for the area of these domains?

Paillotin: I don't think so; I can derive the number of chlorophyll molecules in the space where the exciton may migrate freely but not the dimension of that space.

Joliot: May I ask the same question in a different way? Let us suppose that each particle contains only one photosynthetic unit. In this case, what distance must we assume between these particles to take into account the efficiency of energy transfer between units?

Paillotin: Maybe 5.0 nm or less.

Joliot: That is smaller than the distance observed by Dr Staehelin.

Anderson: The particles could be moving rapidly and colliding so that they could transfer energy from one to another, if each only has a single reaction centre.

Paillotin: Did you modify the distance in freeze-etch experiments?

Staehelin: The density slightly increases (about 10%) when I use no glycerol and freeze the preparation within about 5 ms. On average one finds 1200–1400 large (13.0–16.0 nm) particles/μm², and these have a centre-to-centre spacing of about 25 nm.

Paillotin: But 5.0 nm is much less than this distance.

References

BRETON, J. & GEACINTOV, N. E. (1979) Chlorophyll orientation and exciton migration in the photosynthetic membrane, in *This Volume*, pp. 217–233

CAMPILLO, A. J., SHAPIRO, S. L., KOLLMAN, V. H., WINN, & HYER, R. C. (1976) Picosecond exciton annihilation in photosynthetic systems. *Biophys. J. 16*, 93–97

FLEMING, G. R., MILLAR, D. P., MORRIS, G. C., MORRIS, J. M. & ROBINSON, G. W. (1977) Exciton fusion and annihilation in crystalline tetracene. *Aust. J. Chem. 30*, 2353–2360

GEACINTOV, N. E. & BRETON, J. (1977) Exciton annihilation in the two photosystems in chloroplasts a 100 K. *Biophys. J. 17*, 1–15

GEACINTOV, N. E., BRETON, J., SWENBERG, C. E. & PAILLOTIN, G. (1977) A single pulse pico-second laser study of exciton dynamics in chloroplasts. *Photochem. Photobiol.* 26, 629–638

MAUZERALL, D. (1976) Multiple excitations in photosynthetic systems. *Biophys. J.* 16, 87–91

PORTER, G., SYNOWIEC, J. A. & TREDWELL, C. J. (1977) Intensity effects on the fluorescence of *in vivo* chlorophyll. *Biochim. Biophys. Acta* 459, 329–336

SWENBERG, C. E., GEACINTOV, N. E. & POPE, M. (1976) Bimolecular quenching of excitons and fluorescence in the photosynthetic unit. *Biophys. J.* 16, 1447–1452

Chlorophyll orientation and exciton migration in the photosynthetic membrane

J. BRETON and N. E. GEACINTOV*

Service de Biophysique, Département de Biologie, Centre d'Études Nucléaires de Saclay, Gif-sur-Yvette, France

Abstract Measurements of the linear dichroism and of the polarization of the fluorescence with oriented chloroplasts have revealed a definite orientation of the pigment molecules with respect to the membrane plane. The Q_y transition moments of the chlorophyll a molecules are more closely inclined with respect to this plane for the forms absorbing at longer wavelengths than for those absorbing at shorter wavelengths. The fluorescence depolarization by energy transfer, determined with magnetically-oriented chloroplasts, indicates that the degree of local order increases with wavelength for the different (absorption wavelength) forms of chlorophyll a *in vivo*.

Laser pulses of either picosecond or microsecond duration have been used to probe the emission spectrum, lifetime and quantum yield of fluorescence of chloroplasts at various temperatures. With single picosecond pulses, singlet–singlet annihilations occur within the light-harvesting chlorophyll molecules. In the case of microsecond pulses, triplet excitons act as efficient quenchers of the singlets. By monitoring both the yield of carotenoid triplets and of the fluorescence during and after a laser flash, one can show that the carotenoid triplets account for part of but not all the fluorescence quenching.

In photosynthetic systems, most of the pigment molecules funnel the light energy towards photochemical reaction centres where this energy is trapped, giving rise to an electron-transfer process. Such a specific function of energy transfer suggests that most of the pigment molecules are organized in a precise two- or three-dimensional arrangement in which intermolecular distances and orientations are well defined. Recently, information has been obtained on the organization of the antenna chlorophylls in the photosynthetic membrane (Sauer 1975). In green plants, at least five different spectral forms of chlorophyll a have been revealed, indicating differences in their local environment (French

---

*On leave from the Department of Chemistry, New York University, New York

et al. 1972; Lutz 1977). Strong evidence has been given that the chlorophylls are specifically (but not covalently) bound to hydrophobic proteins (Thornber 1975; Anderson 1975).

Our purpose in this paper is to describe two aspects of chlorophyll organization and energy transfer in photosynthetic systems. First, we shall focus on the orientation of chlorophyll *in vivo*. Linear dichroism and polarized fluorescence spectroscopy on oriented chloroplasts have revealed a considerable degree of orientation of the photosynthetic pigments with respect to the plane of the membrane. Flash-spectroscopy experiments with a polarized measuring beam on oriented chloroplasts have demonstrated that the reaction centres are also oriented with respect to this plane. Studies of fluorescence depolarization by energy transfer with oriented chloroplasts reveal the existence of a certain degree of local order for forms of chlorophyll *a* absorbing at long wavelength.

Secondly, we shall deal with energy-transfer processes and more precisely with the fate of singlet excitons after excitation by laser pulses of various durations and intensities. It has been recently shown with laser pulses of picosecond (Campillo *et al.* 1976*a*), nanosecond (Mauzerall 1976*a,b*) and microsecond duration (Geacintov & Breton 1977) that the quantum yield of fluorescence decreases when the intensity of the pulses increases. This quenching is attributed to the mutual bimolecular annihilation of excitons (Swenberg *et al.* 1976). When trains of picosecond pulses from a mode-locked laser or intense laser flashes of microsecond duration are used, the quenching of singlets by long-lived triplet excitons becomes an important process. We have characterized some of these quenching channels by (i) analysing the quantum yield of fluorescence as a function of the emission wavelength at different temperatures, (ii) measuring the recovery time of the fluorescence yield and (iii) monitoring the kinetics of the build-up and decay of the carotenoid triplets.

ORIENTATION OF THE PHOTOSYNTHETIC PIGMENTS WITH RESPECT TO THE MEMBRANE PLANE

Photosynthetic membranes can be oriented by various techniques such as flow or air-drying (Breton & Roux 1971), electric fields (Gagliano *et al.* 1977) and magnetic fields (Geacintov *et al.* 1972; Breton *et al.* 1973*a*). The last technique is of particular interest for orienting intact chloroplasts in physiological conditions because an oriented ensemble of particles is produced in which the planes of the membranes are lined up perpendicular to the direction of the applied magnetic field.

With such oriented samples, measurement of the linear dichroism (difference in the absorbance of light polarized either parallel or perpendicular to the

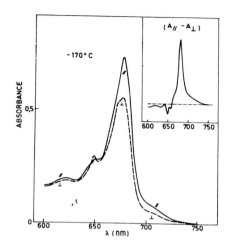
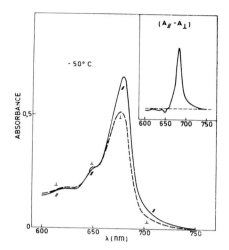

FIG. 1. Polarized absorption spectra of oriented spinach chloroplasts at low temperature (—— $A_{//}$, ---- $A_{\perp}$). The chloroplasts, suspended in glycerol/buffer (2/1), were placed for a few min in a 12 kG electromagnet and slowly cooled to the desired temperature. This procedure allows the trapping of oriented chloroplasts. The insets represent the linear dichroic spectra measured on the same samples with a different instrument recording directly $A_{//}-A_{\perp}$.

membrane plane) as a function of wavelength reveals an anisotropy in the orientation of the various antenna pigments within the photosynthetic membranes. This anisotropy can be in turn related to the tilt angle of the transition dipole moments of the pigments with respect to the membrane plane, provided that the extent of orientation of the membranes is known and that some assumptions about the distribution of the transition moments with respect to the normal to the membrane are made (Breton *et al.* 1973a).

Fig. 1 shows the polarized absorption and linear dichroism spectra of spinach chloroplasts at −50 and −170 °C for the red absorption band of chlorophyll. Relative to the absorption, the dichroism increases towards the longer wavelengths. For the Q_y transition moments of chlorophyll *b*, the angle between the transition moments and the membrane plane is greater than 35° (Vermeglio *et al.* 1976). This angle is close to 35° for the short wavelength forms of chlorophyll *a* and becomes smaller (about 20°) for the longer wavelength forms. A similar orientation close to the membrane plane has also been detected for the Q_y transition moments of bacteriochlorophyll in chromatophores of various photosynthetic bacteria (Morita & Miyazaki 1971; Breton 1974a).

Owing to their small concentration relative to the antenna pigments, the dichroism of the transitions of the chlorophyll molecules in the reaction centres cannot be detected by direct absorption. However, the measurements of specific flash-induced changes in absorbance of magnetically-oriented chloroplasts with a measuring beam polarized either parallel or perpendicular to the membrane plane reveal selectively the orientation of the transitions involved in the photooxidation of these centres. For P700, the primary donor of the photosystem I, both the major band at 700 nm and the radical-cation band of P700$^+$ at 820 nm have their transitions oriented at a small angle with respect to the membrane plane (Junge & Eckhof 1974; Breton et al. 1975; Breton 1977). A similar in-plane orientation of the transition moments has also been detected for the radical-cation band at 820 nm of the photooxidized primary donor of photosystem II (Mathis et al. 1976). Finally, in photosynthetic bacteria, the long-wavelength (870 nm) transition moment of the primary donor bacteriochlorophylls is also oriented close to the plane of the chromatophore membrane (Vermeglio & Clayton 1976).

All these observations suggest that the preferential in-plane orientation of the Q_y transitions of both the antenna absorbing at longer wavelength and the primary donor chlorophyll molecules in the reaction centres is relevant to the final trapping of the excitation energy by the reaction centres. After the absorption of a photon by the light-harvesting antenna-chlorophyll molecules and non-radiative relaxation to the first excited singlet state, the energy migrates preferentially to the lower energy forms of chlorophyll absorbing at longer wavelength. It reaches then an array of molecules whose transition dipole moment vectors are oriented close to or within the membrane planes and it is captured by the specialized chlorophyll molecules of the reaction centres whose transition moments are also located close to or within the membrane plane.

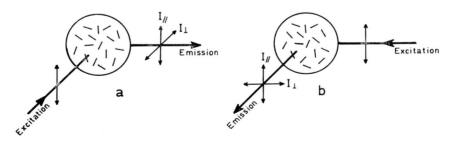

FIG. 2. Examples of two different measurements of polarized fluorescence on an oriented membrane. The small bars represent a homogeneous set of emission dipoles lying at random within a plane. It is assumed that they can transfer the excitation energy. In (a) measurement reveals the orientation of the dipoles with respect to the plane and in (b) the relative order between the dipoles can be determined.

Fluorescence spectroscopy in polarized light has also been used with oriented chloroplasts (Fig. 2*a*) to probe the anisotropy of the emission dipole oscillators with respect to the membrane plane (Geacintov *et al.* 1974; Breton 1974*b*). The results also indicate a preferential in-plane orientation of the long-wavelength transition dipoles involved in the emission and are consistent with the conclusion drawn from the linear dichroism experiments. Furthermore, a heterogeneity in the emission is detected, indicating a higher degree of orientation of the chlorophyll molecules emitting at the longer wavelengths. This is best shown by work at low temperature at which the existence of at least five different fluorescence-emitting forms of chlorophyll has been detected by their different degree of orientation with respect to the membrane plane (Garab & Breton 1976).

FLUORESCENCE DEPOLARIZATION BY ENERGY TRANSFER—LOCAL ORDER BETWEEN CHLOROPHYLL MOLECULES

Both linear dichroism and polarized fluorescence spectroscopy on oriented photosynthetic membranes reveal an orientation of the pigments with respect to the normal to the membrane plane. Attempts to reveal a further orientation with respect to an axis lying within the membrane plane have been so far unsuccessful. Using flow (Breton *et al.* 1973*a*) or electric field (Gagliano *et al.* 1977) orientation, in which the membrane planes orient parallel to the lines of the applied force, we have been unable to detect any long-range order of the pigments extending over the entire area of the membrane. Nevertheless, a degree of local order between pigment molecules is still possible over smaller domains. Such a local order may be significant for energy transfer.

For study of the efficiency of energy transfer and the extent of local order, the determination of the polarization of the fluorescence is a suitable approach. However, in the case of photosynthetic membranes, such a measurement on suspensions of randomly-oriented chloroplasts cannot be interpreted in the same manner as in the case of an isotropic solution of pigments. We have shown (Breton *et al.* 1973*b*; Becker *et al.* 1976) that, for a random suspension of such membranes, the polarization of the fluorescence can be divided into two contributions P_{in} and P_{an}. The latter term arises from the optical anisotropy of the membranes which is due to the forementioned orientation of the pigments and also to various other specific effects (absorption flattening, anisotropic absorption cross-section, light scattering). If only the intrinsic polarization P_{in} (which reflects the energy transfer between the pigments and their degree of mutual orientation) is to be measured, P_{an} must be suppressed. This can be done by using oriented membranes to detect the polarization of the fluorescence propagating along the normal to the membrane plane (axis of symmetry for the membrane). Such a geometry is depicted in Fig. 2b and can be conveniently obtained with a magnetic field.

With such oriented chloroplasts at room temperature we have found (Becker *et al.* 1976) some evidence of local order between the carotenoid molecules and the Q_y transition moments of chlorophyll *a*. The polarization of the emission at 720 nm has also been monitored by excitation in the red region of the spectrum where the absorption and the emission dipoles of chlorophyll have the same *Y* polarizations. In this case, we have shown that P_{in} increases from $2.2 \pm 0.8\%$ for 670 nm excitation to $5.4 \pm 1.1\%$ for 680 nm excitation. An increase of P_{in} towards the longer wavelengths has also been detected in the case of the broad emission band appearing around 735 nm at low temperature (Garab & Breton 1976). These results indicate that the degree of local order between the Q_y transition moments of chlorophyll *a* increases towards the longer wavelength. If this local order extends also to the transitions of the primary donors, then this arrangement further improves the efficiency for driving the excitation energy from the antenna molecules towards the reaction centre.

FLUORESCENCE QUENCHING BY SINGLE PICOSECOND LASER PULSES

Both the quantum yield and the lifetime of the fluorescence of chlorophyll *in vivo* decrease when single picosecond pulses of increasing intensities are used for the excitation (Campillo *et al.* 1976a,b; Geacintov *et al.* 1977a,b). These effects have been attributed to the annihilation of singlet excitons which occurs when many of these excitons are simultaneously (within 5–20 ps) created within the light-harvesting array of photosynthetic pigments. This process may be written as equation (1), where S_0, S_1 and S_n denote the ground state, first and

$$S_1 + S_1 \rightarrow S_0 + S_n \tag{1}$$

upper excited states of chlorophyll, respectively. Since this reaction allows us to probe some of the properties related to the mobility of singlet excitons in the photosynthetic membranes, we have been investigating the decrease of the fluorescence yield as a function of the energy of single picosecond pulses for chloroplasts at various temperatures. Lowering the temperature allows us to resolve the emission spectrum into two main bands, one at 685 nm corresponding to the emission of the light-harvesting antenna-chlorophyll molecules and the other at 735 nm which has been assigned to photosystem I (Butler & Kitajima 1974).

The experimental set-up (described in detail by Geacintov & Breton 1977)

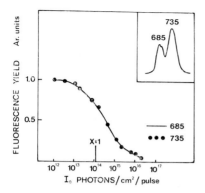

FIG. 3. Relative fluorescence yield of spinach chloroplasts at 100 K as a function of the energy of the single picosecond laser pulse used for the excitation. Note the superposition of the data for the two emission wavelengths of 685 and 735 mm. The mark X = 1 indicates the energy at which 1 photon is absorbed per photosynthetic unit per pulse. Inset: uncorrected emission spectrum of spinach chloroplasts at 100 K as recorded with the OMA–spectrograph combination.

consists mainly of a dye laser operated with a mode-locking dye (output wavelength about 610 nm). Single pulses (or sequences of pulses) can be selected from the entire pulse train by use of a Pöckel cell. After calibration of the energy and proper attenuation, the laser flash is focused onto a small area (about 2 mm²) within the sample. The fluorescence from that area is collected and focused onto the entrance slit of a spectrograph. The emission spectrum is recorded in digital form by use of the array of photodiodes of an Optical Multichannel Analyzer (OMA) coupled to the spectrograph. With this device we could record complete fluorescence spectra with either a single picosecond pulse or any other type of laser flash.

When using picosecond pulses of various energies we observed no difference, within experimental error, in the fluorescence yield measured either at 685 or at 735 nm for spinach chloroplasts at 100 K (Fig. 3). Quenching sets in at about 10^{13} photon cm^{-2} per laser pulse (Geacintov et al. 1977a) and the fluorescence yield decreases by one order of magnitude when the energy reaches 10^{16} photon cm^{-2}. Similar results were also obtained at 200 K and 21 K (Geacintov et al. 1977b). The emissions at these two wavelengths originate from pigment systems which are known to be different in many respects (pigment composition, lifetime of fluorescence, energy distribution). Consequently, if one assumes that singlet–singlet annihilations occur in both pigment systems, it is difficult to understand why the quenching curves are identical at 685 and at 735 nm.

We propose instead that the identical quenching efficiencies observed at these two wavelengths are due to singlet–singlet annihilations within the light-harvesting antenna and that the 735 nm emission derives its energy mainly by energy transfer from the light-harvesting pigments. In this model, the excitons that reach the photosystem I pigments have been created within the light-harvesting system. Annihilation of singlet excitons within the light-harvesting pigments reduces the number of excitons reaching the photosystem I pigments. Such a model for the distribution of energy predicts that the lifetime of the 735 nm emission should stay constant when the energy of the single picosecond pulse is increased, since no exciton is annihilated in this photosystem. On the other hand, a strong variation of the lifetime of the fluorescence at 685 nm is to be expected in the same conditions.

The lifetimes measured at these two wavelengths for various pulse energies are shown in Table 1 (Geacintov *et al.* 1977*a*). The lack of an intensity dependence of the lifetime for the 735 nm emission band in the intensity range studied although a strong variation is observed at 685 nm thus supports the conclusion that the photosystem I pigments responsible for the 735 nm emission band sample the exciton density within the light-harvesting pigment system.

TABLE 1

Dependence of the fluorescence lifetimes measured at 690 and 735 nm on the energy of the single picosecond pulse for spinach chloroplasts at 77 K with excitation at 530 nm (from Geacintov *et al.* 1977*a*).

Energy/pulse (photon cm^{-2})	Fluorescence lifetime at 690 nm (ns)	Fluorescence lifetime at 735 nm (ns)
2.5×10^{14}	0.38	
3.5×10^{14}		1.1
7.5×10^{14}	$\leqslant 0.13$ (instrument-limited)	
2.6×10^{15}		1.1

FLUORESCENCE QUENCHING BY MULTIPLE PICOSECOND PULSES AND BY MICROSECOND LASER PULSES

When laser pulses of picosecond duration are used to excite chloroplasts, mainly singlet–singlet annihilations occur since there is not enough time for a substantial build-up of long-lived quenchers such as triplets. However, quenching of singlet excitons by long-lived excited states has been observed when sequences of picosecond pulses from a mode-locked laser were used for the excitation (Breton & Geacintov 1976; Porter *et al.* 1977) and also when a

TABLE 2

Dependence of the relative fluorescence yield (normalized to 100 at 685 nm) of spinach chloroplasts on the number of exciting picosecond pulses in pulse sequences. The pulses (10^{15} photon cm^{-2} each) were interspaced by 5 ns.

Wavelength (nm)	Temperature (K)	Yield for 1 pulse	4 pulses	22 pulses
685	300	100 ± 10	77 ± 6	56 ± 3
635	100	100 ± 10	34 ± 5	23 ± 1
735	100	192 ± 20	57 ± 8	24 ± 1

non-mode-locked laser pulse of 2 μs duration (Geacintov & Breton 1977) was used.

At room temperature, where only the 685 nm emission is present, we have compared the relative yields of the fluorescence produced either by a single picosecond pulse or by sequences of 4 and 22 individual pulses, each pulse in the sequence bearing the same intensity as the single pulse. In these sequences the pulses were interspaced by dark intervals of 5 ns. In the multiple-pulse cases, the results shown in Table 2 indicate that an additional quenching is taking place as compared to the single-pulse case. At these energy levels the fluorescence lifetime is considerably shortened owing to singlet–singlet annihilations and the population of singlet excitons remaining 5 ns after the first pulse must be negligible. As a consequence the additional quenching which is observed for the following pulses must be due to excited states which survive from one pulse to the next. It has been proposed that triplet excitons (T_1) account for the quenching of singlets according to process (2). Triplet states of chlorophyll

$$S_1 + T_1 \rightarrow S_0 + T_n \tag{2}$$

can be efficient quenchers of singlets (Rahman & Knox 1973). It is also possible that chlorophyll anions (N^-) or cations (N^+) are produced by singlet–singlet fusion; these ions may constitute another quenching channel for the singlets (equation 3).

$$S_1 + N^{(\pm)} \nrightarrow S_0 + N^{(\pm)} \tag{3}$$

At low temperature we have also monitored the fluorescence yield changes at 685 and at 735 nm for various sequences of picosecond pulses. In this case the fluorescence yields at these two wavelengths do not have the same dependence on the number of pulses in the sequences (Breton & Geacintov 1976). Table 2 indicates that the drop of the yield is greater at 735 than at

685 nm. When measuring the fluorescence yields as a function of the energy of the whole output of the mode-locked laser (about 300 pulses), we found that the fluorescence yield at 735 nm starts to decline at an energy of 10^{15} photon cm^{-2}, an intensity which is about 30 times lower than the energy at which quenching at 685 nm sets in. The quenching factors (ratio of the fluorescence yields at the lowest and highest energy used) are also different: 45 at 735 nm and 10 at 685 nm (Geacintov & Breton 1977). These observations confirm that these two emission bands have different origins. Since the quenching efficiency is larger at 735 nm than at 685 nm we conclude that a greater density of long-lived quenchers is built up in photosystem I than in the light-harvesting system. Possible contributions from local heating, photo-chemistry, two-photon absorption and ground-state depletion have been discussed elsewhere (Geacintov & Breton 1977) and shown to be of minor importance in these experiments.

Conventional laser flashes of 2 μs duration have also been used to excite the fluorescence of chloroplasts at 100 K (Fig. 4). By simply removing the mode-locking dye from the laser cavity, we can compare the effect of such flashes with the effect of picosecond-pulse trains using the same sample. The fluorescence-quenching properties were found to be similar for these two types of excitation with only slight differences in the magnitudes of the quenching factors. However, these two types of excitation are different. In a mode-locked train, the photons are delivered in pulses of high peak intensities within about 10 ps bursts separated by dark intervals of 5 ns. In the micro-second flash, the arrival of photons is uniformly spread over a 2 μs interval and the instantaneous density of singlet excitons at any one time is low enough

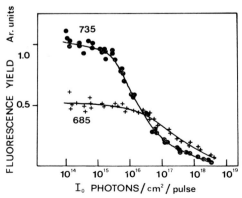

FIG. 4. Relative fluorescence yield at 685 and 735 mm for spinach chloroplasts at 100 K plotted against the energy (I_0) of the microsecond pulse used for the excitation.

so that singlet annihilations are minimized. In the latter conditions, inter-system crossing from the singlet excited states can build up an appreciable concentration of triplet excitons. These triplets will in turn annihilate some of the singlet excitons which are created on the subsequent portion of the pulse.

In view of the above results it appears that single picosecond pulses are ideal for studying singlet–singlet annihilations and thus for obtaining information about singlet exciton dynamics, e.g. diffusion coefficients, annihilation constants, temperature dependence, range of excitons (Geacintov et al. 1977b; Paillotin, this volume). For the study of long-lived triplet excitons, non-mode-locked microsecond flashes are useful. The use of picosecond-pulse trains is not advantageous since $S_1 + S_1$ and $S_1 + T_1$ (or $S_1 + N^{\pm}$) can occur and their respective contributions would be difficult to evaluate.

In the preceding section of the article we have described experiments which provided evidence for the existence of long-lived quenchers of the fluorescence. We have made the reasonable assumption that these long-lived quenchers are triplets and/or ions but no evidence for this conclusion was provided. In the following section we shall provide evidence that triplet excitons are involved in these processes.

TRIPLET EXCITONS GENERATED BY LASER-PULSE EXCITATION AND FLUORESCENCE QUENCHING

We have devised an experimental set-up to measure the kinetics of triplet–triplet absorption, as well as to determine the fluorescence quantum yield of a given sample. The cross-section of the actinic laser flash was adjusted by using slit apertures in front of the sample so that an image 1 mm in height and 3 mm in length was obtained on the sample (about 50 μm thick) which was tilted at an angle of about 45° with respect to the direction of propagation of the laser beam (Fig. 5). A xenon flash of about 3 μs duration was focused onto the sample as a vertical image 1.5–2 mm in width and about 10 mm in height. Both beams were centred on the same point of the sample (see Fig. 5). After traversing the sample the measuring xenon beam was in turn refocused on the array of diodes of the OMA detector. Lenses, interference and/or blocking filters and also neutral density filters were placed in the light paths as indicated in Fig. 5. The planar array of photodiodes of the OMA was placed in such a position that the intensity profile of the light from the xenon flash which was transmitted through the sample could be recorded with a resolution of 25 μm per channel. Furthermore, the OMA was gated with a pulse generator so that the gated OMA acted as a fast photographic shutter allowing the recording of a signal which was proportional to the amount of

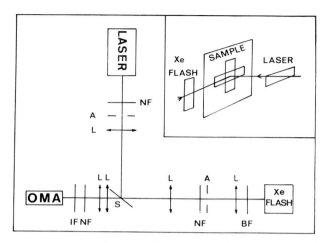

FIG. 5. Schematic diagram of the set-up used to monitor absorbance changes and variations of the fluorescence yield after excitation by various laser pulses: L, lenses; A, aperture; BF, blocking filter; NF, neutral density filter; IF, interference filter; S, sample consisting of a suspension of chloroplasts squeezed between two microscope slide cover-slips (the sample was about 50 μm thick). Inset: directions and shapes of the actinic and probe beams. Note their intersection at the sample plane.

light which reached each of the 500 linear arrays of photodiodes during a given time interval (usually 50 or 200 ns). A triggering system and various delay lines allow for (1) the synchronization of the OMA with the maximum output of the xenon flash, and (2) for adjustable delays between the laser pulse and the xenon flash. When this last delay was fixed at a given value, several (3 to 5) shots were recorded in one of the two memory banks of the OMA with the laser beam blocked. This measurement gave the intensity profile of the transmitted xenon flash light along the vertical axis of the sample. Next, the same number of shots with the laser beam incident on the sample were recorded in the second half of the memory of the OMA. An absorbance increase or decrease in the sample at the wavelength of observation (selected by an interference filter in front of the OMA) appeared as a decrease or increase in the amplitude of the signal at the intersection of the two beams. From these two signals, I and $(I + \Delta I)$ could be obtained. A noise level corresponding to $\Delta I / I = \pm 1\%$ was obtained with this arrangement. Spectra could be recorded point by point by the use of different interference filters and we obtained the kinetics of the decay of the triplets by altering the delay between the actinic and measuring flashes. With the same set-up, variations in the fluorescence quantum yield after the laser flash could be monitored.

In this case, a blue filter was set between the xenon flash and the sample and a complementary red cut-off filter was secured in front of the OMA. The intensity profile of the fluorescence emitted during the excitation by the xenon flash and focused onto the detector could be measured at various delay times between the laser actinic flash and the xenon probe flash.

With this arrangement and an observation wavelength of 820 nm, at which the radical cation of chlorophyll a exhibits an absorption band (Borg et $al.$ 1970) we have not been able to detect any absorbance change increase during or after excitation by the whole laser flash, either mode-locked or non-mode-locked. Owing to the concentration of the sample and the noise level inherent in our measurements, this limit corresponds to an amount of chlorophyll a^+ not larger than 2 per photosynthetic unit. We, therefore, conclude that chlorophyll cations alone cannot be responsible for the large quenching of the fluorescence observed in identical conditions of excitation.

In the same set of conditions but with a 500 ns gate to select a square portion of the excitation pulse, a large absorbance increase can be observed at 515 nm. $\Delta I/I$ signals of 15–20% were routinely observed. When both the excitation energy and the delay were kept constant, the wavelength dependence of the $\Delta I/I$ signal was measured with samples of appropriate optical densities (0.5–1.0) at each wavelength. The spectrum agrees with the one previously reported (Mathis 1970) for the triplet of carotenoids (Car^T) observed in $vivo$. Furthermore, the decay time of the transients in ambient air was decreased or increased when the sample was flushed with either oxygen or nitrogen, respectively. These effects were fully reversible. For a given energy of the laser, the magnitude of the $\Delta I/I$ signals was the same whether the laser was mode-locked or not. A typical kinetic profile for the build-up and decay of Car^T under an atmosphere of air is depicted in Fig. 6, together with the excitation profile of the laser flash. The triplets are still accumulating 500–1000 ns after the end of the flash. This effect will be discussed in detail elsewhere.

The recovery of the fluorescence yield was monitored after the laser flash (Fig. 6). We observed that (1) the recovery of the fluorescence yield was significantly slower than the decay of Car^T and (2) the kinetics were affected by the presence of oxygen or nitrogen in the same way as for Car^T.

The fluorescence quenching was also monitored during the square-excitation laser pulse. In these measurements the experimental arrangement was similar to the one described previously (Geacintov & Breton 1977). The main difference was that we allowed an attenuated portion of the excitation light to enter the spectrograph together with the fluorescence light in order to have an internal measure of the actual energy used to excite the fluorescence

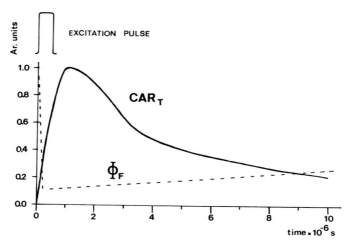

FIG. 6. ——, Kinetics of the build-up and decay of carotenoid triplets observed at 515 nm for spinach chloroplasts at 20 °C in an air atmosphere and excited by a 500 ns square pulse selected from a non-mode-locked dye laser (10^{16} photon cm^{-2}; $\lambda = 610$ nm). The shape of the excitation pulse is represented on the top part of the figure. ----, decrease and recovery of the fluorescence yield observed at 685 nm during and after excitation by the laser pulse.

during the time the gate of the OMA was opened. These data are also included in Fig. 6.

Although the data reported here have been obtained at room temperature, no significant modification of the results was observed for samples at 100 K. A more detailed analysis of these data will be presented elsewhere. However, the major conclusions that we derive from these experiments are:

(1) the quenchers appear to be mainly pigments in their triplet state;

(2) Car^T are formed to the same extent by microsecond flashes as by picosecond-pulse trains. The maximum concentrations of Car^T that we have achieved are about 3 Car^T per photosynthetic unit;

(3) Car^T are responsible for part of, but not all, the quenching of chlorophyll singlets. Other molecules (most probably chlorophyll) in their triplet state must account for the rest of the quenching. Similar conclusions have been reached for photosynthetic bacteria (Monger & Parson 1977).

CONCLUSION

Using polarized light spectroscopy on oriented photosynthetic membranes, we have demonstrated that a high degree of orientation of the pigments exists *in vivo* which is much higher than was previously supposed. Not only most of the antenna molecules, but also the primary donor chlorophylls appear to be oriented with respect to the plane of the membranes. Furthermore, a certain degree of local order between some of the antenna pigments has been detected. The possibility that such an order plays a role in the mechanism of transfer of the excitation energy from the antenna to the reaction centres has been outlined.

We have also described experiments showing that singlet excitons created during an intense laser pulse of microsecond duration are efficiently quenched by pigments (carotenoid and chlorophyll molecules) in their triplet state. In the case of single picosecond pulses, the quenching is produced by singlet–singlet fusions within the light-harvesting chlorophyll–protein complexes.

The energy transfer between the antenna molecules can, in principle, provide a tool for studying the local order between these pigments and to obtain information about the molecular organization of chlorophyll molecules in the photosynthetic membrane. Ideally, the determination of the time dependence of the fluorescence depolarization after excitation by a single picosecond pulse should provide this kind of information. The fluorescence should be excited at several wavelengths within the Q_y band, and the energy should be kept low enough to avoid singlet–singlet annihilations, but this experiment cannot be easily done at present. However, an alternative method can be used taking advantage of the singlet annihilation processes described in this article. In this case single picosecond pulses of polarized light are used to excite the chlorophylls in their red Q_y band; the fluorescence is then analysed by a polarizer, a monochromator and a photodetector that measures only the total amount of fluorescence light emitted from the sample (not the decay kinetics). At the lowest excitation energies, when no singlet–singlet annihilations occur, the fluorescence polarization should be the same as that measured for low-light continuous excitation. It corresponds then to the depolarization of the fluorescence induced by the normal migration of excitons from the antenna towards the reaction centres. When the energy of the single pulse is increased, exciton annihilations will progressively shorten the lifetime and thus the migration range of the singlet exciton. The degree of fluorescence polarization depending on the short- and long-range mutual orientation of chlorophyll molecules may thus change on increasing the excitation energy, so allowing us to probe the mutual orientation of chlorophyll molecules over shorter distances. At the highest energies that are possible without damaging the sample, we achieved about 100 hits per photosynthetic unit within 10 ps (Geacintov *et al.* 1977*a*). Under such circumstances, the excitation energy cannot migrate over more than a few molecules and the polarization should be appreciably higher than at low energy levels of excitation.

References

ANDERSON, J. M. (1975) The molecular organization of chloroplast thylakoids. *Biochim. Biophys. Acta 416*, 191–235

BECKER, J. F., BRETON, J., GEACINTOV, N. E. & TRENTACOSTI, F. (1976) Anisotropy of photosynthetic membranes and the degree of fluorescence polarization. *Biochim. Biophys. Acta 440*, 531–544

BORG, D. C., FAJER, J., FELTON, R. H. & DOLPHIN, D. (1970) The π-cation radical of chlorophyll *a. Proc. Natl. Acad. Sci. U.S.A. 67*, 813–820

BRETON, J. (1974*a*) The state of chlorophyll and carotenoid *in vivo*: II. A linear dichroism study of pigment orientation in photosynthetic bacteria. *Biochem. Biophys. Res. Commun. 59*, 1011–1017

BRETON, J. (1974*b*) Polarized light spectroscopy on oriented spinach chloroplasts: fluorescence emission and excitation spectra, in *Proceedings of the Third International Congress on Photosynthesis* (Avron, M., ed.), pp. 229–234, Elsevier, Amsterdam

BRETON, J. (1977) Dichroism of transient absorbance changes in the red spectral region using oriented chloroplasts. II. *P*-700 absorbance changes. *Biochim. Biophys. Acta 459*, 66–75

BRETON, J. & GEACINTOV, N. E. (1976) Quenching of fluorescence of chlorophyll *in vivo* by long-lived excited states. *FEBS (Fed. Eur. Biochem. Soc.) Lett. 69*, 86–89

BRETON, J. & ROUX, E. (1971) Chlorophylls and carotenoid states *in vivo*: I. A linear dichroism study of pigment orientation in spinach chloroplasts. *Biochem. Biophys. Res. Commun. 45*, 557–563

BRETON, J., MICHEL-VILLAZ, M. & PAILLOTIN, G. (1973*a*) Orientation of pigments and structural proteins in the photosynthetic membrane of spinach chloroplasts: a linear dichroism study. *Biochim. Biophys. Acta 314*, 42–56

BRETON, J., BECKER, J. F. & GEACINTOV, N. E. (1973*b*) Fluorescence depolarization study of randomly oriented and magneto-oriented spinach chloroplasts in suspension. *Biochem. Biophys. Res. Commun. 54*, 1403–1409

BRETON, J., ROUX, E. & WHITMARSH, J. C. (1975) Dichroism of chlorophyll a_{I} absorption change at 700 nm using chloroplasts oriented in a magnetic field. *Biochem. Biophys. Res. Commun. 64*, 1274–1277

BUTLER, W. L. & KITAJIMA, M. (1974) Tripartite model for chloroplast fluorescence, in *Proceedings of the Third International Congress on Photosynthesis* (Avron, M. ed.), pp. 13–24, Elsevier, Amsterdam

CAMPILLO, A. J., SHAPIRO, S. L., KOLLMAN, V. H., WINN, K. R. & HYER, R. C. (1976*a*) Picosecond exciton annihilation in photosynthetic systems. *Biophys. J. 16*, 93–97

CAMPILLO, A. J., KOLLMAN, V. H. & SHAPIRO, S. L. (1976*b*) Intensity dependence of the fluorescence lifetime of *in vivo* chlorophyll excited by a picosecond light pulse. *Science (Wash. D.C.) 193*, 227–229

FRENCH, C. S., BROWN, J. S. & LAWRENCE, M. C. (1972) Four universal forms of chlorophyll *a. Plant. Physiol. 49*, 421–429

GAGLIANO, A. G., GEACINTOV, N. E. & BRETON, J. (1977) Orientation and linear dichroism of chloroplasts and sub-chloroplast fragments oriented in an electric field. *Biochim. Biophys. Acta 461*, 460–474

GARAB, G. & BRETON, J. (1976) Polarized light spectroscopy on oriented spinach chloroplasts. Fluorescence emission at low temperature. *Biochem. Biophys. Res. Commun. 71*, 1095–1102

GEACINTOV, N. E. & BRETON, J. (1977) Exciton annihilation in the two photosystems in chloroplasts at 100°K. *Biophys. J. 17*, 1–15

GEACINTOV, N. E., VAN NOSTRAND, F., BECKER, J. F. & TINKEL, J. B. (1972) Magnetic field induced orientation of photosynthetic systems. *Biochim. Biophys. Acta 267*, 65–79

GEACINTOV, N. E., VAN NOSTRAND, F. & BECKER, J. F. (1974) Polarized light spectroscopy of photosynthetic membranes in magneto-oriented whole cells and chloroplasts: fluorescence and dichroism. *Biochim. Biophys. Acta 347*, 443–463

GEACINTOV, N. E., BRETON, J., SWENBERG, C. E., CAMPILLO, A. J., HYER, R. C. & SHAPIRO, S. L. (1977a) Picosecond and microsecond pulse laser studies of exciton quenching and exciton distribution in spinach chloroplasts at low temperatures. *Biochim. Biophys. Acta 461*, 306–312

GEACINTOV, N. E., BRETON, J., SWENBERG, C. E. & PAILLOTIN, G. (1977b) A single pulse picosecond laser study of exciton dynamics in chloroplasts. *Photochem. Photobiol. 26*, 629–638

JUNGE, W. & ECKHOF, A. (1974) Photoselection studies on the orientation of chlorophyll a_I in the functional membrane of photosynthesis. *Biochim. Biophys. Acta 357*, 103–117

LUTZ, M. (1977) Antenna chlorophyll in photosynthetic membranes. A study by resonance Raman spectroscopy. *Biochim. Biophys. Acta 460*, 408–430

MATHIS, P. (1970) *Étude de Formes Transitoires des Caroténoides*, Thèse doctorale d'État, Orsay

MATHIS, P., BRETON, J., VERMEGIO, A. & YATES, M. (1976) Orientation of the primary donor chlorophyll of photosystem II in chloroplast membranes. *FEBS (Fed. Eur. Biochem. Soc.) Lett. 63*, 171–173

MAUZERALL, D. (1976a) Multiple excitation in photosynthetic systems. *Biophys. J. 16*, 87–91

MAUZERALL, D. (1976b) Fluorescence and multiple excitation in photosynthetic systems. *J. Phys. Chem. 80*, 2306–2309

MONGER, T. G. & PARSON, W. W. (1977) Singlet–triplet fusion in *Rhodopseudomonas sphaeroides* chromatophores. A probe of the organization of the photosynthetic apparatus. *Biochim. Biophys. Acta 460*, 393–407

MORITA, S. & MIYAZAKI, T. (1971) Dichroism of bacteriochlorophyll in cells of the photosynthetic bacterium *Rhodopseudomonas palustris. Biochim. Biophys. Acta 245*, 151–159

PAILLOTIN, G. & SWENBERG, C. E. (1978) Dynamics of excitons created by a single picosecond pulse, in *This Volume*, pp. 201–209

PORTER, G., SYNOWIEC, J. A. & TREDWELL, C. J. (1977) Intensity effects on the fluorescence of *in vivo* chlorophyll. *Biochim. Biophys. Acta 459*, 329–336

RAHMAN, T. S. & KNOX, R. S. (1973) Theory of singlet-triplet exciton fusion. *Phys. Stat. Solidi (B) 58*, 715–721

SAUER, K. (1975) Primary events and the trapping of energy, in *Bioenergetics of Photosynthesis* (Govindjee, ed.), pp. 115–181, Academic Press, New York

SWENBERG, G. E., GEACINTOV, N. E. & POPE, M. (1976) Bimolecular quenching of excitons and fluorescence in the photosynthetic unit. *Biophys. J. 16*, 1447–1452

THORNBER, J. P. (1975) Chlorophyll-proteins: light harvesting and reaction center components of plants. *Annu. Rev. Plant. Physiol. 26*, 127–158

VERMEGLIO, A. & CLAYTON, R. K. (1976) Orientation of chromophores in reaction centers of *Rhodopseudomonas sphaeroides*. Evidence for two absorption bands of the dimeric primary electron donor. *Biochim. Biophys. Acta 449*, 500–515

VERMEGLIO, A., BRETON, J. & MATHIS, P. (1976) Trapping at low temperature of oriented chloroplasts: application to the study of antenna pigments and of the trap of photosystem-I. *J. Supramol. Struct. 5*, 109–117

Discussion

Cogdell: Are more carotenoid triplets formed when the intensity of the laser flash is increased or is the carotenoid triplet state saturated?

Breton: At relatively low intensity the concentration of triplets rises steeply with increasing intensity of the flash but at still higher intensities the carotenoid triplet concentration tends to level off but does not reach a saturation limit.

Cogdell: If triplet–triplet energy transfer from the chlorophyll to the carotenoid is efficient and if one supposes that the low rate of recovery of fluorescence yield is due to some residual chlorophyll triplets, then presumably these triplets must be unable to communicate with carotenoids.

Breton: This is how we tentatively interpret the slow rise of Car^T after the termination of the flash. When only a few chlorophyll triplets are present, the transfer $Chl^T + Car \rightarrow Chl + Car^T$ is efficient. However, this process might be easily saturated at higher excitation energies and some Chl^T may be created in regions of the membrane where there are no nearby available carotenoid molecules (in the ground state) to which they could transfer their energy efficiently. These Chl^T will transfer energy more slowly to less-accessible carotenoid molecules and thus might be efficient quenchers of the fluorescence.

Junge: Your studies on exciton annihilation were designed to extend the information content of studies on fluorescence polarization. Can you give us some preliminary data on time-dependent fluorescence polarization?

Breton: We tried these experiments but not in the way we have just described, because we do not have picosecond dye-laser excitation at the appropriate wavelength (650–700 nm). Consequently we have tried to use singlet–triplet instead of singlet–singlet annihilation and we made use of a 500 ns dye-laser excitation in this spectral range.

For most of the excitation wavelengths we have not found significant changes in the fluorescence polarization over the range of excitation energies for which we see singlet–triplet annihilations. One possible explanation is that triplets are formed far from the antenna that we excite, close to the reaction centres, for example. Then the polarization of the excitation is 'smeared out' and we cannot expect any large change as a function of the excitation energy.

Joliot: The lifetime of the fluorescence quencher, which is probably the carotenoid triplet, seems to be longer than that measured by Wolff & Witt (11969) on illumination by a short flash. Is that owing to the intense excitation puse?

Breton: The decay of the Car^T seems to be independent of intensity although a precise measurement at very low intensity of the laser flash was not possible with our experimental set-up. The 5 μs lifetime that we have measured in air is probably due to the presence of glycerol in our suspension medium (P. Mathis, personal communication).

Duysens: Could the slow fluorescence rise be caused by the disappearance of a carotenoid triplet?

Breton: At low energy of the laser pulse, the kinetics of recovery of the fluorescence and of the decay of the carotenoid triplet are similar. However, at high energy the recovery of the fluorescence is slower than the decay of Car^T and we believe that another triplet (probably Chl^T) is operative.

Knox: You were investigating whether the dipoles neighbouring the dipole that is excited are parallel to it. Our results indicate that at least two of them are not parallel.

Breton: This reduces the maximum polarization that we can expect to find in our experiments, unless we excite at longer wavelengths.

Knox: Also, excitation at 677 nm and at lower temperature will prevent transfer into the 670 nm group of states. Another point is that, when exciting at 650 nm, we got almost zero polarization in emission at 730 nm; that means that excitation at 650 nm might enable us to find out what is going on in the reaction centre complex.

Clayton: Are there technical difficulties in the use of a ruby laser to obtain short enough pulses?

Breton: I do not think so; it is relatively easy to do it with a neodymium laser.

Clayton: Would 694 nm be a good wavelength?

Breton: Yes; you can also shift this frequency by using either a dye or stimulated Raman emission.

Porter: With regard to the orientation in the magnetic field, you are orienting the whole chloroplast. Could you orient parts of the chloroplast?

Breton: We have no evidence that such a re-orientation, of chlorophyll–protein complexes for example, occurs to a detectable extent. This can be shown by the complete absence of magnetically induced linear dichroism with French-press treated chloroplasts or with isolated chromatophores (Breton 1974).

Amesz: As you stated, the identical intensity dependences of the fluorescence at 685 and 735 nm indicate that the pigment emitting at 735 nm 'samples' the exciton density in the 685 nm pigment. However, as is well known, this does not appear to be the case when one measures the initial and variable fluorescence at 'normal' light intensities; the variable fluorescence is relatively less at the latter wavelength. Does that mean that the 685 nm emission comes from an inhomogeneous pigment population?

Breton: Briefly, we propose that in our picosecond experiments the excitons pass directly into photosystems I and II from the light-harvesting antenna pigments (Geacintov *et al.* 1977). In the induction experiments, on the other hand, the excitation is steady-state and thus corresponds to an entirely different time regime in which it is possible for excitons to flow from photosystem II to photosystem I, either directly or *via* the photosystem II → light-harvesting →

photosystem I pathway. Thus the experimental conditions are totally different in the picosecond excitation case.

Beddard: One might expect the decay time at 735 nm to be long (independent of excitation wavelength), but at 685 nm it should be quenched.

Breton: This is what we have observed. In the intensity range $\sim 10^{14}$–10^{16} photon cm^{-2} per pulse the lifetime of the 685 nm fluorescence decreases from about 500 ps down to less than 50 ps. Within the same range of excitation energy the fluorescence at 735 nm has a lifetime of 1.5 ± 4 ns.

Katz: What are the energy levels of the triplets of the species fluorescing at 685 nm and at 735 nm?

Knox: The chlorophyll *a* triplet is at 10 420 cm^{-1} in ethanol (Mau & Puza 1977).

Robinson: The transfer might go to higher triplets and then drop down.

Cogdell: Bacteriochlorophyll is about 5100 cm^{-1} (Connolly *et al.* 1973) and the carotenoid must be lower than that because there is efficient energy transfer from bacteriochlorophyll.

Clayton: Was the background light sufficient to keep the variable fluorescence of photosystem II near its maximum level?

Breton: Yes.

Clayton: In that case the correlation between 735 and 685 nm is not a problem because you will be sampling a certain proportion of the total 685 nm.

References

BRETON, J. (1974) The state of chlorophyll and carotenoid *in vivo*. II. A linear dichroism study of pigment orientation in photosynthetic bacteria. *Biochem. Biophys. Res. Commun.* 59, 1011–1017

CONNOLLY, J. S., GORMAN, D. S. & SEELY, G. R. (1973) *Ann. N.Y. Acad. Sci.* 206, 649–669

GEACINTOV, N. E., BRETON, J., SWENBERG, C. E. & PAILLOTIN, G. (1977) A single pulse picosecond laser study of exciton dynamics in chloroplasts. *Photochem. Photobiol.* 26, 629–638

MAU, A. W. H. & PUZA, M. (1977) Phosphorescence of chlorophylls. *Photochem. Photobiol.* 25, 601–604

WOLFF, C. H. & WITT, H. T. (1969) On metastable states of carotenoids in primary events of photosynthesis. *Z. Naturforsch. 24b*, 1031–1037

Tripartite and bipartite models of the photochemical apparatus of photosynthesis

W. L. BUTLER*

Service de Biophysique, Département de Biologie, Centre d'Études Nucléaires de Saclay, Gif-sur-Yvette, France

Abstract Tripartite and bipartite models for the photochemical apparatus of photosynthesis are presented and examined. It is shown that the equations for the yields of fluorescence from the different parts of the photochemical apparatus of the tripartite model transform into the simple equations of the bipartite formulation when the probability for energy transfer from the light-harvesting chlorophyll *a/b* complex to photosystem II is unity. The nature of the 695 and 735 nm fluorescence bands which appear in the emission spectrum of chloroplasts at low temperature is examined. It is proposed that these bands are due to fluorescence from energy-trapping centres which form in the antenna chlorophyll of photosystem II and photosystem I on cooling to low temperature. Even though these fluorescence emissions can be regarded as low temperature artifacts since they are not present at physiological temperatures, they nevertheless are proportional to the excitation energy in the two photosystems and can be used to monitor energy distribution in the photochemical apparatus. However, the question of their artifactual nature is crucial to the interpretation of fluorescence-lifetime measurements at low temperature.

Fluorescence was one of the earliest methods used to explore primary photochemical mechanisms of photosynthesis. Initially, it was hoped that simple competitive relationships between fluorescence, photochemistry and non-radiative decay processes, similar to those found in simple model systems, could be adapted to the study of the primary photochemistry of photosynthesis but, until recently, those expectations have been largely unrealized, primarily because of the absence of an adequate theoretical framework capable of accommodating and explaining the experimental data without making excessive *ad hoc* assumptions. Recently, a comprehensive photochemical model for the photosynthetic apparatus, termed the tripartite model, was developed

---

*On leave from the University of California San Diego.

237

(initially by Butler & Kitajima 1975a and later in a more rigorous form by Butler & Strasser 1977), based on specific competitive relationships between fluorescence, photochemistry and non-radiative decay including energy transfer to different parts of the photochemical apparatus. Those relationships were established through systematic studies of fluorescence and photochemistry in chloroplasts at -196 °C in conditions where energy quenching and energy distribution could be varied (Kitajima & Butler 1975a; Butler & Kitajima 1975b). The model appears to accommodate existing data, including the explanation of otherwise anomalous results, and has been tested and confirmed in several experiments where the results were predicted initially by the model (e.g., Kitajima & Butler 1975b; Satoh *et al.* 1976; Strasser & Butler 1977a, 1978; Ley & Butler 1977a). The work to date on the development and application of the model has been reviewed recently (Butler 1978). My purpose in this paper is to explore some of the interpretations of the equations, especially the correspondence between the simple formulation (Butler & Kitajima 1975a) and the more rigorous and more complex formulation (Butler & Strasser 1977), to examine the available evidence on the low temperature fluorescence bands to determine if those emissions are consistent with certain basic assumptions made in order to develop and evaluate the model, and to discuss some implications of the model on fluorescence-lifetime measurements.

THE TRIPARTITE MODEL

I shall initially describe briefly the model with emphasis on basic assumptions and definitions to establish the groundwork for subsequent discussions of some of the functional relationships that are inherent in the model. For green plants (Butler & Kitajima 1975a, b) the model assumes three major types of light-harvesting antenna-chlorophyll assemblages which are in close physical association with one another. These are the antenna chlorophyll *a*s associated with the photochemically-active photosystem (PS) I and II units (Chl a_I and Chl a_{II}, respectively) and the light-harvesting chlorophyll a/b complex (Chl LH) which may transfer some energy to PS I but appears to serve largely as additional antenna chlorophyll for PS II. (The model for red algae and cyano-bacteria [Ley & Butler 1977b] is similar except that the chlorophyll a/b complex is replaced by phycobilisomes.) It was also assumed that the three fluorescence bands which appear at about 685, 695 and 735 nm in the emission spectrum of chloroplasts at -196 °C represent fluorescence from the three types of antenna chlorophyll and that those fluorescence bands can be used to monitor the excitation energy in Chl LH, Chl a_{II} and Chl a_I, respectively. It was

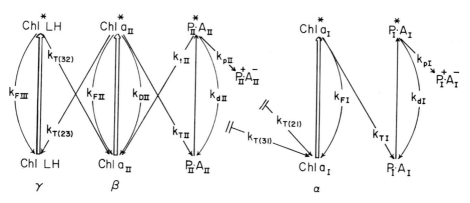

FIG. 1. Rate-constant diagram of excitation and de-excitation processes in the photochemical apparatus of chloroplasts according to the tripartite model (Butler & Strasser 1977): α, β and γ represent the relative optical cross-sections of Chl a_I, Chl a_{II} and Chl LH, respectively. Rate constants for energy transfer between the different beds of antenna chlorophyll are specified by the appropriate subscripts; the subscript 3 is used to represent Chl LH. Processes occurring at the reaction centre chlorophyll are indicated by lower-case subscript letters: p for photochemistry, d for non-radiative decay and t for energy transfer back to the antenna chlorophyll. The arrow k_{FI} indicating the 735 nm fluorescence from Chl a_I includes energy trapping by C-705 and the subsequent fluorescence emission from that species.

apparent from the fluorescence emission spectra of purified preparations of PS I and PS II particles and of the chlorophyll a/b–protein complex at $-196\,°C$ (Satoh & Butler 1978a) and from a deconvolution of the low-temperature emission spectrum of a green leaf into its three constituent components (Strasser & Butler 1977c) that measurements of fluorescence at 680, 695 and 730 nm represent reasonably-pure emissions from each of the three types of antenna chlorophyll.

The various excitation and de-excitation processes which are assumed to play a significant role in the photochemical apparatus are defined in the rate-constant diagram shown in Fig. 1. The initial distribution of quanta absorbed by the photochemical apparatus is specified as the fraction α going to PS I, the fraction β to PS II and the fraction γ to Chl LH, where the sum of α, β and γ is unity. The steady-state distribution of excitation energy also includes energy transfer terms; e.g., $\psi_{T(23)} = k_{T(23)}/\Sigma k_{II}$ is the probability that an exciton in Chl a_{II} will be transferred to Chl LH, where Σk_{II} is the sum of the competing rate constants which function in the antenna chlorophyll of PS II.

The influence of the photochemical activities of PS I and PS II on the overall photochemical properties of the photosynthetic apparatus are markedly different. In green plants no changes in fluorescence yield result from the primary photochemical activity of PS I. (In red algae and cyanobacteria, photooxidation of P700 at -196 °C results in a small increase in the yield of fluorescence from PS I [Ley & Butler 1977b].) It was suggested (Butler & Kitajima 1975a) that the oxidized state of the reaction centre chlorophyll, P_I^+, quenches fluorescence as effectively as the reduced state so that the photooxidation of P_I causes no change in the yield of fluorescence. This suggestion has some precedent in the earlier work (Butler *et al.* 1973) showing that P_{II}^+ quenches the fluorescence of PS II. In PS II, however, the photo-reduction of the primary electron acceptor, A_{II}, causes a marked increase in the yield of fluorescence. In contrast to PS I where the stable closed state of a reaction centre is $P_I^+ \cdot A_I$ (states involving the strong reductant A_I^- turn over too fast even at -196 °C to accumulate to any significant extent), the stable closed state of PS II reaction centres is $P_{II} \cdot A_{II}^-$ (here, the P_{II}^+ states are too reactive to accumulate). It is assumed in Fig. 1 that excitation energy can visit the reaction centre chlorophyll of a closed reaction centre. However, the energy cannot be used for photochemistry but must either be returned to the antenna chlorophyll *via* k_{tII} or dissipated by non-radiative decay *via* k_{dII}. It is assumed for purposes of simplicity that k_{pII} at open reaction centres is much larger than k_{tII} or k_{dII} so that photochemistry is accomplished whenever an exciton is trapped by an open reaction centre. At closed reaction centres the probability that a trapped exciton will be returned to the antenna chlorophyll, where it may reappear as fluorescence, is $\psi_{tII} = k_{tII}(k_{tII} + k_{dII})^{-1}$. In essence, the return of excitation energy from closed PS II reaction centres increases the exciton density in Chl a_{II} so that the fluxes through the PS II de-excitation pathways increase as the PS II reaction centres close. The yields ϕ for the local PS II de-excitation processes can be expressed in terms of a constant probability factor, $\psi_{XII} = k_{XII}/\Sigma k_{II}$, and a variable function of A_{II}, $f(A_{II})$ (equations 1).

$$\phi_{FII} = \psi_{FII} f(A_{II}) \tag{1a}$$

$$\phi_{T(21)} = \psi_{T(21)} f(A_{II}) \tag{1b}$$

$$\phi_{T(23)} = \psi_{T(23)} f(A_{II}) \tag{1c}$$

The particular form of $f(A_{II})$ depends on the degree of energy transfer between PS II units. However, $f(A_{II})$ always varies between the same limiting values regardless of the degree of energy transfer; when the PS II reaction centres are all open ($A_{II} = 1.0$), $f(A_{II}) = 1.0$ and when the centres are all closed ($A_{II} = 0$), $f(A_{II}) = (1 - \psi_{TII}\psi_{tII})^{-1}$. (According to the model, the probability for trapping by the PS II reaction centre chlorophyll is constant,

$\psi_{\text{TII}} = k_{\text{TII}}/\Sigma k_{\text{II}}$, independent of the state of the reaction centre.) Thus, any application of the model which involves measurements made only at the end points, i.e. when $A_{\text{II}} = 1.0$ or 0, should be valid for any degree of energy transfer between PS II units.

The de-excitation processes in Chl a_{II} which depend on the state of A_{II} are differentiated from those in Chl LH and Chl a_{I} which are represented by constant probability terms, ψ, by the use of the symbol ϕ to represent the PS II processes which incorporate the f(A_{II}) function according to equations (1). Following the previous derivation (Butler & Strasser 1977), we can write the overall yields of fluorescence and photochemistry as

$$\Phi_{\text{FIII}} = \frac{F_{\text{III}}}{I_a} = \left(\frac{\gamma + \beta\phi_{\text{T(23)}}}{1 - \psi_{\text{T(32)}}\phi_{\text{T(23)}}} \right) \psi_{\text{FIII}} \qquad (2)$$

$$\Phi_{\text{FII}} = \frac{F_{\text{II}}}{I_a} = \left(\frac{\beta + \gamma\psi_{\text{T(32)}}}{1 - \psi_{\text{T(32)}}\phi_{\text{T(23)}}} \right) \phi_{\text{FII}} \qquad (3)$$

$$\Phi_{\text{FI}} = \frac{F_{\text{I}}}{I_a} = \left(\alpha + \frac{[\beta + \gamma\psi_{\text{T(32)}}]\phi_{\text{T(21)}} + [\gamma + \beta\psi_{\text{T(23)}}]\psi_{\text{T(31)}}}{1 - \psi_{\text{T(32)}}\phi_{\text{T(23)}}} \right) \psi_{\text{FI}} \qquad (4)$$

$$\Phi_{\text{PII}} = \frac{P_{\text{II}}}{I_a} = \left(\frac{\beta + \gamma\psi_{\text{T(32)}}}{1 - \psi_{\text{T(32)}}\phi_{\text{T(23)}}} \right) \psi_{\text{TII}}A_{\text{II}} \qquad (5)$$

$$\Phi_{\text{PI}} = \frac{P_{\text{I}}}{I_a} = \left(\alpha + \frac{[\beta + \gamma\psi_{\text{T(32)}}]\phi_{\text{T(21)}} + [\gamma + \beta\phi_{\text{T(23)}}]\psi_{\text{T(31)}}}{1 - \psi_{\text{T(32)}}\phi_{\text{T(23)}}} \right) \psi_{\text{TI}}A_{\text{I}} \qquad (6)$$

equations (2–6), where I_a is the rate at which quanta are absorbed by the entire photochemical apparatus and F and P are the rates of fluorescence and photochemistry from the different parts of the photochemical apparatus. One can readily show that the sum of the overall yields for fluorescence, non-radiative decay and photochemistry is always unity, which is a necessary test for any such set of equations.

The above equations for fluorescence consist of a probability term for fluorescence from a particular bed of chlorophyll [in (3), ψ_{FII} is a part of ϕ_{FII} according to (1)] multiplied by an expression which represents the probability for excitation in that bed of chlorophyll. For some purposes it is preferable to write the equations for the yield of fluorescence in terms of the rate constant for fluorescence. Thus, (2) can be transformed into (7)

$$\Phi_{\text{FIII}} = \frac{\gamma + \beta\phi_{\text{T(23)}}}{1 - \psi_{\text{T(32)}}\phi_{\text{T(23)}}} \cdot \frac{k_{\text{FIII}}}{\Sigma k_{\text{III}}} \qquad (7)$$

$$\Phi_{\text{FIII}} = \frac{\gamma + \beta\phi_{\text{T(23)}}}{k_{\text{FIII}} + k_{\text{T(31)}} + k_{\text{T(32)}}[1 - \phi_{\text{T(23)}}]} k_{\text{FIII}} \qquad (8)$$

$$\Phi_{FII} = \frac{[\beta + \gamma\psi_{T(32)}]f(A_{II})}{[1-\psi_{T(32)}\psi_{T(23)}f(A_{II})]} \cdot \frac{k_{FII}}{\sum k_{II}} \tag{9}$$

$$\Phi_{FII} = \frac{[\beta + \gamma\psi_{T(32)}]f(A_{II})}{k_{FII} + k_{TII} + k_{T(21)} + k_{23}[1 - \psi_{T(32)}f(A_{II})]} k_{FII} \tag{10}$$

and (8) and (3) into (9) and (10), where for simplicity k_D terms have been included as a part of the k_F terms. In these latter equations (9) and (10), the expression which multiplies the rate constant for fluorescence represents the lifetime of the excitation energy in that bed of chlorophyll.

The tripartite model incorporates the concept of energy coupling between Chl LH and Chl a_{II}. The product $\psi_{T(32)} \phi_{T(23)}$ is the probability that an excitation in Chl LH will be transferred to Chl a_{II} and back to Chl LH or that an exciton in Chl a_{II} will be transferred to Chl LH and back to Chl a_{II}. It is apparent that this cycling of excitation energy back and forth between Chl LH and Chl a_{II} will increase as the PS II reaction centres close owing to the increase of $\phi_{T(23)}$. However, any changes in the physical state of the thylakoid membranes which alters the association between Chl LH and Chl a_{II} should also be reflected in the energy coupling between Chl LH and Chl a_{II}. The product $\psi_{T(32)}\psi_{T(23)}$, which is independent of the state of the PS II reaction centres, will be used to indicate the degree of that coupling. The equations of the model allow an experimental evaluation of energy coupling between Chl LH and Chl a_{II} from measurements of fluorescence from chloroplasts at -196 °C. The measurements were made at the minimum (F_0) and maximum (F_M) levels of fluorescence at 680 and 695 nm in the absence and presence of bivalent cations (Butler & Strasser 1978). The product $\psi_{T(32)}\psi_{T(23)}$ was found to increase from 0.66 to 0.74 on the addition of Mg^{2+}. When the PS II reaction centres were all closed $\psi_{T(32)}\phi_{T(23)(M)}$ was 0.84 in the absence of Mg^{2+} and 0.91 in the presence of Mg^{2+}. It is apparent from these values that energy coupling between Chl LH and Chl a_{II} is tight (i.e., that there is a great deal of energy cycling back and forth between these two beds of chlorophyll) and that the degree of coupling can be influenced by the ionic environment of the thylakoid membranes. It was also possible to calculate the overall yield of photochemistry of PS II from the measurements of fluorescence. Φ_{PII} was estimated to be 0.43 in the absence of Mg^{2+} and 0.50 in the presence of Mg^{2+}, an increase of about 15% due to the addition of bivalent cations. Earlier work showed a similar decrease in the distribution of excitation energy to PS I in the presence of Mg^{2+} (Butler & Kitajima 1975b).

The regulation of the distribution of excitation energy at the level of the primary photochemical apparatus via the ion pumps associated with the electron transport system is one of the feedback-control mechanisms which has been built into the overall process of photosynthesis. The effective cross-sections of PS I and PS II are such that, when a plant is first illuminated, there is an excess of excitation energy distributed to PS II. As photosynthetic electron transport begins, however, protons are pumped into the thylakoids and Mg^{2+} comes out as a counter-ion movement. The expulsion of Mg^{2+} from the membranes and from the inner space of the thylakoids causes conformational changes in the thylakoid membranes which result in an increase in energy distribution to PS I, at the expense of PS II, thereby tending to overcome the initial imbalance in energy distribution.

THE BIPARTITE MODEL

Our earlier derivations (Butler & Kitajima 1975*a, b*) started with the same physical model involving three beds of antenna chlorophyll but we assumed, on the basis of measurements of the ratio of F_M/F_0 at 680 and 695 nm at −196 °C, that the energy coupling between Chl LH and Chl a_{II} was sufficiently tight that, to a first approximation, Chl LH could be considered as part of the antenna chlorophyll of PS II. That simplifying assumption, in essence, reduced the tripartite system to a bipartite system; γ was incorporated into β and the sum of α and β was taken as unity. The equations of the bipartite model, although somewhat less rigorous, are much simpler. According to that formulation, we obtain (11) and (12):

$$\Phi_{FII} = \beta\phi_{FII} = \beta\psi_{FII}f(A_{II}) \tag{11}$$

$$\Phi_{FI} = \left(\alpha + \beta\phi_{T(II\to 1)}\right)\psi_{FI} \tag{12}$$

where

$$\phi_{T(II\to 1)} = \psi_{T(II\to 1)}f(A_{II}) \tag{13}$$

We applied the equations to experimental situations by assuming that the fluorescence measured at 730 nm at −196 °C was representative of fluorescence from PS I and that measured at 695 nm was representative of PS II. The yield of energy transfer from PS II to PS I has the same dependence on the state of the PS II reaction centres, $f(A_{II})$, as does the yield of fluorescence from PS II, ϕ_{FII}. The only source of variable yield fluorescence from PS I is energy transfer from PS II. This formulation has been used extensively to examine energy transfer from PS II to PS I from measurements of fluorescence at −196 °C (e.g., Butler & Kitajima 1975*b*; Kitajima & Butler 1975*b*; Strasser & Butler 1976, 1977*a, b*; Ley & Butler 1976, 1977*b*). In addition, energy transfer from PS II to PS I has been confirmed by more direct measurements on the rate of photooxidation of P700 at −196 °C (Satoh *et al.*

1976; Strasser & Butler 1976, 1977b; Ley & Butler 1977a) and of electron-transport activities of PS I at room temperature (Satoh *et al.* 1976).

It is shown in the Appendix that the tripartite set of equations is transformed into the much simpler bipartite formulation if the probability for energy transfer from Chl LH to Chl a_{II}, the $\psi_{T(32)}$ term, is allowed to go to unity. This was our initial assumption (Butler & Kitajima 1975a). The essential validity of that assumption was confirmed in the later work (Butler & Strasser 1978) on the effects of Mg^{2+} on the coupling between Chl LH and Chl a_{II}. We estimated from the experimentally determined values of the product $\psi_{T(32)}$-$\phi_{T(23)(M)}$ (0.84 in the absence of Mg^{2+} and 0.91 in the presence of Mg^{2+}) that $\psi_{T(32)}$ was 0.93 in the absence of Mg^{2+} and 0.97 in the presence of Mg^{2+}, values which are indeed close to unity. Thus, questions involving energy transfer from PS II to PS I can be examined in the context of the bipartite formulation from measurements of fluorescence at 695 and 730 nm at $-196\,°C$. Studies involving changes in the degree of energy coupling between Chl LH and Chl a_{II} can be approached through measurements of fluorescence at 680 and 695 nm in the context of the tripartite model.

THE ORIGIN OF THE FLUORESCENCE OF VARIABLE YIELD

It was envisaged in the model of PS II (Butler & Kitajima 1975c) which is used in the bipartite and tripartite formulations that excitation energy can be transferred back and forth between the antenna chlorophyll and the reaction centre chlorophyll of PS II units quite readily in either direction. It was assumed that k_{pII} is much larger than k_{tII} or k_{dII} at the reaction centre chlorophyll (see Fig. 1) so that when an exciton visits an open reaction centre ($P_{II} \cdot A_{II}$) there is a high probability for charge separation. However, when an exciton visits the reaction centre chlorophyll of a closed centre ($P_{II} \cdot A_{II}^-$) charge separation cannot be effected but the excitation energy may be transferred back to the antenna chlorophyll or dissipated by non-radiative decay at the reaction centre chlorophyll. According to this model all fluorescence of variable yield is due to excitation energy which was returned to the antenna chlorophyll from closed PS II reaction centres. Recently, however, Klimov *et al.* (1977) found that a pheophytin molecule, I, functions in the reaction centre complex of PS II as an electron acceptor prior to A_{II} (i.e., $P_{II} \cdot I \cdot A_{II}$) and that photochemical charge separation can occur between P_{II} and I when A_{II} is reduced. Furthermore, they suggested, in analogy with earlier work on reaction centre preparations of *Chromatium* (Shuvalov & Klimov 1976), that the increase of fluorescence above the minimum F_0 level which occurs as A_{II} is reduced is a luminescence generated by a back-reaction between P_{II}^+ and I^-. This proposal, which at first glance appears novel and revolutionary in comparison with generally accepted views of fluorescence, merits further consideration.

The reaction centre complex suggested by Klimov et al. for PS II is similar to that which had been proposed for reaction centres of photosynthetic bacteria, generally denoted $P \cdot I \cdot X$ (Parsons et al. 1975; Dutton et al. 1977). When X is reduced, irradiation produces primary charge separation to form $P^+ \cdot I^- \cdot X^-$, which Parsons et al. (1975) found to have a lifetime of about 10 ns before the back-reaction between P^+ and I^- generated the triplet state of the bacteriochlorophyll. Shuvalov & Klimov (1976) found with reaction centre preparations of Chromatium that the back-reaction also produced luminescence from the bacteriochlorophyll. This luminescence was clearly distinguished from normal fluorescence in that it showed on an activation energy of 0.12 eV, presumably because the energy available from the back-reaction was 0.12 eV less than that of the emitted photon (about 1.4 eV). Although it seems clear that luminescence can be generated by the back-reaction in reaction centre preparations of photosynthetic bacteria, it is not so clear that an analogous luminescence in PS II accounts for the fluorescence of a variable yield in green plants. The F_v component of PS II appears to be a real fluorescence which shows no activation energy and appears at very low temperatures. Also, fluorescence quenching experiments (Kitajima & Butler 1975a) and measurements of fluorescence lifetime as a function of relative yield in the region between F_0 and F_M (Moya et al. 1977) indicate that F_v of PS II emanates from the same bed of antenna chlorophyll that gives rise to F_0 when the PS II reaction centres are all open. (Even if the lifetime of the luminescence happened to be the same as that which would be predicted for the F_M level from the lifetime at the F_0 level and the ratio F_M/F_0, the plot of lifetime against relative yield would still not be consistent with luminescence from the reaction centre chlorophyll.) Thus, although it seems likely that charge separations and back-reactions may occur between P_{II} and I in closed PS II reaction centres, it also seems clear that F_v is a normal fluorescence rather than a luminescence.

The apparent dichotomy can be resolved if it is assumed that the singlet excitation energy generated by the back-reaction between P_{II}^+ and I^- is transferred from P_{II} back to the antenna chlorophyll where the fluorescence is emitted. Such a mechanism requires that the charge separation and recombination occur rapidly in times short compared to the lifetime of fluorescence. The 10 ns lifetime of the $P^+ \cdot I^- \cdot X^-$ intermediate in isolated reaction centre preparations from bacteria is much too long to be consistent with fluorescence from PS II. However, the analogous intermediate in PS II units where excitation energy can presumably be transferred from P_{II} to the antenna chlorophyll may have a much shorter lifetime, perhaps in the order of tens of picoseconds. Also, the free energy available from the back-reaction cannot be much less than the energy of the photon of fluorescence or a temperature-dependent luminescence would replace the normal fluorescence. It may be that a negative charge on A_{II} inhibits the stabilization of charge separation between P_{II} and I and induces the back-reaction before the forward reaction has been fully consumated. Klimov *et al.* (1977) reported that the quantum yield for the reduction of I in closed PS II reaction centres was 500 times less than the yield for the reduction of A_{II} in open reaction centres. The picture which emerges as the photochemical model of PS II is essentially unchanged from that presented originally by Butler & Kitajima except that a rapid reversible exchange between excitation energy and charge separation may intervene in the excitation of the fluorescence of variable yield.

THE NATURE OF THE FLUORESCENCE FROM PS I AND PS II

It was assumed in the development of the bipartite and tripartite models that the fluorescence from chloroplasts at 695 and 730 nm at -196 °C represented fluorescence from the antenna chlorophylls associated with PS II and PS I, respectively. These assignments were not without question, however. The reasons for the strong temperature dependence of both of these bands had not been determined. (In contrast to the 685 nm fluorescence which shows a small dependence on temperature, the 735 nm emission band increases markedly at temperatures below -100 °C and the 695 nm band which appears only at temperatures below -160 °C shows an even stronger temperature dependence near -196 °C; Satoh & Butler 1978*b*.) Also, it had been proposed much earlier (Butler 1961) that the 735 nm fluorescence at -196 °C is due to small amounts of a long wavelength form of chlorophyll, C-705, which traps excitation energy in PS I. The relationship of the 735 nm fluorescence to PS I was investigated further by measurement of the fluorescence and the rate of photooxidation of P700 as a function of temperature (Satoh & Butler 1978*b*). The results showed that the 735 nm fluorescence competes with the primary photochemistry of PS I: the rate of photooxidation of P700 decreased by a factor of about 2 on cooling from -78 to -196 °C while the intensity of the 735 nm emission band increased by a factor of about 10. Satoh & Butler (1978*b*) proposed that either the C-705 species of chlorophyll only formed on cooling to low temperatures in which case the temperature dependence of the 735 fluorescence would be the temperature dependence for the formation of C-705 or that the C-705 species was present at higher temperatures and only became fluorescent at low temperature. In the latter case it was suggested that lowering the temperature might cause the absorption maximum of C-705 to shift to longer wavelength thereby inhibiting the transfer of energy from C-705 to P700 and causing C-705 to become a terminal energy trap which dissipated its excitation energy by fluorescence. In either case the effect of lowering the temperature would be to form 735 nm fluorescing energy traps which would compete with P700 for the excitation energy in the antenna chlorophyll of PS I.

By analogy with the 735 nm fluorescence from PS I, we proposed (Satoh & Butler 1978*b*) that the 695 nm fluorescence from PS II is due to fluorescent energy-trapping centres which appear in PS II at temperatures below −160 °C. It was apparent from curves for the temperature dependence of the 685, 695

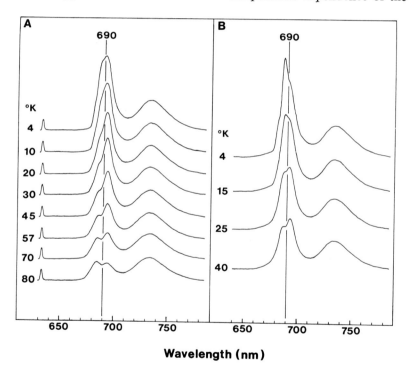

FIG. 2. Low-temperature fluorescence-emission spectra of spinach chloroplasts. Samples were frozen in a 1 mm cuvette in a cryostat with a cold helium-gas flow system. Fluorescence was excited at 633 nm with a He–Ne laser and measured with a monochromator with a 2 nm half-width passband: A, chloroplasts (0.04 absorbance in the 1 mm cuvette at room temperature) frozen in buffer; B, chloroplasts (0.07 absorbance) frozen in 50% glycerol. The absorbance of the frozen samples was sufficiently low that self-absorption of the fluorescence was negligible.

and 735 nm emission bands down to −196 °C that the 695 nm emission should continue to increase relative to the other bands as the temperature was lowered below −196 °C. Indeed, the earlier work of Cho *et al.* (1966) on *Chlorella* had shown that, as the temperature was lowered from that of liquid nitrogen to that of liquid helium, the short wavelength emission increased dramatically and became much larger than the 735 nm band. However, their spectra indicated that the increase occurred primarily in the short wavelength (687 nm) band and the middle wavelength (697 nm) band showed a much smaller temperature dependence. The fluorescence emission spectra of a dilute suspension of chloroplasts in the absence or presence of glycerol at temperatures down to that of liquid helium are shown in Fig. 2. In the absence of glycerol it is apparent that the 695 nm band increases most sharply as the temperature is lowered and shifts to a maximum at about 692 nm. At the very low temperatures, however, a shoulder on the short wavelength side of the 692 nm emission band is increasing. For the chloroplasts frozen in 50% glycerol the short wavelength band at 687 nm becomes the dominant emission at liquid helium temperatures. The reason for the effect of glycerol is not known. It is clear, however, that the 695 nm emission band of PS II is increasing steeply with decreasing temperature in the region of −200 °C, presumably owing to the appearance of fluorescent energy-trapping centres in PS II.

These studies on the temperature dependence of fluorescence indicate that the conditions which prevail at −196 °C are probably optimal for the study energy distribution in chloroplasts. The three emission bands which monitor the excitation energy in Chl LH, Chl a_{II} and Chl a_I are well distinguished from one another but the fluorescence emission at 695 nm, which functions in competition with the other PS II deexcitation pathways, does not represent a major energy drain. At lower temperatures, the 695 nm fluorescence probably does compete significantly with other PS II processes such as photochemistry and energy transfer to PS I. The fact that the 735 nm fluorescence does represent a significant energy drain at −196 °C is of little consequence (except for studies of the photochemistry of PS I) since there are no fluorescence yield changes due to PS I nor any energy transfer from PS I back to the rest of the photochemical apparatus. This temperature dependence is especially fortuitous in view of the relative ease with which measurements can be made at liquid-nitrogen temperature compared to operations at lower temperature.

FLUORESCENCE LIFETIMES

The tripartite model has important implications for the interpretation of fluorescence lifetime data. It should be recognized, however, that even though the intensity of the 735 nm fluorescence of C-705 molecules may be a valid monitor of the density of excitation energy in PS I, the lifetime of that fluorescence may not be a useful index of the lifetime of the excitation energy in PS I. Recently Campillo *et al.* (1977) reported that the risetime of the 735 nm fluorescence, after a 30 ps light flash, was about 140 ps and the lifetime of the fluorescence was about 1.5 ns. They interpreted the 140 ps risetime to be the time for energy transfer from an auxiliary bed of antenna chlorophyll associated with PS I to the PS I units. If the proposed identification of the 735 nm fluorescence with C-705 trapping centres is correct, a more likely interpretation is that the risetime represents the time for energy transfer from Chl a_I to C-705. The actual lifetime of the 735 nm fluorescence may be of little significance to the process of photosynthesis. Nevertheless, a clear understanding of the nature of the fluorescing species is crucial for the interpretation of fluorescence lifetime data at low temperatures.

Given the observation that the 735 nm fluorescence at low temperature is in competition with the photooxidation of P700 (Satoh & Butler 1978*b*), it should be possible to obtain additional information on the manner in which C-705 functions from measurements of the lifetime and the relative yield of the 735 nm fluorescence as a function of temperature over a range (e.g. between -100 and $-200\ °C$) where the yield varies strongly. If C-705 only forms on cooling to low temperatures, the lifetime of the fluorescence should be independent of the yield since, to a first approximation, the yield of fluorescence should be determined solely by the concentration of the C-705 energy trapping centres. On the other hand, if C-705 were present over the entire temperature range, the yield of fluorescence would be determined by the competition of fluorescence with the other deexcitation processes which were competing for the excitation energy in C-705. In the latter case the lifetime of fluorescence should be proportional to the yield. It is possible of course, that the results of such an experiment will be intermediate between these two sets of predictions. However, if either one of the two cases does prevail, the distinction between them should be clear-cut since the yield of the 735 nm fluorescence can be changed by approximately an order of magnitude by changing the temperature.

ACKNOWLEDGEMENT

I gratefully acknowledge the John Simon Guggenheim Foundation for a fellowship. The work was partially supported by a National Sciences Foundation grant, PCM 76–07111.

APPENDIX

The purpose of the appendix is to show that the tripartite equations transform into the bipartite equations if $\psi_{T(32)}$ is assumed to be unity. This requires the demonstration that Φ_{FII} is identical in both formulations. If $\psi_{T(32)}$ is unity, then ψ_{FIII} must be zero and so Φ_{FIII} is also zero (the 685 nm band is then due to PS II). It will also be apparent that Φ_{FI} is identical in both formulations since $\psi_{T(31)}$ is zero. Φ_{FII} will be evaluated at the end points, i.e., at $A_{II} = 1.0$ where $f(A_{II}) = 1.0$ and at $A_{II} = 0$ where $f(A_{II}) = (1-\psi_{TII}\psi_{tII})^{-1}$ for all $f(A_{II})$.

The bipartite formulation

$$\sum k_{II} = k_{FII} + k_{T(II\rightarrow1)} + k_{TII}$$

(k_{DII} is included in k_{FII})

$$\Phi_{FII} = \beta\phi_{FII} = \beta\psi_{FII}f(A_{II}) = \frac{\beta k_{FII}}{\sum k_{II}} f(A_{II})$$

at $A_{II} = 1.0$

$$\Phi_{FII(0)} = \frac{\beta k_{FII}}{k_{FII} + k_{T(II\rightarrow1)} + k_{TII}} \tag{1}$$

at $A_{II} = 0$

$$\Phi_{FII(M)} = \frac{\beta k_{FII}}{(1 - \psi_{TII}\psi_{tII})\sum k_{II}} = \frac{\beta k_{FII}}{(1 - \frac{k_{TII}}{\sum k_{II}} \psi_{tII})\sum k_{II}}$$

$$= \frac{\beta k_{FII}}{k_{FII} + k_{T(II\rightarrow1)} + k_{TII}(1 - \psi_{tII})} \tag{2}$$

The tripartite formulation

$$\sum k_{II} = k_{FII} + k_{T(21)} + k_{TII} + k_{T(23)}$$

$$\psi_{FII} + \psi_{T(21)} + \psi_{TII} + \psi_{T(23)} = 1.0$$

$$\Phi_{FII} = \frac{\beta + \gamma\psi_{T(32)}}{1 - \psi_{T(32)}\phi_{T(23)}} \phi_{FII}$$

Let $\psi_{T(32)} = 1.0$ and replace $(\beta + \gamma)$ by β, then

$$\Phi_{FII} = \frac{\beta\psi_{FII}f(A_{II})}{1 - \psi_{T(23)}f(A_{II})}$$

at $A_{II} = 1.0$

$$\Phi_{FII(0)} = \frac{\beta}{1 - \psi_{T(23)}} \cdot \frac{k_{FII}}{\sum k_{II}} = \frac{\beta k_{FII}}{\left(1 - \frac{k_{T(23)}}{\sum k_{II}}\right) \sum k_{II}}$$

$$= \frac{\beta k_{FII}}{k_{FII} + k_{T(21)} + k_{TII}} \qquad \text{(see eq. 1)}$$

at $A_{II} = 0$

$$\Phi_{FII(M)} = \frac{\beta k_{FII}}{\left(1 - \frac{\psi_{T(23)}}{(1 - \psi_{TII}\psi_{tII})}\right)(1 - \psi_{TII}\psi_{tII})\sum k_{II}}$$

$$= \frac{\beta k_{FII}}{(1 - \psi_{T(23)} - \psi_{TII}\psi_{tII})\sum k_{II}}$$

$$= \frac{\beta k_{FII}}{(\psi_{FII} + \psi_{T(21)} + \psi_{TII} - \psi_{TII}\psi_{tII})\sum k_{II}}$$

$$= \frac{\beta k_{FII}}{k_{FII} + k_{T(21)} + k_{TII}(1 - \psi_{tII})} \qquad \text{(see eq. 2)}$$

References

BUTLER, W. L. (1961) A far-red absorbing form of chlorophyll *in vivo*. *Arch. Biochem. Biophys. 92*, 413–422

BUTLER, W. L. (1978) Energy distribution in the photochemical apparatus of photosynthesis. *Annu. Rev. Plant Physiol. 29*, 345–378

BUTLER, W. L. & KITAJIMA, M. (1975a) A tripartite model for chlorophyll fluorescence, in *Proceedings of the Third International Congress on Photosynthesis* (Avron, M., ed.), pp. 13–14, Elsevier, Amsterdam

BUTLER, W. L. & KITAJIMA, M. (1975b) Energy transfer between photosystem II and photosystem I in chloroplasts. *Biochim. Biophys. Acta 396*, 72–85

BUTLER, W. L. & KITAJIMA, M. (1975c) Fluorescence quenching in photosystem II of chloroplasts. *Biochim. Biophys. Acta 376*, 116–125

BUTLER, W. L. & STRASSER, R. J. (1977) Tripartite model for the photochemical apparatus of green plant photosynthesis. *Proc. Natl. Acad. Sci. U.S.A. 74*, 3382–3385

BUTLER, W. L. & STRASSER, R. J. (1978) Effect of divalent cations on energy coupling between the light-harvesting chlorophyll *a/b* complex and photosystem II, in *Photosynthesis 77: Proceedings of the Fourth International Congress on Photosynthesis* (Hall, D. O., Coombs, J. & Goodwin, T. W., eds.), pp. 9–20, The Biochemical Society, London

BUTLER, W. L., VISSER, J. W. M. & SIMONS, H. L. (1973) The kinetics of light-induced changes of C-550 cytochrome b_{559} and fluorescence yield in chloroplasts at low temperature. *Biochim. Biophys. Acta 292*, 140–151

CAMPILLO, A. J., SHAPIRO, S. L., GEACINTOV, N. E. & SWENBERG, C. E. (1977) Single pulse picosecond determination of 735 nm fluorescence risetime in spinach chloroplasts. *FEBS (Fed. Eur. Biochem. Soc.) Lett. 83*, 316–320

CHO, F., SPENCER, J. & GOVINDJEE (1966) Emission spectra of *Chlorella* at very low temperatures (−269° to −196°). *Biochim. Biophys. Acta 126*, 174–176

DUTTON, L. P., PRINCE, R. D., TIEDE, D. E., PETTY, K. M., KAUFMANN, K. J., NETZEL, T. L. & RENTZEPIS, P. M. (1977) Electron transfer in the photosynthetic reaction center. *Brookhaven Symp. Biol. 28*, 213–237

KITAJIMA, M. & BUTLER, W. L. (1975*a*) Quenching of chlorophyll fluorescence and primary photochemistry in chloroplasts by dibromothymoquinone. *Biochim. Biophys. Acta 376*, 105–115

KITAJIMA, M. & BUTLER, W. L. (1975*b*) Excitation spectra for photosystem I and photosystem II in chloroplasts and the spectral characteristics of the distribution of quanta between the two photosystems. *Biochim. Biophys. Acta 408*, 297–305

KLIMOV, V. V., KLEVANIK, V. A., SHUVALOV, V. A. & KRAVSNOVSKY, A. A. (1977) Reduction of pheophytin in the primary light reaction of photosystem II. *FEBS (Fed. Eur. Biochem. Soc.) Lett. 82*, 183–186

LEY, A. C. & BUTLER, W. L. (1976) The efficiency of energy transfer from photosystem II to photosystem I in *Porphyridium cruentum*. *Proc. Natl. Acad. Sci. U.S.A. 73*, 3957–3960

LEY, A. C. & BUTLER, W. L. (1977*a*) Energy transfer from photosystem II to photosystem I in *Porphyridium cruentum*. *Biochim. Biophys. Acta 462*, 290–294

LEY, A. C. & BUTLER, W. L. (1977*b*) The distribution of excitation energy between photosystem II and photosystem I in *Porphyridium cruentum*, in *Photosynthetic Organelles* (Miyachi, S., Katoh, S., Fujita, Y. & Shibata, K., eds.) (special issue of *Plant and Cell Physiology 3*), pp. 33–46, Japanese Society for Plant Physiology, Tokyo, Japan

MOYA, I., GOVINDJEE, VERNOTTI, C. & BRIANTAIS, J. M. (1977) Antagonistic effect of mono- and di-valent cations on lifetime and quantum yield of fluorescence in isolated chloroplasts. *FEBS (Fed. Eur. Biochem. Soc.) Lett. 75*, 13–18

PARSON, W. W., CLAYTON, R. K. & COGDELL, R. J. (1975) Excited states of photosynthetic reaction centres at low redox potentials. *Biochim. Biophys. Acta 387*, 265–278

SATOH, K. & BUTLER, W. L. (1978*a*) Low temperature spectral properties of subchloroplast fractions purified from spinach. *Plant. Physiol. 61*, 373–379

SATOH, K. & BUTLER, W. L. (1978*b*) Competition between the 735 nm fluorescence and the photochemistry of photosystem I in chloroplasts at low temperature. *Biochim. Biophys. Acta 502*, 103–110

SATOH, K., STRASSER, R. J. & BUTLER, W. L. (1976) A demonstration of energy transfer from photosystem II to photosystem I in chloroplasts. *Biochim. Biophys. Acta 440*, 337–345

SHUVALOV, V. A. & KLIMOV, V. V. (1976) The primary photoreactions in the complex cytochrome-P-890–P-760 (Bacteriopheophytin$_{760}$) of *Chromatium minutissimum* at low potentials. *Biochim. Biophys. Acta 440*, 587–599

STRASSER, R. J. & BUTLER, W. L. (1976) Energy transfer in the photochemical apparatus of flashed bean leaves. *Biochim. Biophys. Acta 449*, 412–419

STRASSER, R. J. & BUTLER, W. L. (1977*a*) Energy transfer and the distribution of excitation energy in the photosynthetic apparatus of spinach chloroplasts. *Biochim. Biophys. Acta 460*, 230–238

STRASSER, R. J. & BUTLER, W. L. (1977*b*) The yield of energy transfer and the spectral distribution of excitation energy in the photochemical apparatus of flashed bean leaves, *Biochim. Biophys. Acta 462*, 295–306

STRASSER, R. J. & BUTLER, W. L. (1977*c*) Fluorescence emission spectra of photosystem I, photosystem II and the light-harvesting chlorophyll *a/b* complex of higher plants. *Biochim. Biophys. Acta 462*, 307–313

STRASSER, R. J. & BUTLER, W. L. (1978) Energy coupling in the photosynthetic apparatus during development, in *Photosynthesis 77: Proceedings of the Fourth International Congress on Photosynthesis* (Hall, D. O., Coombs, J. & Goodwin, T. W., eds.), pp. 527–536, The Biochemical Society, London

Discussion

Amesz: Does the emission spectrum of a preparation that does not have the light-harvesting protein (say, a greening leaf or a mutant) lack the 685 nm emission?

Butler: The emission spectrum of a flashed bean leaf at $-196\ °C$ shows only two emission bands at 695 and 730 nm. On further greening in continuous light, however, the chlorophyll a/b–protein begins to form and accumulate; then the 685 nm emission band appears (Strasser & Butler 1977).

Amesz: The blue-green and red algae also have an emission band at 685 nm at low temperature.

Butler: That band is due to allophycocyanin B. This emission also shows a fluorescence of variable yield which depends on the state of the photosystem II reaction centre. Thus, we conclude that there is energy transfer back and forth between allophycocyanin B (which serves as an energy trap for the phycobilisome) and the antenna chlorophyll of photosystem II (Ley & Butler 1977).

Barber: The barley mutant which lacks chlorophyll has a 685 nm emission (Boardman & Thorne 1968).

Thornber: The barley mutant, which lacks the chlorophyll a/b–protein, poses a problem! With figures of 0.3 and 0.2 for the cross-section of capture for chlorophyll a_1 and a_2, respectively, in the wild type, how does the barley mutant live efficiently? Is there spillover from photosystem I to photosystem II to maximize energy distribution to the two photosystems?

Butler: I don't know what the relative cross-sections of photosystems I and II are in the barley mutant. We could certainly determine those but as yet we have not.

I see no evidence for any energy transfer from photosystem I to II nor any reason to assume such energy transfer.

Barber: Even though transfer from photosystem II to I is efficient, you say that there is little transfer from the light-harvesting protein to photosystem I but, on the other hand, the addition of magnesium ions somehow increases the transfer from the light-harvesting protein to photosystem II. There seems to be a conflict.

Butler: No; we conclude that the ion-induced membrane conformational change which increases energy coupling between the light-harvesting chlorophyll a/b–protein complex and photosystem II also decreases the connection from photosystem II to I. We cannot account for the effects of Mg^{2+} on fluorescences solely on the basis of the increased energy coupling between the light-harvesting chlorophyll and a_{II}. We have to assume that at least one other photochemical rate constant in the system is altered as well and it is simplest to assume that the rate constant for spillover is altered.

Barber: One change in transfer may be the consequence of another. That is, if transfer between photosystems II and I is reduced, there must be more extensive energy migration between photosystem II units.

Staehelin: Maybe our results and yours are not so disparate, especially in relation to energy transfer from II to I in a low concentration of magnesium ions. We find that most of the photosystem II particles diffuse out of the stacked regions and everything is intermixed; each photosystem II unit is now surrounded by many more photosystem I units than if it were concentrated in a stacked region. The other relationship—of light-harvesting protein to photosystem II—is probably on a much finer scale and will not be seen by electron microscopy.

Butler: I cannot exclude that possibility. However, I still feel that the changes we observe in energy transfer from photosystem II to I reflect membrane changes which are much more subtle than those you describe.

Staehelin: Have you looked at the proportion of light-harvesting components to the photosystem II and I components in winter and in summer spinach?

Butler: No; we have not systematically examined summer spinach.

Searle: You reported the quantum yield of photochemistry in photosystem II as 0.5 or slightly lower. Why is this quantum yield so low?

Butler: An overall quantum yield of 0.5 for the photochemistry photosystem II is high. I calculate the overall quantum yields (Φ) on the basis of the energy absorbed by the entire photochemical apparatus. Since some of the energy is absorbed by, or transferred to, photosystem I the optimum yield of photochemistry for photosystem II should be somewhat less than 0.5.

Searle: Are you saying then that the trapping at photosystem II reaction centres is inefficient?

Butler: No. The probability for trapping the excitons which are in chlorophyll a_{II} and light-harvesting chlorophyll by the photosystem II reaction centres is about 0.75 when the reaction centres are all open (in the presence of Mg^{2+}) and some of those excitons that are not trapped are transferred on to photosystem I.

Joliot: You admit that in your model energy is transferred between photosystem II units through the complex CP II. I am surprised that in *Cyanidium*, which lacks CP II, the probability of energy transfer between photosystem II units is the same as in *Chlorella* (Diner & Wollman 1978).

Butler: We have used our model to analyse the photochemical apparatus of the red alga *Porphyridium cruentum* on the assumption that the phycobilisomes play a role similar to that of the chlorophyll *a/b*–protein complex of green plants. We concluded that the photosystem II units are small with about 5% of the chlorophyll, the photosystem I units much larger with 95% of the chlorophyll and that the phycobilisomes transferred excitation energy exclusively to the small photosystem II units *via* allophycocyanin B (Ley & Butler 1977). We also found that the transfer of energy from photosystem II to photosystem I was much more efficient in *Porphyridium* than in green plants (Ley & Butler 1976).

Joliot: But what about transfer between system II units in this case?

Butler: We have not examined energy transfer between photosystem II units in *Porphyridium*.

Joliot: Since the fluorescence-induction curves are the same for *Cyanidium* and *Chlorella*, I am surprised that the light-harvesting protein is so important for the energy transfer. We do not have yet the same type of information for *Porphyridium*.

Butler: We examined the influence of the light-harvesting chlorophyll *a/b*–protein complex on energy transfer between photosystem II units in flashed bean leaves in which none of the chlorophyll *a/b* complex is present initially. Energy transfer between these units was inferred from the shape of the fluorescence-induction curve measured at room temperature in the presence of DCMU. Initially when there is no chlorophyll *a/b* complex present, the fluorescence yield increases exponentially during irradiation, a fact that indicates no energy transfer between photosystem II units. With further greening in continuous light the leaves accumulate the chlorophyll *a/b*–protein complex and the fluorescence-induction curves develop the initial lag phase which you (Joliot *et al.* 1973) have interpreted to indicate energy transfer between photosystem II units. These observations do not prove that energy transfer between these units must pass through the light-harvesting chlorophyll *a/b*–protein complex but they are consistent with that interpretation.

Anderson: Why is the half-band-width of the fluorescence emission at 735 nm greater than that of the 680 and 695 nm fluorescence bands?

Butler: I don't know.

Barber: Is the C-705 competing with oxidation of P700? If it were possible to change the yield of fluorescence by changing the redox state of the photosystem I reduction centre, one would see changes at 730 nm. Could there be variable changes in fluorescence yield which are not due to spillover alone?

Butler: We have looked carefully for changes in fluorescence yield that accompany the photooxidation of P700 at low temperature. In red and blue-green algae the photooxidation of P700 causes a small increase (about 15%)

in the yield of photosystem I fluorescence at $-196\,°C$ (Ley & Butler 1977) but we have found no such change in higher plant chloroplasts. In chloroplasts all the fluorescence of variable yield at 730 nm is controlled by the closure of photosystem II reaction centres.

Seely: In the systems we made with the particles the 720 nm band often develops at lower temperatures. Perhaps it also arises from a species of chlorophyll absorbing near 705 nm. Presumably there is no competition between fluorescence at 720 nm and transfer to P700.

Clayton: If fluorescence from C-705 competes with radiative transfer to P700, the state of the P700 should affect the fluorescence. If it does not, there must be a buffer in the form of another physical state or another molecule.

Butler: We have proposed that the oxidized form of P700 quenches fluorescence as efficiently as the reduced form so that there is no change of fluorescence yield accompanying the photooxidation of P700 (Butler & Kitajima 1975).

Clayton: That implies a radiation acceptor that is not the 700 nm absorption transition.

Butler: When P700 is oxidized an absorption band at 690 nm increases as the 700 nm band is bleached (Lozier & Butler 1974). Possibly the 690 nm band is related to a quenching species which forms on the photooxidation of P700.

References

BOARDMAN, N. K. & THORNE, S. W. (1968) Studies on a barley mutant lacking chlorophyll *b*. *II*. Fluorescence properties of isolated chloroplasts. *Biochim. Biophys. Acta 153*, 448–458
BUTLER, W. L. & KITAJIMA, M. (1975) Energy transfer between photosystem II and photosystem I in chloroplasts. *Biochim. Biophys. Acta 396*, 72–85
DINER, B. & WOLLMAN, F. A. (1978) A functional comparison of the photosystem II center–antenna complex of a phycocyanin-less mutant of *Cyanidium caldarium* with that of *Chlorella pyrenoidosa*. *Plant Physiol.*, in press
JOLIOT, P., BENNOUN, P. & JOLIOT, A. (1973) New evidence supporting energy transfer between photosynthetic units. *Biochim. Biophys. Acta 305*, 317–328
LEY, A. C. & BUTLER, W. L. (1976) The efficiency of energy transfer from photosystem II to photosystem I in *Porphyridium cruentum*. *Proc. Natl. Acad. Sci. U.S.A. 73*, 3957–3960
LEY, A. C. & BUTLER, W. L. (1977) The distribution of excitation energy between photosystem I and photosystem II in *Porphyridium cruentum*, in *Photosynthetic Organelles* (Myachi, S., Katoh, S., Fujita, Y. & Shibata, K., eds), (special issue of *Plant and Cell Physiology 3*), pp. 33–46, Society of Plant Physiology, Tokyo
LOZIER, R. H. & BUTLER, W. L. (1974) Light-induced absorbance changes in chloroplasts mediated by photosystem I and photosystem II. *Biochim. Biophys. Acta 333*, 465–480
STRASSER, R. J. & BUTLER, W. L. (1977) Fluorescence emission spectra of photosystem I, photosystem II and the light-harvesting chlorophyll *a/b* complex of higher plants. *Biochim. Biophys. Acta 462*, 307–313

Picosecond fluorescence from photosynthetic systems *in vivo*

G. F. W. SEARLE and C. J. TREDWELL*

Department of Botany, Imperial College, and **Davy Farady Research Laboratory of the Royal Institution, London*

Abstract Picosecond time-resolved fluorescence emission from the pigments of intact photosynthetic systems and isolated pigment–protein fractions has been used to probe the mechanism of energy transfer and the organization of the pigments. The fluorescence kinetics of chlorophyll and the phycobilins of the red alga, *Porphyridium cruentum*, are governed by time-dependent kinetics, but the observed time dependence of the chlorophyll *a* fluorescence decay from dark-adapted *Chlorella pyrenoidosa* and spinach sub-chloroplast fractions is still open to conjecture. In contrast to the green plants containing only chlorophyll and carotenoids, *Porphyridium* shows distinct emission bands for each of the pigments in the transfer sequence. The rate of energy transfer *in vivo* has the empirical form: $\mathrm{d}S^*/\mathrm{d}t = -\tfrac{1}{2}S^*At^{-1/2}$, where S^* is the excited-state population of the donor pigment and A is the overall rate of energy transfer to the acceptor pigment. This kinetic analysis can describe closely the observed fluorescence risetimes and lifetimes of the photosynthetic pigments of *Porphyridium*. The extremely rapid rates of energy transfer, determined by this treatment, imply that exciton migration within each pigment bed of the phycobilisome is less extensive than in the chlorophyll-antenna systems. Changes in the fluorescence yield and decay kinetics of chlorophyll *a* and allophycocyanin *in vivo* can be induced at high excitation intensities by exciton–exciton annihilation.

Picosecond spectroscopy of the fluorescence from photosynthetic pigments *in vivo* can be used to gain an insight into the mechanism of excitation energy transfer and the organization of these pigments. Since excitation energy can be dissipated by intramolecular processes, for example fluorescence or intersystem crossing, or by intermolecular processes such as energy transfer to another pigment or to a photochemical reaction centre, the fluorescence kinetics of a particular pigment provide a direct means of probing these processes. We have used a picosecond laser streak-camera system to study the fluorescence kinetics of the phycobiliproteins and chlorophylls in the alga, *Porphyridium cruentum* (Porter *et al.* 1978; Searle *et al.* 1978), and of chlorophyll *a* in both the alga *Chlorella pyrenoidosa* (Porter *et al.* 1977; Tredwell *et al.* 1977) and spinach chloroplast particles (Searle *et al.* 1977*a*). We have shown that the changes in fluorescence quantum yield, observed on either closure of the photosystem II reaction centre or cooling to 77 K, can be correlated with changes in the fluorescence lifetime of chlorophyll *a* *in vivo* (Tredwell *et al.* 1977).

As a result of the low fluorescence quantum yield of these pigments, single picosecond pulses with intensities in the range of 10^{13}–10^{16} photon cm^{-2} are often used to produce a detectable fluorescence signal. The first picosecond measurements of the chlorophyll *a* fluorescence emission, from dark-adapted photosystem II (683 nm) in green photosynthetic systems, gave extremely short lifetimes of between 50 and 100 ps (Beddard *et al.* 1975); fluorescence lifetimes of the order of 600 ps had previously been recorded with low excitation intensity techniques such as single-photon counting and phase fluorimetry (Mar *et al.* 1972). The subsequent use of high-intensity single picosecond pulses ($>10^{14}$ photon cm^{-2}) or complete pulse trains of lower intensity gave rise to singlet–singlet and singlet–triplet exciton annihilation processes owing to the large excited-state population generated by the pulse (Porter *et al.* 1977). These quenching processes produce a marked reduction in the fluorescence lifetime and quantum yield of chlorophyll *a* *in vivo* in various experimental conditions. Low excitation intensity picosecond laser measurements of the fluorescence lifetimes of chlorophyll *a* ($\leqslant 5 \times 10^{13}$ photon cm^{-2}) have since been reported (Tredwell *et al.* 1977); these results agree well with those recorded with the other techniques and will be fully discussed below.

The accessory pigments of *P. cruentum*, the phycobiliproteins, are contained within prolate structures known as phycobilisomes which are attached to both surfaces of the thylakoid membrane (Gantt & Conti 1966). Three major phycobiliproteins have been characterized: B-phycoerythrin, R-phycocyanin and allophycocyanin; another pigment, allophycocyanin B, has also been isolated (Ley *et al.* 1977). Steady-state fluorescence studies indicated that the phycobilisomes preferentially serve photosystem II (Ley & Butler 1977) and that the energy transfer sequence probably is:

$$\text{B-phycoerythrin} \rightarrow \text{R-phycocyanin} \rightarrow \text{allophycocyanin} \begin{array}{c} \nearrow \text{allophycocyanin B} \\ \updownarrow\uparrow? \\ \searrow \text{chlorophyll } a \end{array}$$

$$\text{575 nm} \qquad\qquad \text{636 nm} \qquad\qquad \text{660 nm} \qquad\qquad \text{685 nm}$$

Since the fluorescence maxima (shown in the scheme) are well defined, we have been able to wavelength-resolve the fluorescence kinetics of each

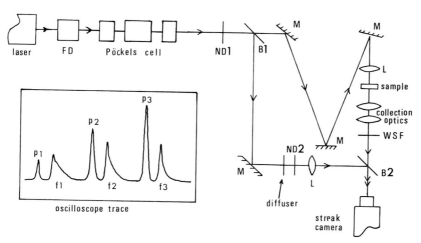

FIG. 1. Diagram of the optical set-up for measurements of relative quantum yield with the picosecond-laser apparatus: FD, frequency-doubling crystal of caesium dihydrogen arsenate; ND 1 and 2, non-saturable neutral optical-density filters; B1 and 2, 10% beam splitters; M, 100% front surface mirror; L, lens; WSF, wavelength-selection filter. The inset shows a typical trace for the first three pulses of a mini-pulse train (p1–3) and the fluorescence emission excited by those pulses (f1–3).

pigment by picosecond spectroscopy. We shall discuss a kinetic model describing energy transfer between these accessory pigments and chlorophyll *a* and describe the effects of preventing energy transfer from allophycocyanin to chlorophyll *a*, by separating the phycobilisomes from the thylakoid membrane.

MATERIAL AND METHODS

Fig. 1 is a simplified diagram of the picosecond laser system (see Archer *et al.* 1977 for more details). The mode-locked neodymium-glass laser generates a train of light pulses (at 1060 nm and of 6 ps duration) with an interpulse separation time of 6.9 ns. A temperature-tuned crystal of caesium dihydrogen arsenate converts the 1060 nm pulse train into one of 530 nm with 10–15% efficiency. After frequency doubling, a single pulse or group of 10–20 pulses is selected from the centre of the pulse train by a Pöckels cell electro-optic shutter; the average energy per pulse after selection is 50 μJ. Fluorescence from the sample is collected by two short-focal-length lenses and focused through a wavelength-selection filter onto the slit of an S20 Imacon 600-streak camera (John Hadland [P.I.] Ltd.). The resulting streak trace is detected and digitized by an optical multichannel analyser (OMA 1205 A and B, Princeton Applied Research). Data stored in the 500-channel memory of the OMA can be displayed on an x, y-oscilloscope or transferred to computer punch tape for analysis. The linearity of the response of the detection system to incident light intensity is better than $\pm 3\%$ between 30 and 3000 counts in any channel of the OMA memory. In general, the streak velocity is selected to give the best resolution of the complete decay curve over 1.5–2.0 orders of magnitude, unless the initial decay kinetics are to be studied, in which case the fastest streak speeds are used (5–10 ps resolution).

A small portion of the excitation pulse (8%) is directed along a shorter optical path to the streak-camera slit so that it arrives just before the fluorescence from the sample. When calibrated in terms of the energy of the excitation pulse, this reference pulse can be used for calculation of the relative quantum yield of the fluorescence from the integrated intensities of the excitation pulse and the fluorescence decay curve (OMA integrate function); it can also be used to provide a 'zero-time' reference point on the OMA traces.

The unfocused excitation pulse, incident on the sample, covers an area of 0.28 cm² which corresponds to a peak excitation intensity of 5×10^{14} photon cm⁻²; we attained higher excitation intensities by inserting a lens before the sample to reduce the excitation area to 0.005 cm² and to increase the intensity to 3×10^{16} photon cm⁻². Neutral optical-density filters, inserted into the main laser beam, were used to vary the excitation intensity over the desired range. Unless stated to the contrary, the excitation intensity at 530 nm was kept below 10^{14} photon cm⁻².

We have described the culture conditions for *P. cruentum* and the isolation of the phycobilisomes using Triton X-100 and high-speed centrifugation elsewhere (Porter *et al.* 1978; Searle *et al.* 1978). Samples of the intact alga and phycobilisomes were diluted immediately before the experiment to give a transmission of 50% and 30%, respectively, at 530 nm in a cuvette with a 1 mm pathlength. Fluorescence-kinetics experiments with *P. cruentum* were all done at room temperature. The following interference filters were used to wavelength-resolve the pigment emissions:

     B-phycoerythrin: 576 nm, 9 nm bandwidth (Balzer).
     R-phycocyanin: 640 nm, 13 nm bandwidth (Balzer).
     allophycocyanin: 661 nm, 14 nm bandwidth (MTO Intervex A).
     chlorophyll *a*: 685 nm, 11 nm bandwidth (Balzer).

Since allophycocyanin fluorescence dominates above 645 nm with isolated phycobilisomes, an RG 645 (Schott) cut-off filter could be used in these experiments. All fluorescence measurements on phycobilisomes were made on concentrated (non-scattering) suspensions, as in dilute suspensions they disaggregate with time even in the presence of 0.5M-sodium phosphate.

The samples of *Chlorella* (Porter *et al.* 1977) and spinach sub-chloroplast fractions were prepared as described below. The chloroplast fractions were of the following types:

(*a*) a stromal lamellae vesicle fraction, prepared with 0.2% digitonin (SLV), which contains the photosystem I reaction centre and the associated antenna system (Searle *et al.* 1977a);

(*b*) a fraction (designated F_I) prepared with 1.3% digitonin which contains the photosystem I reaction centre and antenna system in a small particle (Searle *et al.* 1977a);

(*c*) a digitonin-extracted fragment fraction (DEF) prepared with 1.3% digitonin in the presence of high concentrations of salt, containing the photosystem II reaction centre and antenna system together with light-harvesting chlorophyll *a/b* (Searle *et al.* 1977a);

(*d*) a fraction (designated F_{III}), which is a 1.3% digitonin-prepared light-harvesting chlorophyll a/b protein complex of diameter 6 nm (Wessels 1968);

(*e*) a fraction (designated CPI) which is a chlorophyll *a*–protein complex, derived from photosystem I by treatment with 0.25% sodium dodecyl sulphate, but without active reaction centres. It is purified by hydroxyapatite chromatography and precipitation with ammonium sulphate (Kung & Thornber 1971).

Experimental conditions will be numbered as follows for clarity in the text:
(i) dark-adapted at room temperature;
(ii) 3-(3,4-dichlorophenyl)-1,1-dimethylurea (DCMU) was added and the sample was preilluminated with 633 nm light at room temperature;
(iii) temperature lowered to 77 K.

At room temperature, the chlorophyll fluorescence in conditions (i) and (ii) was monitored at wavelengths above 665 nm (RG 665 filter, Schott). At 77 K, the DEF fragments were monitored above 665 nm, and the photosystem I emission from *Chlorella* and the photosystem I preparations (SLV and F_I) were monitored above 715 nm (RG 715 filter, Schott). The transmission of the samples at 530 nm was adjusted to 50% in a cuvette of 1 mm pathlength.

RESULTS AND DISCUSSION

Energy transfer between the accessory pigments and chlorophyll a *of* P. cruentum

The red alga, *P. cruentum*, has a series of accessory pigments that absorb light in the region between 500 and 650 nm; the spectroscopic characteristics

TABLE 1

The spectroscopic characteristics of the phycobiliproteins of *Porphyridium cruentum*

Phycobiliprotein	Chromophore[a]	Absorption maximum (nm)	Emission maximum (nm)	Relative phycobiliprotein content of the isolated phycobilisome (%)[b]
B-Phycoerythrin	PUB	498	575	84
	PEB	545, 563		
R-Phycocyanin	PEB	555	636	10
	PCB	617		
Allophycocyanin	PCB	650	660	8

[a] PUB, phycourobilin; PEB, phycoerythrobilin; PCB, phycocyanobilin.
[b] Calculated from the absorption spectrum (Searle *et al.* 1978).

and composition of the phycobiliproteins in the phycobilisomes of *P. cruentum* are summarized in Table 1. Absorption and fluorescence emission spectra of *P. cruentum* are shown in Fig. 2. Additional absorption bands at 436 and 678 nm can be ascribed to chlorophyll *a*, and at 470 and 500 nm to the carotenoids; the absorption band at 625 nm is attributable to both chlorophyll *a* and R-phycocyanin. Changes in the fluorescence yield of chlorophyll *a* on the addition of DCMU and preillumination with light at 633 nm were relatively low in 10–12-day-old cultures of *P. cruentum* (see Fig. 2B), whereas a much larger increase was observed in the 2–3-day-old cultures (Fig. 2C). This result may be due to the higher chlorophyll *a*/phycobilin ratio observed in these young cultures. Measurement of chlorophyll *a* kinetics have been made on young cell cultures showing a pronounced peak at 685 nm on addition of DCMU. In older cells the chlorophyll fluorescence is superimposed on a large background of phycobilin fluorescence when 530 nm excitation is used (however, with 430 nm excitation the chlorophyll fluorescence is dominant and the expected 2–3-fold increase in yield is seen on addition of DCMU). The lower chlorophyll fluorescence in older cells was probably not due to a reduced photosystem II activity because addition of $10mM-NH_2OH$ did not increase the level significantly above that seen with DCMU alone. Excitation into the B-phycoerythrin absorption band (545–565 nm) results in transfer of excitation energy to the succeeding pigments in the sequence as shown by the fluorescence excitation spectra (Fig. 2D–G). The large amount of chlorophyll *a* observed in the absorption spectrum does not contribute significantly to the fluorescence emission.

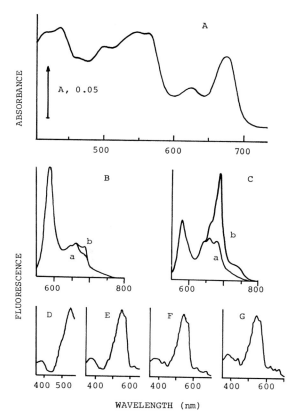

FIG. 2. Spectroscopic characteristics of *Porphyridium cruentum*: A, the absorption spectrum taken on an Aminco DW2 spectrophotometer with diffusing plate to reduce scattering artifacts (pathlength 10 mm); B and C, the fluorescence emission spectrum on excitation at 530 nm of (B) 12-day-old and (C) 2-day-old cultures in the (*a*) absence and (*b*) presence of 10^{-4}M-DCMU; D–G, the fluorescence excitation spectra for emission at (D) 576 nm, (E) 661 nm, (F) 730 nm and (G) 730 nm after addition of 10^{-4}M-DCMU. The fluorescence spectra (B–G) are uncorrected for photodetector sensitivity or xenon-lamp emission, and intensity is expressed on an arbitrary linear scale.

When the phycobilisomes are isolated from the thylakoid membrane, the absorption spectrum no longer shows bands attributable to chlorophyll *a* and the carotenoids (Fig. 3). The fluorescence excitation spectrum monitored at 670 nm is similar to that of the intact alga, whereas the fluorescence emission spectrum (530 nm excitation) exhibits a new intense band centred near 672 nm. It is possible that the main band is comprised of two overlapping emissions from allophycocyanin and allophycocyanin B, respectively. Since the allo-

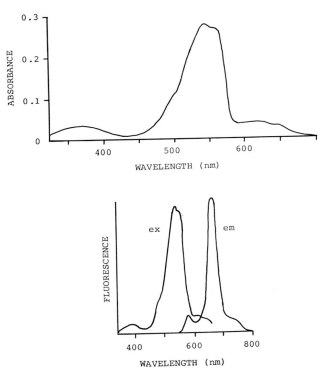

FIG. 3. Spectroscopic characteristics of phycobilisomes from *Porphyridium cruentum*. The absorption spectrum was measured on a diluted suspension. The fluorescence emission spectrum (em) on excitation at 530 nm and the fluorescence excitation spectrum (ex) for emission at 670 nm were measured on a concentrated suspension by reflection off the front surface of the cuvette, and are not corrected for photodetector sensitivity or xenon-lamp emission. Fluorescence intensity is expressed on an arbitrary linear scale.

phycocyanins have a negligible absorption at 530 nm, it is clear that energy is being transferred from B-phycoerythrin toward allophycocyanin which, in the absence of energy transfer to chlorophyll *a*, must dissipate this energy by intersystem crossing and fluorescence emission.

The fluorescence decay characteristics for the accessory pigments in the intact alga and the isolated phycobilisomes are summarized in Table 2. The fluorescence decay of B-phycoerythrin obeys an $\exp(-At^{1/2})$ law, whereas the fluorescence emissions from R-phycocyanin, allophycocyanin and chlorophyll *a* (dark-adapted 2–3-day-old cultures) all appeared to be governed by an $\exp(-kt)$ decay law. It also appeared that the latter pigments exhibited risetimes in their fluorescence emission which could not be explained by the

TABLE 2

Fluorescence risetime τ_{rise} and decay kinetics (see equation 2) of the photosynthetic pigments of dark-adapted *Porphyridium cruentum*

Pigment	$\tau_{1/e}$ (ps)	τ_{rise} (ps)	A (ps$^{-1/2}$)	k^c (ps^{-1})	$\tau_{1/e}{}^a$ (ps)	$\tau_{\text{rise}}{}^a$ (ps)
B-Phycoerythrin	70 ± 5^b	0	0.26^c		70 ± 5^b	0
R-Phycocyanin	90 ± 10	12	0.48^a	0.0110	85 ± 5	12
Allophycocyanin	118 ± 8	24	0.52^a	0.0085	115 ± 8	22
Chlorophyll a	175 ± 10	50	0.40^a	0.0057	176 ± 8	52

[a] Calculated from kinetic model (Porter *et al*. 1978).
[b] Mean lifetime.
[c] Experimentally recorded fluorescence decay rates.
A and k are fluorescence decay constants

resolution of the detection system; the risetimes given in Table 2 correspond to the time elapsed between the maximum of the B-phycoerythrin emission and the maximum emission of the pigment. These risetimes to maximum fluorescence emission are illustrated by the initial portion of the decay curves shown in Fig. 4.

If we consider energy transfer between the excited-state population of B-phycoerythrin (J^*) and the ground-state population of R-phycocyanin (K), a simple exponential transfer equation predicts that the form of the R-phycocyanin fluorescence decay will be equation (1)

$$K^*(t) = \frac{J^*(0)\, k_j}{(k_k - k_j)} (e^{-k_j t} - e^{-k_k t}) \qquad (1)$$

where k_j and k_k are the fluorescence decay rates of the excited-state populations of B-phycoerythrin (J^*) and R-phycocyanin (K^*), respectively, in the presence of energy transfer. Using this equation and the mean fluorescence lifetime of B-phycoerythrin $(k_j^{-1} = 70$ ps$)$, we found it impossible to match simultaneously both the observed risetime and the fluorescence decay of R-phycocyanin by varying k_k. Consequently, we derived a series of kinetic equations based upon the empirical rate equation (2) where S^* is the excited-

$$\frac{dS^*}{dt} = -\tfrac{1}{2} S^* A t^{-1/2} \qquad (2)$$

state population of the donor pigment and A is the overall rate of energy transfer to the acceptor pigment. The final form of the fluorescence-decay

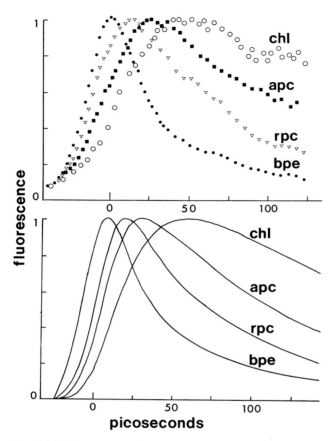

FIG. 4. The fluorescence risetimes of the phycobiliproteins and chlorophyll *a* in dark-adapted *Porphyridium cruentum*. The upper set of curves shows the experimental results obtained with the picosecond laser for the four wavelength-resolved emissions, after normalization of the fluorescence intensity and alignment on the time axis. The experimental points represent individual channels of the OMA memory. The lower set of theoretical curves were generated from the kinetic analysis published previously for direct excitation of B-phycoerythrin and energy transfer to the other pigments (Porter *et al.* 1978): Chl, chlorophyll *a*; apc, allo-phycocyanin; rpc, R-phycocyanin; bpe, B-phycoerythrin. The zero time of the experimental curves is arbitrarily set at the fluorescence maximum of B-phycoerythrin but that for the theoretical curves is a calculated fixed-point zero. Fluorescence is on a normalized linear scale.

equations is similar to that derived by Tomita & Rabinowitch (1962), except that the exponential terms all contain $t^{1/2}$ instead of t. From these equations, the transfer rate constants *(A)* were sequentially evaluated for each pigment, and values of A and the calculated fluorescence rise and decay times are

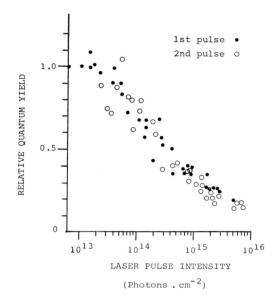

FIG. 5. The dependence of the fluorescence relative quantum yield of phycobilisomes, isolated from *Porphyridium cruentum*, on the laser-pulse intensity. The allophycocyanin fluorescence was monitored above 645 nm. The integrated fluorescence intensity relative to that of the excitation pulse is taken to be unity at the lowest photon densities used.

given in Table 2. After convolution with the resolution function of the streak camera, the calculated curves gave good agreement with those observed experimentally, as Fig. 4 illustrates.

The fluorescence decay of B-phycoerythrin in the isolated phycobilisomes was the same as that in the intact alga but the fluorescence lifetime of allophycocyanin increased to 4.2 ns (at 2×10^{13} photon cm^{-2}) and followed an $\exp(-kt)$ decay law. At higher excitation intensities, the allophycocyanin fluorescence lifetime (see Searle *et al.* 1978) and the quantum yield (Fig. 5) both decreased. Up to an excitation intensity of 5×10^{14} photon cm^{-2} the decay law remained exponential (with a fluorescence lifetime shortened to 1.96 ns), but above this intensity the decay became increasingly non-exponential. This fluorescence quenching process can be ascribed to singlet–exciton annihilation within the allophycocyanin pigment system. Since the fluorescence lifetime and quantum yield are reduced proportionately, we conclude that exciton annihilation within the preceding pigments is negligible. We also conclude that energy transfer to chlorophyll *a* in the intact alga will reduce the lifetime of the excited state of allophycocyanin sufficiently to preclude exciton

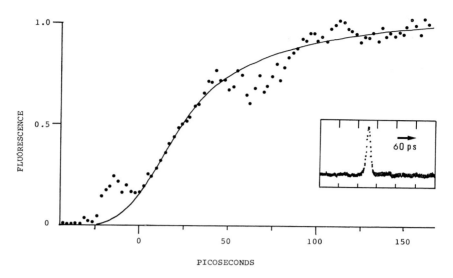

FIG. 6. The risetime of allophycocyanin fluorescence in phycobilisomes isolated from *Porphyridium cruentum*. Fluorescence was monitored above 645 nm and is expressed on a normalized scale, with the asymptote of the rise curve being taken as unity. Each experimental point represents an individual channel of the OMA memory. The solid line represents the theoretical curve derived from the kinetic analysis (see Searle *et al.* 1978). The inset shows the laser pulse profile taken at the same streak rate.

annihilation at the intensities used in the measurements reported above (Fig. 4 and Table 2). In light-harvesting complexes lacking reaction centres, such as phycobilisomes, the quenching centres induced by high photon densities may be used to investigate pigment organization and mechanisms of energy transfer.

To test further the validity of the kinetic equations derived for the intact alga, we examined the risetime of the fluorescence emission from allophyco-cyanin in the isolated phycobilisomes. Simply by inserting the fluorescence lifetime found in the absence of energy transfer to chlorophyll *a* (4.2 ns) into the equation derived for allophycocyanin, we obtained the convoluted risetime shown in Fig. 6. Although the experimental data had a relatively low signal-to-noise ratio, the theoretical curve is seen to give a good fit.

The extremely rapid rates of energy transfer, observed in the intact alga imply that few migrations are made within each pigment bed before the energy is transferred to the next pigment. Also, comparison of the rate constants *A* (Table 2) and the ratio of the pigments in the phycobilisomes (Table 1) indicates that the excitation energy resides longer in the large B-phycoerythrin pigment bed (84%, 0.26 ps$^{-1/2}$) than in R-phycocyanin (10%, 0.48 ps$^{-1/2}$) or

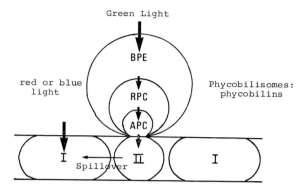

FIG. 7. A model of the pathways of energy transfer in *Porphyridium cruentum*. BPE, B-phycoerythrin; RPC, R-phycocyanin; APC, allophycocyanin; I, photosystem I; II, photosystem II.

allophycocyanin (8%, 0.52 ps$^{-1/2}$). The highly-efficient energy transfer observed in these systems is consistent with the layered structure of the phycobilisomes proposed by Gantt *et al.* (1977) (illustrated in Fig. 7), which would certainly allow the rapid transfer of energy between the pigments. This type of structure would also tend to limit the distance over which excitons could migrate in any given pigment; the probability of singlet-exciton annihilation will, therefore, be much less than in a comparable chlorophyll-antenna system. Only when energy transfer is prevented, as in the case of the isolated phycobilisomes, will the singlet-exciton lifetime be long enough to allow such quenching processes.

Chlorophyll *a* in *P. cruentum* is located in the thylakoid membrane; most of the chlorophyll acts as antenna for photosystem I, and only about 10% is associated with photosystem II. The energy absorbed by the phycobilisomes is transferred to photosystem II, and the chlorophylls in photosystem I absorb light directly or receive energy by 'spillover' transfer from photosystem II (Ley & Butler 1977). We have found that the fluorescence lifetime of chlorophyll *a* in dark-adapted *P. cruentum* is 175 ps in contrast to a value of about 500 ps recorded for green plants (Tredwell *et al.* 1977). How can this difference be explained? If we accept that energy transfer and not trapping is the rate-determining step, then the smaller size of the antenna system of photosystem II might be expected to increase the rate of migration to the reaction centre. It is also possible that in *P. cruentum* energy is being transferred from photosystem II to photosystem I even in the dark-adapted state, also leading to a shortened fluorescence lifetime, and some evidence has been presented for this (Ley & Butler 1977). If we follow Paillotin's (1976) proposal that energy transfer is extremely rapid but trapping by the reaction centre of photosystem II is not 100% efficient, then the shorter lifetime would suggest that trapping at the photosystem II reaction centre in *P. cruentum* is more efficient compared to green plants.

Finally, we have noted in a preliminary experiment that when *P. cruentum* cells were preilluminated with 633 nm light (12.5 W/m²) for several minutes, the chlorophyll-fluorescence lifetime showed a marked reduction but slowly reverted to the original value after a dark period of several minutes. This is consistent with the rate of spillover from photosystem II being increased on illumination, perhaps reflecting a State 1-to-State 2 transition (Ried & Reinhardt 1977).

The fluorescence kinetics of chlorophyll a *in* Chlorella *and chloroplasts*

Fig. 8 shows a model of the photosynthetic unit of higher plant chloroplasts based on the tripartite model of Butler & Strasser (1977) and adopted by us as a framework for discussion. The chlorophyll light-harvesting system of green photosynthetic species is a particularly difficult structure to study, as the fluorescence emissions from the chlorophyll molecules in slightly different environments occur over such a relatively narrow spectroscopic range. Consequently we have not been able to wavelength-resolve the fluorescence kinetics of these various components by picosecond spectroscopy. The fluorescence lifetimes and decay kinetics reported for *Chlorella* and for spinach chloroplast fragments, therefore, represent an average over all these components

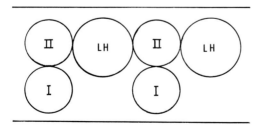

L H   Light Harvesting, chl a/b
 I   Photosystem I, chl a
II   Photosystem II, chl a

FIG. 8. A tripartite model of the photosynthetic unit of green photosynthetic systems.

of chlorophyll emission. Only in the case of the photosystem I emission at 77 K (715–730 nm) has it been possible to investigate a wavelength-resolved fraction of the emission. After our initial report of energy transfer affecting the risetime of fluorescence emission in *Porphyridium* (Searle *et al.* 1977*b*) a study appeared of the risetime of photosystem I emission (735 nm) at 77 K in chloroplasts (Campillo *et al.* 1977). The difficulties Campillo *et al.* found in explaining the kinetics of the observed risetime could possibly be resolved using the time-dependent kinetics of energy transfer which we have proposed.

Table 3 summarizes the fluorescence lifetimes and decay kinetics recorded in our investigation of green plants at various excitation intensities. Time-dependent kinetics are only noted when quenching centres are present. At the lowest intensities used, the fluorescence lifetimes and quantum yields are in close agreement with values recorded by single-photon counting and phase-shift fluorimetry (Mar *et al.* 1972).

Unlike the situation in the pigments of *P. cruentum*, energy migration within the chlorophyll-antenna systems probably proceeds by many transfer steps before the energy is trapped by the reaction centre or is dissipated by fluorescence emission or radiation-less intramolecular relaxation. The observed kinetics of fluorescence decay should, therefore, reflect the overall kinetics of migration to the reaction centre. However, at high excitation intensities, the singlet-exciton population becomes so large that they interact, producing a spurious quenching process (3):

$$S_1 + S_1 \rightarrow S_0 + S_n \tag{3}$$

where S_1 is the first excited singlet state, S_n is an upper excited singlet state, and S_0 is the ground state. Harris *et al.* (1976) found that an empirical decay law of the form $\exp(-At^{1/2})$ could describe the decay curves recorded at high intensities without recourse to a combination of exponential components. This decay law was applicable in various experimental conditions in which high excitation intensities were used (see Table 3). Several workers also reported that a long-lived quencher, the triplet or ion of chlorophyll *a*, was generated when low-intensity pulse trains were employed. Porter *et al.* (1977) have suggested that this quencher is formed within the exciton population generated by the first pulse and, as a consequence, singlet–triplet annihilation probably contributes to the decay kinetics even at this stage. It has been suggested that stimulated emission could cause the reduction in fluorescence lifetimes (Hindman *et al.* 1977), but if this were true the quantum yield measured at increasing excitation intensities should remain constant and not decrease in the manner that we have observed. Recently, Geacintov *et al.* (1977) reported that the photosystem I emission at 735 nm from spinach chloroplasts

TABLE 3

The fluorescence decay kinetics of chlorophyll a in green plants[a]

System	Conditions[b]	Intensity (photon cm^{-2}) ×10^{-13}	$\tau_{1/e}$ (ps)	A (ps$^{-1/2}$)	k (ps^{-1})	$\phi_{calc.}$
Chlorella						
	(i)	3	450	0.047		0.052
(PS II)	(i)	50	280	0.060		0.032
	(i)	800	50	0.093		0.013
	(ii)	3	1800		0.00056	0.104
(PS II)	(ii)	50	660	0.039		0.076
	(ii)	800	220	0.068		0.025
(PS I)	(iii)	3	2800		0.00036	0.162
	(iii)	800	430	0.048		0.050
DEF						
	(i)	5	500	0.045		0.057
	(i)	800	71	0.119		0.008
(PS II)						
	(iii)	5	2500		0.00040	0.145
	(iii)	800	430	0.048		0.050
SLV						
	(i)	50	100			0.006
(PS I)						
	(iii)	5	1800		0.00055	0.105
	(iii)	800	450	0.047		0.052
F_I	(i)	50	100			0.006
(PS I)	(iii)	5	1900		0.00053	0.110
F_{III}	(i)	10	4000		0.00025	0.232
CP 1	(i)	10	4500–5000		0.00021	0.275

[a] PS I, photosystem I; PS II, photosystem II; see Material and Methods for a description of the various sub-chloroplast fractions DEF, SLV, F_I, F_{III} and CP 1; A and k are fluorescence decay constants.
[b] (i) Dark-adapted; (ii) DCMU and preillumination; (iii) 77 K.

at 77 K does not exhibit an intensity-dependent decay rate at excitation intensities below 10^{15} photon cm^{-2}. This is contrary to our findings given in Table 3; in particular, the lifetime of 1.1 ns they reported is much shorter than the value of 1.8 ns obtained from photosystem I fractions of spinach chloroplasts and the value of 2.8 ns obtained from photosystem I of *Chlorella* at 77 K, suggesting an intensity-dependent shortening of the lifetime in their conditions. In general, it appears that exciton annihilation occurs in green photosynthetic systems, in various experimental conditions, at excitation intensities in excess of 1×10^{14}–5×10^{14} photon cm^{-2}.

Of the low excitation intensity measurements reported in Table 3, only the dark-adapted cases still appear to be governed by an $\exp(-At^{1/2})$ decay law. If this was the result of exciton-annihilation processes, it is surprising that the decay curve obtained in condition (ii) at the same excitation intensity does not exhibit this type of decay law. We, therefore, conclude that the $\exp(-At^{1/2})$ fluorescence decay law, in this particular instance, is related to the overall migration of energy towards an open photosystem II reaction centre. As yet we have had only limited success in the study of photosystem I at room temperature, a lifetime of 100 ps has been measured but we have been unable to time-resolve the kinetics of the fluorescence decay.

The antenna chlorophyll *a* of the two photosystems forms discrete units which can be separated from each other with detergents: treatment of chloroplasts with digitonin can separate photosystem I (F_I) from the light-harvesting–photosystem II complex (DEF). Table 3 illustrates the marked difference between the fluorescence lifetime of photosystem I and that of photosystem II at room temperature. On cooling to 77 K the chlorophyll-fluorescence lifetimes of both photosystems increase and appear to approach a similar value. It is possible that this is related to the formation of lower energy chlorophyll forms both in photosystem I (F730) and photosystem II (F695) on freezing.

The fluorescence lifetimes recorded for the larger digitonin-derived chlorophyll *a/b*–protein complex (F_{III}, 4 ns) and the chlorophyll *a*–protein complex prepared with sodium dodecyl sulphate (CP 1, 4.5–5 ns) at room temperature (Table 3) both indicate that concentration quenching is a minor deactivation pathway for singlet excitons in these chlorophyll–proteins. Fig. 9 shows that F_{III} fluorescence decays exponentially both at low excitation intensity (single-photon counting, 3.9 ns) and high intensity (picosecond laser at 10^{15} photon cm^{-2}, 2.3 ns). At low photon densities (1×10^{14} photon cm^{-2}) the lifetime of 4.0 ns was in good agreement with the single-photon-counting result. These preparations were characterized by their absorption and fluorescence emission spectra and shown not to contain detergent-solubilized chlorophyll (F_{III}:

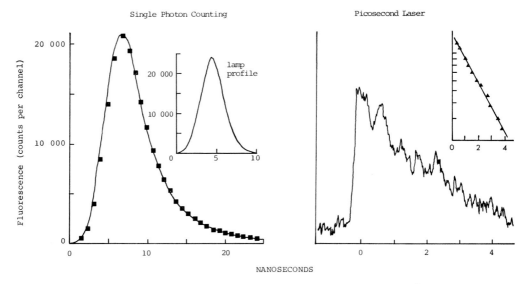

FIG. 9. Chlorophyll-fluorescence kinetics of F_{III} at room temperature: in the single-photon-counting experiment the solid squares represent experimental points, and the line is the theoretical curve of an exponential decay with a lifetime of 4.0 ns convoluted with the excitation-lamp profile (see inset). The sample was excited at 435 nm ($A_{435}^{10mm} = 0.3$) and the emission monitored by a Philips 56 TUVP protected by a Balzer B-40 685 nm interference filter. The picosecond laser fluorescence decay curve was obtained at a high photon density of 10^{15} photon cm^{-2} and was exponential (see inset).

fluorescence maximum at 681 nm at both 295 and 77 K, absorbance maximum 676 nm; CP1: fluorescence maximum at 678 nm at 295 K and 683 nm at 77 K, absorbance maximum at 674 nm).

On closing the photosystem II reaction centres of *Chlorella* in condition (ii) (see Table 3), the chlorophyll lifetime only increases to 1.8 ns, corresponding to about 50% quenching compared to F_{III}. Since concentration quenching does not appear to be a likely cause, it could be possible that the closed reaction centre can still act as a quencher, although other quenching species may also be involved.

Finally, we have started to investigate the organization of the constituent chlorophyll–proteins of chloroplasts and the rates of energy transfer between them by varying the cation composition of the suspending medium (Barber *et al.* 1978). In the presence of DCMU in the low-fluorescence-yield state ($-MgCl_2$) the chlorophyll fluorescence could be described closely by an $\exp(-At^{1/2})$ decay law. Addition of $MgCl_2$ produced the high fluorescing state and caused the decay to become predominantly exponential. In both cases the final part of the decay had a lifetime of 1.6 ns. These results support the hypothesis that energy transfer from photosystem II to photosystem I ('spillover') is active in the low-fluorescing state and that photosystem I acts as a quenching centre for excitons in the light-harvesting–photosystem II pigment bed (see Fig. 8); MgCl₂ inhibits this energy transfer. Changes in partitioning of incoming quanta can be ruled out as the cause of the fluorescence yield changes in the absence and presence of MgCl₂.

CONCLUSIONS

Our investigation of energy transfer between the phycobiliproteins of *P. cruentum* has shown that the observed fluorescence risetimes and lifetimes can be described by a series of time-dependent rate equations (Porter *et al.* 1978). Since the calculated rates of energy transfer are extremely rapid, it seems probable that only a few migration jumps occur in each pigment before the energy is transferred to the next pigment in the sequence. If this were the case, then the time-dependent nature of the kinetics could be explained in terms of a Förster-type mechanism for resonance-energy transfer (Harris *et al.* 1976). The model of the phycobilisome proposed by Gantt *et al.* (1977) appears to be consistent with the highly-efficient energy transfer observed in these experiments.

The time-dependent behaviour of the fluorescence emission from dark-adapted photosystem II in green plants is still open to discussion. It could be that the time dependence of a single migration jump is lost as many jumps are required to reach the reaction centre, although this has not yet been demonstrated experimentally for photosynthetic systems. As well as a Förster-type mechanism, a diffusion mechanism for energy migration to the reaction centre could also lead to the time-dependent kinetics reported.

ACKNOWLEDGEMENTS

This work was supported by the Science Research Council and the EEC Solar Energy Research and Development Programme, and by the award of a Ministry of Defence Post-Doctoral fellowship to C.J.T. We also thank Prof. Sir George Porter, Dr J. Barber, Dr L. Harris and Mr. J. A. Synowiec for their invaluable help during the course of this work. The single-photon counting of F_{III} was carried out by Dr G. Beddard.

References

ARCHER, M. D., FERREIRA, M. I. C., PORTER, G. & TREDWELL, C. J. (1977) Picosecond study of Stern–Volmer quenching of thionine by ferrous ions. *Nouv. J. Chim. 1*, 9–12

BARBER, J., SEARLE, G. F. W. & TREDWELL, C. J. (1978) Picosecond time-resolved study of MgCl₂-induced chlorophyll fluorescence yield changes from chloroplasts. *Biochim. Biophys. Acta 501*, 174–182

BEDDARD, G. S., PORTER, G., TREDWELL, C. J. & BARBER, J. (1975) Fluorescence lifetimes in the photosynthetic unit. *Nature (Lond.) 258*, 166–168

BUTLER, W. L. & STRASSER, R. J. (1977) Tripartite model for the photochemical apparatus of green plant photosynthesis. *Proc. Natl. Acad. Sci. U.S.A. 74*, 3382–3385

CAMPILLO, A. J., SHAPIRO, S. L., GEACINTOV, N. E. & SWENBERG, C. E. (1977) Single pulse picosecond determination of 735 nm fluorescence risetime in spinach chloroplasts. *FEBS (Fed. Eur. Biochem. Soc.) Lett. 83*, 316–320

GANTT, E. & CONTI, S. F. (1966) Granules associated with the chloroplast lamellae of *Porphyridium cruentum. J. Cell Biol. 29*, 423–434

GANTT, E., LIPSCHULTZ, C. A. & ZILINSKAS, B. A. (1977) Phycobilisomes in relation to thylakoid membranes. *Brookhaven Symp. Biol. 28*, 347–357

GEACINTOV, N. E., BRETON, J., SWENBERG, C., CAMPILLO, A. J., HYER, R. C. & SHAPIRO, S. L. (1977) Picosecond and microsecond pulse laser studies of exciton quenching and exciton distribution in spinach chloroplasts at low temperatures. *Biochim. Biophys. Acta 461*, 306–312

HARRIS, L., PORTER, G., SYNOWIEC, J. A., TREDWELL, C. J. & BARBER, J. (1976) Fluorescence lifetimes of *Chlorella pyrenoidosa. Biochim. Biophys. Acta 449*, 329–339

HINDMAN, J. C., KUGEL, R., SVIRMICKAS, A. & KATZ, J. J. (1977) Chlorophyll lasers: stimulated light emission by chlorophylls and Mg-free chlorophyll derivatives. *Proc. Natl. Acad. Sci. U.S.A. 74*, 5–9

KUNG, S. D. & THORNBER, J. P. (1971) Photosystem I and II chlorophyll-protein complexes of higher plant chloroplasts. *Biochim. Biophys. Acta 253*, 285–289

LEY, A. C. & BUTLER, W. L. (1977) The distribution of excitation energy between photosystem I and photosystem II in *Porphyridium cruentum*, in *Photosynthetic Organelles* (Miyachi, S., Katoh, S., Fujita, Y. & Shibata, K., eds.), pp. 33–46 (special issue of *Plant and Cell Physiology 3*), Japanese Society of Plant Physiology, Tokyo

LEY, A. C., BUTLER, W. L., BRYANT, D. A. & GLAZER, A. N. (1977) Isolation and function of allophycocyanin B of *Porphyridium cruentum. Plant Physiol. 59*, 974–980

MAR, T., GOVINDJEE, SINGHAL, G. S. & MERKELO, H. (1972) Lifetime of the excited state *in vivo*. I. Chlorophyll *a* in algae, at room and at liquid nitrogen temperatures; rate constants of radiationless deactivation and trapping. *Biophys. J. 12*, 797–808

PAILLOTIN, G. (1976) Capture frequency of excitations and energy transfer between photosynthetic units in the photosystem II. *J. Theor. Biol. 58*, 219–235

PORTER, G., SYNOWIEC, J. A. & TREDWELL, C. J. (1977) Intensity effects on the fluorescence of *in vivo* chlorophyll. *Biochim. Biophys. Acta 459*, 329–336

PORTER, G., TREDWELL, C. J., SEARLE, G. F. W. & BARBER, J. (1978) Picosecond time-resolved energy transfer in *Porphyridium cruentum*, Part 1. In the intact alga. *Biochim. Biophys. Acta 501*, 232–245

RIED, A. & REINHARDT, B. (1977) Distribution of excitation energy between photosystem I and photosystem II in red algae. II Kinetics of the transition between State 1 and State 2. *Biochim. Biophys. Acta 460*, 25–35

SEARLE, G. F. W., BARBER, J., HARRIS, L., PORTER, G. & TREDWELL, C. J. (1977a) Picosecond laser study of fluorescence lifetimes in spinach chloroplast photosystem I and photosystem II preparations. *Biochim. Biophys. Acta 459*, 390–401

SEARLE, G. F. W., BARBER, J., PORTER, G. & TREDWELL, C. J. (1977*b*) A study of energy transfer in *Porphyridium cruentum* using picosecond laser techniques, in *Fourth International Congress on Photosynthesis, Abstracts*, p. 340, U.K.I.S.E.S., London

SEARLE, G. F. W., BARBER, J., PORTER, G. & TREDWELL, C. J. (1978) Picosecond time resolved energy transfer in *Porphyridium cruentum* Part 2. In the isolated light-harvesting complex (phycobilisomes). *Biochim. Biophys. Acta 501*, 245–256

TOMITA, G. & RABINOWITCH, E. (1962) Excitation energy transfer between pigments in photosynthetic cells. *Biophys. J. 2*, 483–499

TREDWELL, C. J., PORTER, G., SYNOWIEC, J. A., BARBER, J., SEARLE, G. F. W. & HARRIS, L. (1977) Picosecond laser spectroscopy of the photosynthetic unit, in *Lasers in Chemistry* (West, M. A., ed.), pp. 304–310, Elsevier, Amsterdam

WESSELS, J. S. C. (1968) Isolation and properties of two digitonin-soluble pigment–protein complexes from spinach. *Biochim. Biophys. Acta 153*, 497–500

Discussion

Junge: Where is the $t^{1/2}$ law derived from?

Tredwell: The $\exp(-At^{1/2})$ fluorescence decay law is purely empirical. Although a sum of exponentials could have been used instead, this decay law minimizes the number of variables and gives a good fit to the experimental decay curves up to 1.5 ns after excitation.

Porter: That is one answer; the other is that it is the form of the decay law of Förster transfer.

Tredwell: The speed of energy transfer within the phycobilins of *Porphyridium cruentum* suggests that transfer is close to a single-step process between each pigment. A Förster-type energy transfer mechanism may be applicable in this case.

Knox: But Förster never wrote a time dependence into a rate constant— the $t^{1/2}$ is a result of an average over the distribution of the acceptors with respect to the donors.

Tredwell: That is correct; this particular treatment was taken from Birks (1968) and extended to a multiple pigment system.

Knox: At one point you implied that exciton annihilation was unreal. I dispute that! May I, therefore, put Dr Junge's question differently: what is the reality behind that time-dependent rate constant? How do you explain the physics of $At^{-1/2}$ as a rate constant?

Tredwell: Exciton annihilation is obviously a real process, as can be seen from the marked reduction in the fluorescence quantum yield at high excitation intensities. However, I suggest that the effect of exciton fusion is negligible at excitation intensities less than 5×10^{13} photon cm^{-2}; this appears to be substantiated by the published fluorescence quantum yield curves (Geacintov & Breton 1977; Campillo *et al.* 1976; Mauzerall 1976).

As I have stated before, the $\exp(-At^{1/2})$ fluorescence decay law is purely empirical and is not intended to imply any form of physical process. Since the kinetics of exciton fusion and energy migration within the photosynthetic unit are still open to debate, time-dependent kinetics cannot be ruled out.

Beddard: With respect to Förster energy transfer, it is single-step transfer that gives the $t^{1/2}$ term whereas presumably there is here multi-step transfer between each of the different pigments.

Porter: Let us leave the single step for a moment. Dr Knox says that Förster did not put a $t^{1/2}$ term into a rate constant, but the Förster treatment gives a $t^{1/2}$ dependence of the rate.

Knox: There is a $t^{1/2}$ term in the result—but not in the rate constant.

Junge: It is an integrated result.

Porter: No; the yield gives an error function.

Knox: The integration is over the distribution of molecules, not over time.

Porter: The dependence becomes an $\exp(-t^{1/2})$ as Drs Searle & Tredwell use and, as Dr Beddard said, for a single transfer to a random distribution of acceptors. A diffusional process of many jumps is a problem that nobody has been able to solve. On the other hand, the purely diffusional process also gives an $\exp(-kt^{1/2})$ term, which at short times predominates. So either a single-step transfer or a diffusion gives a $t^{1/2}$ term at short times.

Paillotin: Maybe that $t^{1/2}$ term comes from mere statistics.

Porter: What do you mean, it comes from mere statistics? It is a rigorous treatment.

Paillotin: The $t^{1/2}$ dependence also applies to the rise of the fluorescence curve. The situation now differs from what I discussed earlier. Here only one part of the pigment array is excited and since there is some heterogeneity in the photosynthetic membrane one can hope to observe the time taken to go from the subsystem that was excited initially to the other one which was not. One can expect to see, for instance, the exciton transfer from system II to system I. Campillo *et al.* (1977) did this experiment and observed an S-shaped lag time which can be related to a diffusion process. When one excites one part but looks at another part which was not excited, such a lag must be observed. It is more difficult to explain the $t^{1/2}$ term for the decrease of fluorescence.

Tredwell: We attempted to fit the various fluorescence risetimes and decay times with simple exponential functions that might be expected for a diffusion-like process but we could not fit the experimental data; I might also point out that Campillo *et al.* (1977) could not fit their data by this treatment either.

Paillotin: The rate constant depends on time but you have a linear equation for the concentration of the excitons. You do not take into account the fusion of excitons, the rate of which depends on the square of the exciton concentration. If one wants to introduce some diffusion term in the fusion process one must consider the time-dependence of the constant of fusion, i.e. a rate of fusion of the form: $\int S^*(t_1)\gamma(t-t_1)S^*(t)dt_1$.

Tredwell: But are we considering exciton fusion? All the evidence suggests that exciton annihilation within the phycobilins of *Porphyridium cruentum* occurs at much higher intensities than those we used. Exciton fusion only occurs within the allophycocyanin pigment bed when energy transfer to chlorophyll *a* is prevented; this can be ascribed to the 40-fold increase in the lifetime of the exciton population in these conditions.

Knox: One of the philosophical problems posed by a time-dependent rate constant is how does the system know when the clock starts? That is not too difficult with only one equation but you propose a chain of pigments (J, K, L, M), all of which seem to know when the clock starts at the same time. How do they do that?

Tredwell: Surely there is always a finite possibility of finding excited-state energy in any one pigment bed immediately after excitation?

Knox: They all start at $t = 0$. Presumably when the excitation reaches the third molecule it finds that the $t^{1/2}$ appearing in the third equation is referred to the same time-scale as that of the first equation.

Duysens: Does the fluorescence from dark-adapted *Chlorella* decay as $\exp(-kt)$?

Tredwell: Measurements for dark-adapted *Chlorella* with a continuous-wave mode-locked dye laser (10^9 photon cm^{-2}) and single-photon-counting techniques give a fluorescence lifetime of 490 ps with a time resolution of 300 ps (G. R. Fleming, G. S. Beddard, G. Porter & J. A. Synowiec, unpublished work, 1978). Streak-camera measurements indicate a fast component, within 100 ps of excitation (3×10^{13} photon cm^{-2}).

Duysens: That is rather small; what kind of exponential does the main part of the decay show?

Tredwell: After the first 100 ps, the fluorescence decay is governed by an exponential decay law, i.e. $\exp(-kt)$. The lifetime of this component is 500 ps when measured with the streak camera.

Duysens: Doesn't that suggest a more regular distribution of the antenna in *Chlorella* than in other species which show an $\exp(-kt^{1/2})$-type decay?

Tredwell: All the photosynthetic species we have studied exhibit this initial fast component at excitation intensities of less than 5×10^{13} photon cm^{-2}. To be consistent with the measurements at higher intensities we used an $\exp(-At^{1/2})$ decay law. The results suggest that the antenna systems of all the photosynthetic species are equally heterogeneous.

Knox: I am not one of those who totally disbelieves the $t^{1/2}$ dependence. I believe it for solutions and I accept Förster's derivation of it but, whenever it appears in a rate constant, and I even include the case of the Smolukowski theory where $k(t) = A + Bt^{-1/2}$, something is being swept under the rug, because in a rate constant it is clearly describing something unphysical at $t = 0$.

Tredwell: We can never determine what happens at $t = 0$, since we use a 6 ps laser pulse. We have shown that the Förster equation is valid for dye systems in solution to within 10 ps of excitation.

Paillotin: It is possible to solve this problem. Let us suppose there is a diffusion-limited process between all the parts of the system and that each part is a plane-parallel layer. Then we can write a diffusion equation and calculate the time-dependence of the yield of fluorescence. That is a simple model without rate constants.

Tredwell: There must still be a rate constant.

Porter: Yes, but it is a diffusion-controlled rate constant.

Paillotin: If exchange between two systems is not diffusion-limited, one can introduce a rate constant but if it is diffusion-limited one cannot use a rate constant.

Porter: If it is diffusion-limited, the rate constant is the rate of diffusion.

Paillotin: In the diffusion-limited case it is difficult to introduce, for instance, the concentration of the excitation in one system; one ought to consider the concentration at the boundary between two systems.

Porter: The Smolukowski–Debye equation gives the diffusion-controlled rate constant.

Paillotin: Yes, but doing that one supposes that the concentration at the boundary is proportional to the total concentration, that is to say, some kind of equilibrium is reached before the transfer. If not, one has to introduce the true concentration at the boundary, which is not directly proportional to the total concentration.

Porter: If you are saying that, if diffusion is rapid compared with reaction, the rate is reaction-limited, then I agree, of course.

Tredwell: We intend to excite the pigments of *Porphyridium cruentum* directly, so that we can obtain data on the transfer mechanism to a subsequent pigment in the energy-transfer sequence. We may then be able to derive a more satisfactory model from a theoretical point of view.

Wessels: The fluorescence emission spectrum of the light-harvesting complex F_{III} at 77 K showed a maximum at 681 nm and a weak shoulder at 695 nm but the fluorescence spectrum shows a major peak at 695 nm.

Searle: The latter spectrum was for the fraction containing photosystem II plus the light-harvesting chlorophyll, and not the isolated F_{III}. The F_{III}

particle has a maximum at 681 nm both at room temperature and at 77 K and no shoulder at 695 nm.

Anderson: F_{III} has a chlorophyll a/b ratio of 1.3 and has no reaction centre of photosystem II. I also find that the light-harvesting chlorophyll a/b–protein complex contains more chlorophyll a than b, with a chlorophyll a/b ratio of 1.25.

Searle: We have also found the chlorophyll a/b ratio to be greater than one in F_{III}.

References

BIRKS, J. B. (1968) *J. Phys. B. (Proc. Phys. Soc.)*, Ser. 2, *1*, 946–957

CAMPILLO, A. J., SHAPIRO, S. L., KOLLMAN, V. H., WINN, K. R. & HYER, R. C. (1976) Picosecond exciton annihilation in photosynthetic systems. *Biophys. J. 16*, 93–97

CAMPILLO, A. J., SHAPIRO, S. L., GEACINTOV, N. E. & SWENBERG, C. E. (1977) Single pulse picosecond determination of 735 nm fluorescence risetime in spinach chloroplasts. *FEBS (Fed. Eur. Biochem. Soc.) Lett. 83*, 316–320

GEACINTOV, N. E. & BRETON, J. (1977) Exciton annihilation in the two photosystems in chloroplasts at 100 K. *Biophys. J. 17*, 1–15

MAUZERALL, D. (1976) Multiple excitations in photosynthetic systems. *Biophys. J. 16*, 87–91

Energy transfer and its dependence on membrane properties

J. BARBER

Department of Botany, Imperial College, London

Abstract With isolated chloroplasts variations in the degree of energy transfer between light-harvesting chlorophyll–protein complexes can be induced by changing the cation content of the suspending medium. The changes can be observed by measuring chlorophyll-fluorescence yields and lifetimes and are probably brought about by conformatial changes in the thylakoid membrane. Detailed studies of the properties of cation-induced changes in chlorophyll fluorescence indicated that the alterations in pigment organization are due to variations in the density of positive charges immediately adjacent to the surface of the thylakoid membrane, being in qualitative agreement with predictions based on the Gouy–Chapman theory of diffuse double layers. Possible mechanisms for the membrane structural changes controlling energy transfer are given.

To gain a full understanding of the mechanisms involved in the capture and transfer of energy between photosynthetic pigments one must understand the properties of the membranes with which they are intimately related. Several papers in this volume emphasize this point but perhaps the most striking case of membrane control of energy transfer is that of the regulation of spillover from photosystem II (PS II) to photosystem I (PS I) as discussed by Professor Butler. Although the early work of Bonaventura & Myers (1969) and Murata (1969*a*, *b*, 1971) demonstrated the existence of this regulatory mechanism *in vivo* and the involvement of changes in ionic levels within the chloroplast, there have been few attempts to explain the physical processes involved. Vague arguments that cationic binding to the thylakoid membrane brings about specific conformational changes in its structure have been put forward. However, in this paper I shall show that the properties of the cation-induced spillover are indicative of electrical effects immediately adjacent to the membrane surface and suggest how these effects can influence energy-transfer processes between the pigment–protein complexes of the two photosystems.

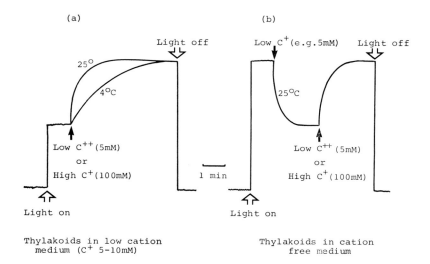

FIG. 1. Typical examples of cation-induced changes in the yield of chlorophyll fluorescence: (a) thylakoids suspended in a medium having a background concentration of univalent salts of 10 mmol/l; the effect of temperature on the cation-induced rise; (b) the antagonistic effect of low concentrations of univalent cations with either low concentrations of bivalent or high concentrations of univalent cations when unwashed thylakoids are initially suspended in a cation-free medium.

PROPERTIES OF CATION-INDUCED CHANGES IN CHLOROPHYLL-FLUORESCENCE YIELD

Homann (1969) and Murata (1969b, 1971) found that, when salts were added to isolated thylakoids treated with 3-(3,4-dichlorophenyl)-1,1-dimethylurea (DCMU) so that the PS II traps were fully closed, the yield of variable chlorophyll fluorescence increased (see Fig. 1). Murata (1969b, 1971) and later many others (see Barber 1976; Butler 1978) presented evidence that the increase in PS II fluorescence yield was associated with a decrease in spillover of energy from PS II to PS I and could be correlated with a State 2 → State 1 change observed in intact systems (Myers 1971) and with changes in chlorophyll-fluorescence lifetime (Barber et al. 1978).

The properties of the salt-induced fluorescence increase (see Mills & Barber 1978) are:

(i) it depends on the cation used but is independent of the anion;

(ii) bivalent cations are far more effective than univalent cations (concentrations necessary for maximum change are: bivalent, 0.5–5 mmol/l; univalent, 100 mmol/l);

(iii) there is no specificity between a range of bivalent cations: Mg^{2+}, Ca^{2+}, Ba^{2+}, Sr^{2+}, Mn^{2+}, including organic cations e.g. lysyl-lysine;

(iv) there is no specificity between a range of univalent cations: Na^+, K^+, Rb^+, Cs^+, including organic cations, e.g. choline and lysine;

(v) the bivalent cation concentration necessary to induce maximum rise depends on the background concentration of univalent cations in the suspending medium: with background concentration of 1 mmol/l, 0.5mM-bivalent cations are needed; with 3 mmol/l background, 1mM-bivalent cations; with background of 10 mmol/l, 2–3mM-bivalent cations;

(vi) as shown in Fig. 1a the fluorescence rise depends on temperature, follows an Arrhenius plot from 4 to 30 °C corresponding to an activation energy of about 12 kcal/mol (Barber, unpublished results).

Another important property of cation-controlled changes in fluorescence was discovered recently by Gross & Hess (1973) who showed that, when freshly-isolated DCMU-treated thylakoids were suspended in a medium totally free of cations, the chlorophyll fluorescence was already at a maximum and was unaffected by additions of low concentrations of bivalent cations. However, addition of 1–10mM-univalent cations decreased the fluorescence to the low level, comparable with the maximum spillover level originally found by Murata (see Fig. 1b). (In the earlier experiments by Homann [1969], Murata [1969b, 1971] and others, there were always some univalent cations present in the media used.) Properties of this type of cation-induced change in fluorescence can be added to the above list:

(vii) there is no specificity among univalent cations in their ability to decrease the fluorescence yield (see [ii]);

(viii) the univalent cation concentration required for the decrease depends on the background concentration of bivalent cations;

(ix) the initial level of fluorescence recorded when suspending thylakoids in cation-free medium depends on their pretreatment. When the membranes are obtained by osmotically shocking the intact chloroplasts during their suspension in the cation-free medium, the fluorescence is maximum. When, however, the membranes are washed or exposed to univalent cations before suspension in the cation-free medium, the fluorescence is not at the maximum; in these conditions one must add a small amount of bivalent cation to the suspension to restore the initial high-fluorescing state;

(x) After adding 1–10mM-univalent cation to the medium, the cation-induced changes in fluorescence have identical properties to those listed in (i)–(vi);

(xi) none of the above changes in fluorescence is due to a change in the osmotic condition of the suspension since they cannot be induced by non-charged solutes;

(xii) more-chemically-reactive cations like Zn^{2+}, Hg^{2+} and La^{3+} do not cause the above reversible changes in fluorescence; they bind to the membrane and generally lower the fluorescence yield (Mills & Barber 1978).

THE EXISTENCE OF NEGATIVE CHARGES ON THE THYLAKOID MEMBRANE

The differential effect of univalent and bivalent cations and the lack of specificity between species of a particular charged group indicate an electrical rather than a chemical mechanism for the cationic control of the chlorophyll fluorescence. These properties are similar to those described by the Schulze–Hardy rule for colloidal aggregation (see Overbeek 1977). Any mechanism to explain the cationic control of spillover must also take into account the intriguing observation of an antagonism between low concentrations of univalent and bivalent cations and between low and high concentrations of univalent cations. The explanation of these effects lies in the fact that the thylakoid membranes are negatively charged. For the sake of electro-neutrality, cations will be drawn close to the membrane and form a diffuse layer, the thickness of which depends on the ionic composition of the suspending medium. Quantitative treatment of this type of electrical double layer stems back to Gouy & Chapman (see Delahay 1965) whose classical theory is frequently applied to biological membranes (Haydon 1964).

The negative charge of the thylakoid membrane has been realized for some years (Nobel & Mel 1966; Berg et al. 1974), but recently we have done a detailed particle electrophoretic study of this membrane system (H. Y. Nakatani, J. Barber & J. A. Forrester, unpublished results). Fig. 2 shows that the thylakoids move towards the positive electrode, thereby indicating that they carry a net negative charge. In contrast to earlier work (Nobel & Mel 1966) we found no detectable effect of light on the mobility. Various chemical treatments of the thylakoids have shown that neither the sugar residues of the galactolipids in the membrane nor the basic groups of the membrane protein are exposed at the surface. However, treatment with water-soluble carbodiimides together with glycine methyl ester neutralized the surface negative charges, a result which implicates the carboxy groups which, because of their pH sensitivity, are likely to be the carboxy groups of aspartic and glutamic acid residues. Experiments with thylakoids isolated from a barley mutant lacking the light-harvesting chlorophyll a/b pigment–protein indicated that the negative charges are not associated with this complex. Fig. 3 shows

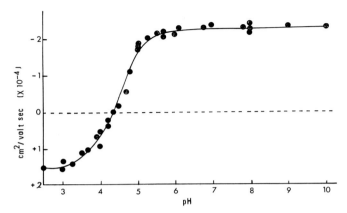

FIG. 2. Profile of pH electrophoretic mobility of pea chloroplast thylakoids suspended in 0.33M-sorbitol and 20mM-KCl at 20 °C; 10–20 measurements were made for each point and the voltages were adjusted to keep measurement times between 7 and 10 s (H. Nakatani, J. Barber & J. A. Forrester, unpublished results).

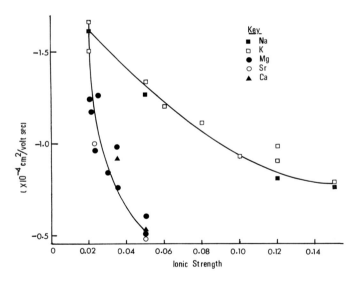

FIG. 3. Effect of univalent and bivalent cations on electrophoretic mobility of pea chloroplast thylakoids suspended in 0.33M-sorbitol–Tris, pH 7.6; bivalent cation concentration added to background univalent cation concentration of 20 mmol/l.

that the electrophoretic mobility decreases with increasing ionic strength. As with chlorophyll fluorescence, the effect is independent of the anion but bivalent cations are far more effective than univalent cations.

DIFFUSE DOUBLE-LAYER THEORY

The basis of the Gouy–Chapman theory is as follows. Consider a charged membrane surface in a solution containing diffusible ions. The distribution of ions in the solution at equilibrium will be such that the electrochemical potential ($\bar{\mu}$) of an ion (i) at any point is the same. For example, if $\bar{\mu}_{ib}$ is the electrochemical potential of i in the bulk solution and $\bar{\mu}_{ix}$ is its electrochemical potential near the surface (distance x from the membrane), then $\bar{\mu}_{ix} = \bar{\mu}_{ib}$.

If $\bar{\mu}$ is described only by chemical and electrical terms and if pressure and activity terms are ignored, the concentration of species i at any point from the membrane surface (C_{ix}) is given by the Boltzmann equation (1),

$$C_{ix} = C_{ib}\exp(-ZF\psi_x/RT) \tag{1}$$

where C_{ib} is the bulk concentration (at infinite distance from the membrane surface, ψ_x is the electrical potential difference at point x relative to bulk solution ($\psi_b = 0$) and the other terms have their usual meanings.

Assuming the membrane surface to be infinitely flat and with a uniformly spread charge density of σ μC/cm², the electrical potential difference at any point x, relative to the bulk solution, is given by the Gauss equation (2),

$$\frac{\mathrm{d}\psi_x}{\mathrm{d}x} = -\frac{4\pi}{\varepsilon}\sigma \tag{2}$$

where ε is the permittivity of water.

For a membrane suspended in aqueous solution at equilibrium, the charge density σ is balanced by an equal diffuse charge density of opposite polarity in the layer of solution adjacent to the membrane surface, equation (3),

$$\sigma = \int\limits_{x=0}^{x=\infty} -\varrho\,\mathrm{d}x \tag{3}$$

where ϱ is the space charge density of ions in solution in a plane parallel to the membrane surface at distance x. This charge density ϱ is given by equation (4),

$$\varrho = \sum_i Z_i F C_i \tag{4}$$

where C_i is concentration of species i in the plane.

Combining equations (2) and (3) we obtain the Poisson equation (5).

$$\frac{\mathrm{d}^2\psi_x}{\mathrm{d}x^2} = -\frac{4\pi}{\varepsilon}\rho \tag{5}$$

The Poisson equation can be combined with equations (1) and (4) to yield the Poisson–Boltzmann expression (6). This can be integrated to give (7).

$$\frac{d^2\psi_x}{dx^2} = -\frac{4\pi}{\varepsilon} \sum_i Z_i F C_{ib} \exp\left(\frac{-Z_i F \psi_x}{RT}\right) \tag{6}$$

$$\frac{d\psi_x}{dx} = \pm 2 \left(\frac{2\pi RT}{\varepsilon}\right)^{\frac{1}{2}} \left\{ \sum_i C_{ib} \left[\exp\left(\frac{-Z_i F \psi_x}{RT}\right) - 1 \right] \right\}^{\frac{1}{2}} \tag{7}$$

The field strength $(d\psi/dx)$ can be related at any point with the potential difference at that point (ψ_x) relative to the bulk solution. Since the Gauss equation also relates the field strength at the membrane $(x = 0)$ to the surface charge density on the membrane σ, combination of equations (2) and (7) gives (8), where ψ_0 is the surface potential $(x = 0)$.

$$\sigma = \pm \left\{ \frac{RT\varepsilon}{2\pi} \sum_i C_{ib} \left[\exp\left(\frac{-Z_i F \psi_0}{RT}\right) - 1 \right] \right\}^{\frac{1}{2}} \tag{8}$$

In deriving equations (7) and (8) several assumptions have been made.

(a) The membrane is assumed to be infinitely flat with a fixed charge 'smeared-out' uniformly over the surface. In practice the membrane can be considered infinitely flat if the radius of curvature of the surface is 30-times the perpendicular distance of the ion from the centre of the membrane. Since the radius of a single thylakoid is about 250 nm for the major surface area and the electrical double-layer can be normally considered to have little influence on ions at a distance greater than 10–20 nm from the surface, this assumption appears to be reasonably valid. More serious is the fact that it is unlikely that the charge is either smeared-out or homogeneous on the membrane surface.

(b) It is assumed in all the integration steps that the dielectric constant term, ε, is independent of ψ. Grahame (1953) has shown that this assumption is valid for a field strength $(d\psi/dx)$ not exceeding 10^6 V/cm. In the following analyses the field strength never exceeds this value.

(c) In the Gouy–Chapman treatment of the double-layer ions in solution are considered to be point charges which can approach the membrane to any distance including $x = 0$. In practice, however, the closest plane of approach will be defined by the radius of the hydrated ion, which for small metal cations is of the order of 0.2–0.4 nm. This error may be accounted for by dividing the double-layer into two regions: (i) the compact double-layer between the charged surface $(x = 0)$ and the plane of closest approach of the ion $(x = x_1)$ and (ii) the diffuse double-layer which extends out from the plane of closest approach. Thus the potential ψ_0 ascribed to the point $x = 0$

should probably be redefined as the potential at the plane of closest approach. However, when the surface charge density is relatively small this effect may be neglected.

(*d*) Ion-binding and changes in the activity coefficients relative to the values of the bulk ions have not been allowed for. Although such effects almost certainly will occur, they are difficult to quantify and introduce into the classical approach.

DIFFUSE DOUBLE-LAYER CONCEPT AND CATION-INDUCED CHANGES IN CHLOROPHYLL-FLUORESCENCE YIELD

When thylakoid membranes are carefully isolated from leaves and suspended in a cation-free medium they will retain their physiological surface cations. Studies involving neutron activation analysis (Barber 1977; H. Y. Nakatani, J. Barber & M. Minski, unpublished results) have demonstrated that, within the intact chloroplast, Mg^{2+} is likely to be the cation normally at the membrane surface. This finding is consistent with the fluorescence yield being at a maximum when unwashed thylakoids are placed in a cation-free medium. Why then does the addition of small amounts of univalent cations to the thylakoids in this condition bring about a decrease in fluorescence and why does subsequent addition of high concentrations of univalent cations mimic the effect of adding back small amounts of bivalent cations? To answer these questions it is necessary to understand what happens when cations exchange in the diffuse layer adjacent to the thylakoid membranes. For this reason we re-write equations (7) and (8) to take account of solutions of mixed electrolytes.

If we define C_b' and C_b'' as the bulk concentrations of symmetrical univalent and bivalent cations, equation (7) can be rewritten as equation (9) and,

$$\frac{d\psi_x}{dx} = \pm \left(\frac{8\pi RT}{\varepsilon}\right)^{\frac{1}{2}} \left[4C_b'\sinh^2\left(\frac{F\psi_x}{2RT}\right) + 4C_b''\sinh^2\left(\frac{F\psi_x}{RT}\right)\right]^{1/2} \qquad (9)$$

after rearrangement, equation (8) becomes equation (10).

$$2C_b''\cosh^2\left(\frac{F\psi_0}{RT}\right) + C_b'\cosh\left(\frac{F\psi_0}{RT}\right) - \left(2C_b'' + C_b' + \frac{\sigma^2}{2A^2}\right) = 0 \qquad (10)$$

where $A = (RT\varepsilon/2\pi)^{1/2}$.

Equation (10) can be used to calculate how the surface potential ψ_0 varies as the bulk concentration of ions is changed. Moreover by using equation (1) and the calculated values of ψ_0, one can find out the concentration of ions at the surface for various electrolyte mixtures. Above it has been argued

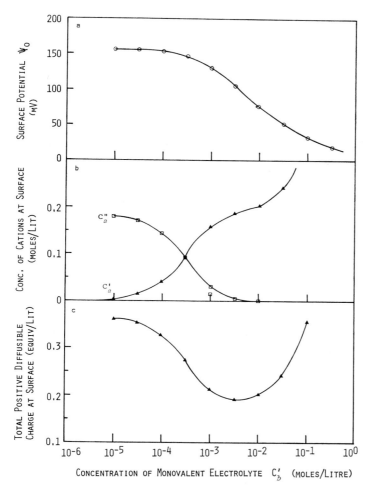

FIG. 4. (a) Surface potential ψ_0 calculated from equation (10) assuming a fixed charge density
(σ) of 2.5 $\mu C/cm^2$ on a membrane surface bathed in a solution containing Z-Z type univalent
and bivalent salts (C_b' and C_b'', respectively): C_b'' was kept constant at 10^{-6} mol/l while C_b' was
varied from 10^{-5} to 10^{-1} mol/l. (b) Concentrations of univalent C_s' and bivalent C_s'' cations
at the membrane surface were calculated with the values of ψ_0 in (a) and equation (1). (c)
Positive space charge density at the membrane surface in equiv./l.

that the high-fluorescence state observed in 'cation-free' medium occurs when
bivalent cations are the major counter ions at the membrane surface. For
the sake of the argument, let us assume that the 'cation-free' medium contains
low concentrations of univalent (10^{-5} mol/l) and bivalent (10^{-6} mol/l) salts

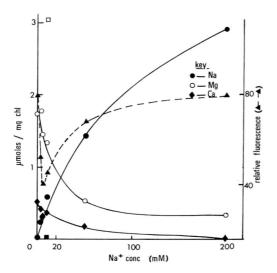

FIG. 5. Simultaneous determination of chlorophyll fluorescence and cationic concentration for pea thylakoids on increasing the external Na^+ concentration: unwashed thylakoids were initially suspended in a cation-free medium consisting of 0.1M-sorbitol brought to pH 7.0 with a small quantity of Tris. The concentrations of Na^+, Mg^{2+} and Ca^{2+} associated with membrane surfaces were measured, after washing with the cation-free medium, by neutron activation analysis (irradiation flux 1.5×10^{12} thermal neutron s^{-1} cm^{-2} for 8 min) (H. Nakatani, J. Barber & M. Minski, unpublished results).

and the surface-charge density σ is 2.5 $\mu C/cm^2$ on the thylakoids. As can be seen in Fig. 4a and 4b, application of equations (1) and (10) shows that in this medium ψ_0 is large and bivalent cations are predominantly attracted to the surface. As the concentration of univalent salt is increased, the value of ψ_0 decreases and surface bivalent cations are replaced by univalent cations. In these same conditions chlorophyll fluorescence would have passed through a minimum before returning back to its maximum value. This is emphasized in Fig. 5 where direct measurements of chlorophyll fluorescence and the bivalent (Mg^{2+}, Ca^{2+}) and univalent (Na^+) cation concentrations associated with thylakoids have been measured on the same sample.

At first sight the reduction of ψ_0 and the cation-exchange curves do not explain the associated changes in chlorophyll fluorescence. To find an explanation for the fluorescence effects one must consider the total positive charge at the membrane surface for the various electrolyte mixtures. This has been done (Fig. 4c); the positive charge at the surface goes through a minimum similar to that seen with fluorescence. The position of the minimum is sensitive to the values of surface-charge density used and we chose the value of 2.5 $\mu C/cm^2$ to give a minimum corresponding to a univalent cation concentration necessary to induce the low-fluorescence state.

In the above we have estimated the positive-charge density immediately adjacent to the membrane surface, that is where $x = 0$. A more satisfactory approach is to obtain values of the space charge density ($d^2\psi_x/dx^2$) for various distances from the membrane surface. This is possible: integration of equation (9) by numerical methods with a computer gives values of ψ_x as a function of distance and, after the appropriate differentiation steps, the plot shown in Fig. 6 can be made. The minimum observed in the space charge density profile is not apparent more than about 0.5 nm from the membrane surface.

The concept that the high- and low-fluorescence states correspond to high- and low-diffusible positive-charge density immediately adjacent to the membrane surface can also explain the differential and antagonistic effects between univalent and bivalent cations (see Barber *et al.* 1977).

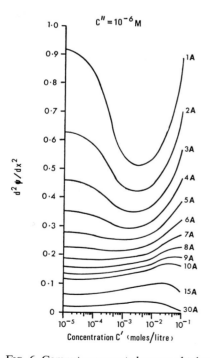

FIG. 6. Computer-generated curves obtained by numerical integration of equation (9) showing how the space charge density ($d^2 \psi/dx^2$) varies as a function of distance from the membrane surface for different univalent salt concentrations (C_b') and with C_b'' kept constant at 10^{-6} mol/l: 1 Å = 0.1 nm.

DISCUSSION OF RESULTS

The foregoing consideration of the cation-induced changes in chlorophyll-fluorescence yield strongly suggests that they are due to electrical effects at the membrane/solution interface. The controlling parameter seems to be the density of diffusible positive charge adjacent to the membrane surface. Just how far this effect extends from the membrane surface is not clear but the estimate of less than 1.0 nm is probably unrealistic and comes about because the theory used does not allow for ionic size or for the existence of a compact double layer. The estimated surface-charge density of 2.5 $\mu C/cm^2$ is equivalent to 1 electronic charge per 6.25 nm^2 and seems a reasonable figure when compared with those calculated from the particle electrophoresis measurements. As can be seen in Fig. 2, at neutral pH the electrophoretic mobility (u) of thylakoids was 2.3×10^{-4} cm^2/(V s). Using equation (11) we can calculate the zeta potential (ζ) which is the potential at the plane of shear relative to the bulk solution; Ω is the viscosity of the suspension medium.

$$\zeta = \frac{4\pi\Omega u}{\varepsilon} \tag{11}$$

By assuming $\zeta = \psi_0$ and using equation (8) we can calculate the charge density at the shear plane. Using the data in Fig. 1, we find that the charge density is 1 electronic charge per 12.25 nm^2. Since the plane of shear does not correspond with the membrane surface this calculation will underestimate the true surface charge (Haydon 1961), making the value assumed above an acceptable estimate.

By what mechanism can the transfer of energy between different types of chlorophyll–protein complexes be controlled by changes in positive-charge density at the surface? The explanation for this lies in the established theories for the aggregation of charged colloids (Overbeek 1952, 1977). Essentially, charged surfaces are exposed to two forces: attractive van der Waals' forces and repulsive electrostatic forces. Thus the spacing between two similarly charged surfaces or particles depends on the balance of these two forces. When there is good screening of the charged surface (for negatively-charged surfaces this occurs when the adjacent positive-charge density is high) particles or surfaces may be drawn together (see Fig. 7a and Fig. 7b). With poor screening the opposite is true. Thus, according to this theory, unwashed thylakoids may be expected to stack in cation-free media and unstack when the positive-charge density adjacent to the membrane surfaces is lowered by adding small amounts of univalent cations; they should restack when small amounts of bivalent cations or large amounts of univalent cations are added back. Such changes in stacking do occur (Gross & Prasher 1974).

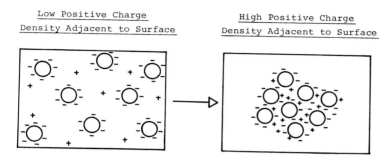

(a) Particle aggregation in membrane

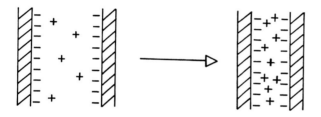

(b) Membrane stacking

(c) Change in repulsive force between fixed surface
charges.

FIG. 7. Three posibilities of how an increase in the positive space charge density adjacent to membrane surfaces can induce changes in their organization.

Although the cation-induced changes in spillover have been implicated with thylakoid stacking (see Williams 1977) there is some evidence which suggests that this is not the case (Telfer *et al.* 1976). Another explanation based on the same principle is that charged pigment–protein complexes aggregate in the membrane when electrostatic screening is at a maximum (see Fig. 7a). As originally observed by Goodenough & Staehelin (1971) and described by Staehelin & Arntzen in this volume, membrane protein complexes do aggregate in the thylakoids in those conditions when the positive space charge density at the surface is maximum. The ability of membrane protein complexes to segregate and aggregate has recently been considered by Chapman *et al.* (1977) who argue that lipid phase-changes control such events. However, most structural lipids of the thylakoids are electrically neutral and, owing to the degree of unsaturation of their acyl groups, give rise to a fluid membrane at normal temperature (Anderson 1975). Thus it seems reasonable to conclude that electrostatic effects of the type described above control the lateral movement of intrinsic protein complexes of the thylakoid membrane. For completeness, I should point out that any changes in the density of diffusible positive charge adjacent to the membrane surface could also change the forces acting between the fixed negative charges, as depicted in Fig. 7c. This latter effect could also induce conformational changes in the membrane.

How can these proposed changes in membrane organization explain changes in energy transfer and take account of the fact that thylakoid stacking and particle aggregation are associated with specific regions of the membrane surface (Staehelin 1976)? There is evidence that thylakoid stacking only occurs when the light-harvesting chlorophyll *a/b* pigment–protein complex is present (see Anderson 1975; Arntzen *et al.* 1976). Moreover, of the two basic types of particles observed on freeze-fracturing of thylakoid membranes, this light-harvesting complex is associated with larger particles which have diameters ranging from 11.0 nm to 17.0 nm and are preferentially distributed in the stacked portions of the membrane (see Staehelin & Arntzen, this volume). However, as I mentioned, electrophoretic mobility studies indicate that the overall level of negative charge on the thylakoid membrane is not changed by the presence or absence of the chlorophyll *a/b*–protein complex. Thus the cation-induced segregation of particles into distinct regions of the membrane could be explained if smaller particles, of about 8.0 nm diameter, also seen in freeze-fracture studies (Staehelin 1976), are negatively charged. The shielding of these negative charges, by increasing the concentration of diffusible positive charge at the surface, may possibly encourage separation of the two types of particles to different regions of the membrane, with membrane fusion occurring at the most hydrophobic areas. If the smaller particles have only PS I activity, which is often assumed, then the change from a random to a segregated distribution of the two particle types could lead to an overall decrease in energy transfer from the light-harvesting chlorophyll a/b complex to the PS I chlorophylls which make up the whole lamellae system.

Whether this dynamic mechanism for controlling spillover to PS I is near to the truth awaits further experimental evidence. Nevertheless, a detailed analysis of the related cation-induced changes in chlorophyll fluorescence emphasizes that the mechanism involves electrical rather than chemical events and predicts ultrastructural membrane reorganization which can be observed in electron microscopy.

ACKNOWLEDGEMENTS

I acknowledge the participation of my colleagues, H. Nakatani, G. F. W. Searle, A. Telfer, J. Mills and A. Love in the development of the ideas and experimental evidence presented in this paper. I also thank the Science Research Council and the EEC Solar Energy Research Programme for their financial support.

References

ANDERSON, J. M. (1975) The molecular organisation of chloroplast thylakoids. *Biochim. Biophys. Acta 416*, 191–235

ARNTZEN, C. J., ARMOND, P. A., BRIANTAIS, J. M., BURKE, J. J. & NOVITZKY, W. P. (1976) Dynamic interaction among structural components of the chloroplast membrane in chlorophyll-protein, reaction centers and photosynthetic membranes. *Brookhaven Sym. Biol. 28*, 316–336

BARBER, J. (1976) Ionic regulation in intact chloroplasts and its effect on primary photosynthetic processes, in '*The Intact Chloroplast*', vol. 1 *Topics in Photosynthesis* (Barber, J., ed.), pp. 89–134, Elsevier, Amsterdam

BARBER, J. (1977) Energy conversion and ion fluxes in chloroplasts, in *Fertiliser Use and Production of Carbohydrates and Lipids (13th Coll. Int. Potash Inst.)*, pp. 67–77

BARBER, J., MILLS, J. & LOVE, A. (1977) Electrical diffuse layers and their influence on photosynthetic processes. *FEBS (Fed. Eur. Biochem. Soc.) Lett. 74*, 174–181

BARBER, J., SEARLE, G. F. W. & TREDWELL, C. J. (1978) Picosecond time resolved study of $MgCl_2$ induced chlorophyll fluorescence yield changes from chloroplasts. *Biochim. Biophys. Acta 501*, 232–245

BERG, S., DODGE, S., KROGMANN, D. W. & DILLEY, R. A. (1974) Chloroplast grana membrane carboxyl groups. Their involvement in membrane association. *Plant Physiol. (Bethesda) 53*, 619–627

BONAVENTURA, C. & MYERS, J. (1969) Fluorescence and oxygen evolution from *Chlorella pyrenoidosa. Biochim. Biophys. Acta 189*, 366–383

BUTLER, W. L. (1978) Energy distribution in the photochemical apparatus of photosynthesis. *Annu. Rev. Plant Physiol. 29*, 345–378

CHAPMAN, D., CORNELL, B. A. & QUINN, P. J. (1977) Phase transitions, protein aggregation and a new method for modulating membrane fluidity, in *Biochemistry of Membrane Transport (FEBS Symp. 42)* (Semenza, G. & Carafoli, E., eds.), pp. 72–85, Springer-Verlag, Berlin

DELAHAY, P. (1965) *Double Layer and Electrode Kinetics*, Wiley, New York

GOODENOUGH, U. W. & STAEHELIN, L. A. (1971) Structural differentiation of stacked and unstacked chloroplast membranes. *J. Cell Biol. 48*, 594–619

GRAHAME, D. C. (1953) Diffuse double layer theory for electrolytes of unsymmetrical valence types. *J. Chem. Phys. 21*, 1054–1064

GROSS, E. L. & HESS, S. (1973) Monovalent cation induced inhibition of chlorophyll *a* fluorescence: antagonism by divalent ions. *Arch. Biochem. Biophys. 159*, 832–836

GROSS, E. L. & PRASHER, S. H. (1974) Correlation between monovalent cation induced decreases in chlorophyll *a* fluorescence and chloroplast structural changes. *Arch. Biochem. Biophys. 164*, 460–468

HAYDON, D. A. (1961) The surface charge of cells and some other small particles as indicated by electrophoresis. 1. The zeta potential surface charge relationship. *Biochim. Biophys. Acta 50*, 450–457

HAYDON, D. A. (1964) Electrical double layers and electrokinetics, in *Recent Progress in Surface Science*, vol. 1 (Danielli, J. F., Pankhurst, K. G. A. & Riddiford, A. C., eds.), pp. 94–158, Academic Press, New York

HOMANN, P. (1969) Cation effects on fluorescence of isolated chloroplasts *Plant Physiol. (Bethesda) 44*, 932–936

MILLS, J. D. & BARBER, J. (1978) Fluorescence changes in isolated broken chloroplasts and the involvement of the electrical double layer. *Biophys. J. 21*, 257–272

MURATA, N. (1969*a*) Control of excitation transfer in photosynthesis. I. Light induced change of chlorophyll *a* fluorescence in *Porphyridium cruentum*. *Biochim. Biophys. Acta 172*, 242–251

MURATA, N. (1969*b*) Control of excitation transfer in photosynthesis. II. Magnesium ion dependent distribution of excitation energy between two pigment systems in spinach chloroplasts. *Biochim. Biophys. Acta 189*, 171–181

MURATA, N. (1971) Effects of monovalent cations on light energy distribution between two pigment systems of photosynthesis in isolated spinach chloroplasts. *Biochim. Biophys. Acta 226*, 422–432

MYERS, J. (1971) Enhancement studies in photosynthesis. *Annu. Rev. Plant Physiol. 22*, 289–312

NOBEL, P. S. & MEL, H. C. (1966) Electrophoretic studies of light induced charge in spinach chloroplasts. *Arch. Biochem. Biophys. 113*, 695–702

OVERBEEK, J. TH. G. (1952) Electrochemistry of the double layer, in *Colloid Science* Vol. 1, (Kruyt, H. R., ed.), pp. 115–193, Elsevier, Amsterdam

OVERBEEK, J. TH. G. (1977) Recent developments in understanding of colloid stability. *J. Colloid Interface Sci. 58*, 408–422

STAEHELIN, L. A. (1976) Reversible particle movements associated with unstacking and restacking of chloroplast membranes *in vitro*. *J. Cell Biol. 71*, 136–158

TELFER, A., NICOLSON, J. & BARBER, J. (1976) Cation control of chloroplast structure and chlorophyll *a* fluorescence yield and its relevance to the intact chloroplast. *FEBS (Fed. Eur. Biochem. Soc.) Lett. 65*, 77–83

WILLIAMS, W. P. (1977) The two photosystems and their interactions, in *Primary Processes of Photosynthesis, Topics in Photosynthesis*, vol. 2 (Barber, J., ed.), pp. 99–147, Elsevier, Amsterdam

Discussion

Amesz: Why do you think that the stacking process is not important in determining the fluorescence yield?

Barber: It is possible to change the fluorescence yield without affecting thylakoid stacking. For example, when Mg^{2+} ions are added to a suspension of chloroplasts, the fluorescence yield does not change until one darkens the samples whereupon the fluorescence yield rises to the maximum and quenches down again.   This is more akin to the physiological control of spillover. To raise the yield in the light one has to add DCMU or an uncoupler (see Fig. 3.14 in Barber 1976).   Electron micrographs of the chloroplasts before and after addition of the Mg^{2+} ions in the light showed unstacked and stacked membranes, respectively,—i.e. the stacking changed but the fluorescence yield did not (see Telfer *et al.* 1976).

Cogdell: How quickly do the particles seen by Dr Staehelin aggregate compared with the change in fluorescence that you see on adding the cations?

Barber: According to Dr Staehelin's published results, reaggregation takes a long time but I am not certain that the gross changes in organization that he reports reflect the more subtle changes that give rise to the chlorophyll-fluorescence effects.

Staehelin: The gross changes in the membrane structure—the stacking and the aggregation of the particles—are too slow to account for the relatively rapid response to bivalent cations that you find in the fluorescence measurements. How long after you add Mg^{2+} ions do the fluorescence properties change?

Barber: It takes a couple of minutes for the fluorescence yield to reach a maximum.

Staehelin: That is fast compared to the structural reorganization that I observe.

Porter: A movement of 5.0 nm is all that is needed for energy transfer. That is much less than you see, Dr Staehelin.

Staehelin: Yes.   The fluorescence changes that we see are not directly related to membrane stacking and to the aggregation of the particles. I agree with Dr Butler that the fluorescence changes are probably related to changes within these larger particles within the complexes themselves.

Porter: But may it not just be the first 5.0 nm movement apart of the membranes that is sufficient for unstacking of the membranes?

Staehelin: Unstacked and unfolded membranes can be over 100 nm apart.

Porter: Yes; even if it takes a minute for them to move about 100 nm apart but only a few seconds for them to move 5.0 nm apart, it may be the same phenomenon.

Staehelin: Two experimentally unstacked membranes may rapidly form several point contacts on addition of cations, and these contact points may also aggregate quickly (see Fig. 17, p. 165), but it takes at least 10–30 min (depending on temperature conditions) for the membranes to adhere together over long distances.   It may take up to 1 h to get the normal 60% stacked membranes of control chloroplasts.

Clayton: If it were the first 5.0 nm of unstacking that caused the changes, the last 5.0 nm of restacking would be important, in which case there would be a long lag between the initiation of restacking and the onset of a change in the fluorescence.

Barber: I predict that, if there are charged particles and the membrane is fluid enough, then in principle these particles should come closer together on addition of salts, owing to the screening effect.

Joliot: In that case, Dr Staehelin, do you observe any change in the average distance between the particles at different salt concentrations?

Staehelin: That is a point I wanted to raise, too. Figs. 7*a* and *b* show low and high positive charge density. Let us suppose that these charged particles are surrounded by many other uncharged particles. When the charge density increases, why should the charged particles aggregate in the centre if one is not using bivalent cations which could cross-link them?

Barber: But this is classical colloid chemistry; charged particles can be moved apart or together by magnesium or calcium ions without cross-linking.

Staehelin: They can come close together, but will they form a stable aggregate?

Barber: Yes.

Joliot: Is there any electron microscopic evidence for this type of behaviour?

Barber: In the intact system we are dealing with subtle changes but one does see reorganization of particles in the membrane in different salt conditions.

Staehelin: After five minutes I do not see any aggregation of particles into the first small stacked regions. Furthermore, any particle aggregation seems to depend on stacking.

Barber: Maybe you are missing the rapid and more subtle reorganization of the complexes or it might be that the reorganization is slow owing to the use of low temperatures. As I described, fluorescence studies indicate a significant activation energy for the reorganization process.

Porter: Are you talking about different particles? Dr Barber, you are describing the interface between photosystem II and photosystem I particles, aren't you?

Barber: In a general sense, yes. However, the mechanism, as emphasized by Dr Butler, probably involves changes in coupling between the light-harvesting chlorophyll *a/b*–protein complex and the photosystem II pigment–protein complex.

Porter: In Dr Staehelin's particles photosystem I and II are stuck together.

Staehelin: No, the ones I described are photosystem II units surrounded by smaller particles.

Porter: We ought to talk about the distance between photosystem I and II particles.

Clayton: For which we have different models from Drs Butler and Staehelin!

Barber: In my analysis I have used changes in chlorophyll fluorescence yield, which are generally agreed to represent changes in energy transfer between different pigment–protein complexes of the photosynthetic light-harvesting system. My analysis of the cation-induced fluorescence changes indicates an electrostatic effect, not a chemical effect with 'cross-linking' or 'rivetting' or whatever. Fig. 7 depicts three possible ways in which electrostatic screening can bring about membrane reorganization. Although they are simply predictions, both stacking and particle reorganization are experimentally observed. I suspect that changes in forces and, therefore, distances between adjacent charged protein complexes in biological membranes due to changes in ionic composition of the diffuse layer may be important in many membrane phenomena.

Porter: I want to take up Dr Clayton's point (p. 299)—how long does one have to wait after unstacking for an increase in fluorescence?

Barber: That is an important point for which I do not have a clear answer. The problem is that, when discussing attractive and repulsive forces, one must remember that attractive forces depend on distances. If the thylakoids are unstacked for a long time then the membranes move so far apart that it is much harder to bring them together. For studies of stacking and unstacking the surfaces must not be allowed to move too far apart.

Porter: In other words one has to wait a long time after unstacking for the fluorescence to return to its original high level?

Barber: It depends on the pretreatment of the membranes.

Clayton: Is there any difference between the lag time for the change from high to low fluorescence and that for the reverse process? Even if the lags differed by only a few ms, that could be interesting because the movements themselves are small.

Barber: Unfortunately this type of experiment has not yet been done.

Staehelin: Incidentally, when we fix the control membranes with glutaral-dehyde before freeze-fracture, we observe the same partitioning of photo-system I and II elements as we do without glutaraldehyde fixation. But when we put the membranes through an unstacking–restacking procedure and then fix with glutaraldehyde before freeze-fracturing, the partitioning of the particles between the two halves of the membrane is changed: 60–70% of the photo-system II particles cling to the side of photosystem I. That fact indicates a truly-internal reorganization of those particles.

Junge: It is difficult to reconcile all the data and their interpretations. Dr Butler, you assume that all photosystem I units are equivalent. If a fraction of these centres were located in a different neighbourhood from the rest, how would that affect your predictions? Would your model then be able to fit Dr Staehelin's data? Is there any experimental evidence for such a heterogeneity?

Butler: We cannot rule that out.

Barber: The charged particles that move in the model shown in Fig. 7 could be photosystem I particles. Evidence to support this view comes from studies on the electrophoretic mobility of chloroplasts with and without the light-harvesting chlorophyll *a/b*–protein complex isolated from the wild-type and a mutant barley (chlorina-f2 mutant). The fact that both types of thylakoids had identical mobilities means that the charges are carried by something other than the chlorophyll *a/b* complex. Thus salt additions may cause some photosystem I pigment–protein complexes to go into the stromal lamellae.

Butler: We have some evidence which indicates that stacking *per se* has little influence on energy distribution. Dark-grown bean leaves which had been partially opened by a repetitive series of brief flashes form thylakoids which align in parallel arrays but do not fuse to any appreciable extent. The adjacent thylakoids fuse, however, with as little as five minutes of continuous light (Strasser & Butler 1976). The fluorescence and energy-distribution properties of these leaves are not altered by the process of fusion (Strasser & Butler 1978). However, on further greening in continuous light these properties do change as the chlorophyll *a/b*–protein complex accumulates.

Paillotin: Can you exclude the possibility that the exchange of energy between the two photosystems is controlled by one particular chlorophyll–protein complex?

One configuration of the complex may allow efficient transfer between the two systems whereas another configuration may suppress the transfer.

Barber: I do not see why not; my argument proposes control by an electrostatic mechanism but it could well include conformational changes within a single type of pigment complex.

Paillotin: One can imagine such control of spillover for *Porphyridium*, with its large photosystem I and small photosystem II.

Joliot: The connection between the 680 nm pigments and P700, and consequently between photosystem II and the traps of photosystem I, passes through a special protein complex that contains a few chlorophyll molecules which absorb at around 690 nm (P. Delepelaire, personal communication, 1977). In a mutant which lacks this protein the photosystem I centres are disconnected from the antenna.

Butler: Is this chlorophyll in photosystem II?

Joliot: No; it is probably the short wavelength part of the CP I complex. In the mutant which lacks the 690 nm protein complex, photosystem I is active but the rate of excitation transfer from the antenna to photosystem I is slow.

Clayton: Can these effects be seen in broken-cell suspensions of *Senedescemus* or *Chlamydomonas*? There may be mutants that lack both C-705 and P700.

Butler: It is difficult to prepare the particles from those mutants. Mutants that lack the chlorophyll *a/b*–protein do not respond to Mg^{2+} ions.

Clayton: That implies that changes in the direct connection between photosystems II and I as described in your model are not involved in the effects caused by Mg^{2+} ions.

Butler: No; photosystems II and I are always connected but how well they are connected could be controlled by bivalent cations.

Clayton: Have you investigated mutants that lack photosystem I? These effects could have nothing to do with spillover.

Butler: As mutants that lack P700 also lack C-705 and the 730 nm fluorescence at low temperatures, it would be difficult to do the experiment.

Clayton: Is the shorter-wavelength fluorescence influenced by salts?

Joliot: Possibly; back excitation transfer from the remaining part of photosystem I antenna to photosystem II can be observed. Bennoun & Jupin (1976) observe an increase in the sensitization of photosystem II by the 680 nm pigment in a mutant which lacks CP I.

Staehelin: With regard to testing the effects of Mg^{2+} ions on thylakoids from blue-green or red algae, I must warn that, when one isolates those membranes, most of the phycobilisomes fall off—nobody yet knows how to keep them on in significant numbers. Another feature that might be useful to take advantage of is the well defined organization of the photosystem II-type particles in algae like *Porphyridium* and certain blue-greens—the photosystem II elements are tightly packed into beaded rows. The same can be said about the associated phycobilisomes. In other algae (e.g. *Spermothamnion*) the photosystem II elements and the phycobilisomes are randomly organized. To study energy transfers between the different elements one could compare these two types of algae.

Duysens: For inducing a shift to pigment state 1 in intact cells, illumination instead of Mg^{2+} ions is used with light that primarily excites photosystem I. From various experiments we concluded that the shift to the pigment state 1 occurs if there are oxidized components in the electron-transfer chain between Q and P700 inclusive (Duysens & Talens 1969). Neither Q^- nor Q has any influence—nor does DCMU. The redox state of one or more of these oxidized components seems to determine the effect. I wonder whether the positive charge on, for instance, P700, cytochrome or plastocyanin may cause this shift of the component you mentioned (it may be attached to a protein) and cause the spillover effect. In chloroplast preparations the influence of charge (perhaps of Mg^{2+} or other cations) may be similar.

Barber: Although a local charge effect of this type cannot be ignored it seems unlikely to me. When we illuminate chloroplasts we detect no change in electrophoretic motility (Nakatani *et al.* 1978) (even though Nobel & Mel [1966] have reported a slight change). This seems to indicate that light-induced charge-separation processes do not contribute significantly to the overall surface charge density of the thylakoid membrane.

Duysens: I am not talking about surface charges but charges on the inside of the thylakoid (perhaps inside the membrane).

Barber: The cation-induced effects reflect the properties of the charges on the outer surface. I have no information about charges on the inner surface or, indeed, within the membrane itself.

References

BARBER, J. (1976) in *The Intact Chloroplast* (Barber, J., ed.) (*Topics in Photosynthesis*, vol. 1), Elsevier, Amsterdam

BENNOUN, P. & JUPIN, H. (1976) Spectral properties of System I-deficient mutants of *Chlamydomonas reinhardi*. Possible occurence of uphill energy transfer. *Biochim. Biophys. Acta* *440*, 122–130

DUYSENS, L. N. M. & TALENS, A. (1969) Reactivation of reaction center II by a product of photoreaction I, in *Progress in Photosynthesis Research (Proceedings of the International Congress on Photosynthesis Research, 1968)* (Metzner, H., ed.), vol. 2, pp. 1073–1081, H. Laupp Jr., Tübingen

NAKATANI, H. Y., BARBER, J. & FORRESTER, J. A. (1978) *Biochim. Biophys. Acta,* in press

NOBEL, P. S. & MEL, H. C. (1966) Electrophoretic studies of light-induced charge in spinach chloroplasts. *Arch. Biochem. Biophys. 113*, 695–702

STRASSER, R. J. & BUTLER, W. L. (1976) The correlation of absorbance changes and thylakoid fusion with the induction of oxygen evolution in bean leaves greened by brief flashes. *Plant Physiol. 58*, 371–376

STRASSER, R. J. & BUTLER, W. L. (1978) Energy coupling in the photosynthetic apparatus during development, in *Proceedings of the Fourth International Congress on Photosynthesis* (Hall, D.O., Coombs, J. & Goodwin, T. W., eds.), pp. 527–536, Biochemical Society, London

TELFER, A., NICHOLSON, J. & BARBER, J. (1976) Cation control of chloroplast structure and chlorophyll-*a* fluorescence yield and its relevance to intact chloroplasts. *FEBS (Fed. Eur. Biochem. Soc.) Lett. 65*, 77–83

Quenching of chlorophyll fluorescence and photochemical activity of chloroplasts at low temperature

C. P. RIJGERSBERG, A. MELIS, J. AMESZ and J. A. SWAGER

Department of Biophysics, Huygens Laboratory, University of Leiden, The Netherlands

Abstract　Fluorescence kinetics and emission spectra of pea and spinach chloroplasts were studied between 294 and 4.2 K. In the presence of $MgCl_2$ the fluorescence-induction curves were sigmoidal between 294 and 180 K but they lost their inflection points at lower temperature. In the absence as well as in the presence of Mg^{2+} ions, analysis of the kinetics of the area over the induction curve revealed two different components, indicating the existence of two different types of reaction centres at all temperatures. The 'rate constants' for these centres were nearly independent of temperature between 294 and 200 K, but showed a sharp decrease on further cooling.

Emission spectra at low temperature showed the previously observed bands at 685, 695 and 735 nm. All three bands showed a considerable increase on cooling between 180 and 4.2 K, but with different temperature-dependence. The amplitude of the 695 band became constant below about 50 K, whereas the 685 emission increased markedly in this region. The relative proportion of the so-called variable fluorescence was almost the same at 685 and 695 nm, both at 80 and at 5 K. The data are discussed in terms of changes of energy-transfer rates on cooling.

The properties of photosynthetic material at sub-zero temperatures have been investigated extensively in many laboratories (for a review see Amesz 1977). These studies include measurements of light-induced electron transport and energy transformation of photosynthetic pigments. A considerable proportion of these investigations deals with the fluorescence properties of chlorophyll and bacteriochlorophyll at or above liquid-nitrogen temperature (for reviews, see Goedheer 1972; Butler 1977), but only few data are available on fluorescence below 77 K (Cho *et al.* 1966; Cho & Govindjee 1970).

We present the results of recent experiments on the fluorescence properties of pea and spinach chloroplasts at temperatures down to 4.2 K. Measurements of emission spectra and fluorescence kinetics indicate substantial changes in the efficiencies of energy transfer between different chlorophyll *a* molecules on cooling. In addition to this the kinetics indicate, in agreement with earlier measurements (Melis & Homann 1976; Rijgersberg & Amesz 1978), the existence of two different types of system II reaction centres with different rates of formation of stable photochemical products.

MATERIALS AND METHODS

Chloroplasts from spinach *(Spinacia oleracea)* or pea leaves *(Pisum sativum)* were obtained as described elsewhere (Visser *et al.* 1974). They were suspended in a medium containing 50mM-Tricine (*N*-tris[hydroxymethyl]methylglycine), pH 7.8, 0.4M-sucrose, 10mM-KCl and 5mM-MgCl$_2$ and stored in the dark on ice until use. For some experiments the MgCl$_2$ was omitted as indicated. Before the measurements the suspension was mixed with a solution of 0.4M-sucrose in glycerol in a ratio 45:55 (v/v) to prevent crystallization on cooling.

Chlorophyll fluorescence at low temperature was measured as described elsewhere (Rijgersberg & Amesz 1978). The emitted light passed a mono-chromator or an interference filter, supplemented by coloured glass filters to absorb scattered excitation light. The monochromator was set at a band-width of 3 nm; the half-width of the interference filters was 10–15 nm. The sample was contained in a perspex vessel of 1 mm thickness. The fluorescence was detected at the illuminated surface of the cuvette. The emission spectra were calibrated by means of a standard ribbon-filament lamp, operated at a known temperature; they are plotted in relative units of energy per wavelength interval.

RESULTS AND INTERPRETATION

Two types of system II reaction centres

The yield of chlorophyll fluorescence in intact cells and isolated chloroplasts increases strongly on lowering the temperature (Goedheer 1972 [review]; Cho *et al.* 1966; Cho & Govindjee 1970; Rijgersberg & Amesz 1978). For spinach chloroplasts the increase in the region 120 to 4.2 K is mainly due to an increase of the initial level of fluorescence on illumination (F_0). The increase of the maximum level (F_{max}) and particularly of the increment, the so-called variable fluorescence (F_v), was less pronounced (Rijgersberg & Amesz 1978). These effects could be explained, at least in part, as a decrease in efficiency of energy transfer to, or of trapping of energy at, the reaction centre.

The kinetics of the increase in fluorescence yield on illumination of spinach chloroplasts can be analysed in two exponential components, both at 80 and

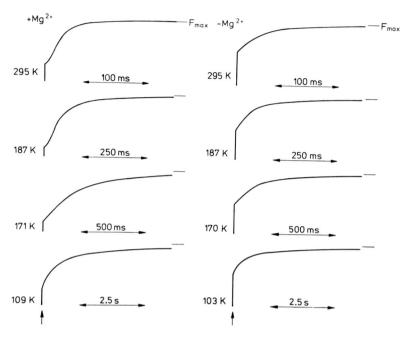

FIG. 1. Time course of fluorescence of pea chloroplasts at different temperatures. Isolated chloroplasts were suspended with ($+ Mg^{2+}$) and without ($- Mg^{2+}$) 5mM-MgCl$_2$: chlorophyll concentration, 50 μg/ml; DCMU [3-(3,4-dichlorophenyl)-1,1-dimethylurea], 10 μmol/l. Excitation was provided in the green region of the spectrum by a combination of a CS 4–96 and a CS 3–96 Corning filter; the incident intensity was 2 mW/cm^2. The fluorescence emission was detected with an AL 686 nm Schott interference filter. All maximal fluorescence values F_{max} have been normalized to the same relative height. Note the absence of an S-shape in the induction kinetics with Mg^{2+} below about 170 K and the high F_v/F_{max} ratio at 187 and 171 K.

at 5 K (Rijgersberg & Amesz 1978). The time constants for the two components differed by a factor of about eight at 5 K and of about 15 at 80 K. These results indicate that there are two different types of reaction centres of system II, with different rates of charge stabilization. They also indicate that the intensity of fluorescence and the concentration of quencher are linearly related at temperatures of 80 K and below (see also Visser 1975), otherwise analysis into exponentials would not have been possible. One may thus conclude that transfer of energy between photosynthetic units of photosystem II (Joliot & Joliot 1964) is negligible at low temperature.

Fig. 1 compares the kinetics of fluorescence of pea chloroplasts at various temperatures. Above about 180 K, the fluorescence increase showed the

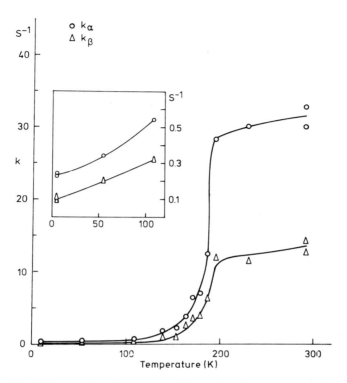

FIG. 2. Temperature-dependence of the system II energy-conversion rate constants k_α ($\bigcirc$) and k_β ($\triangle$) obtained with chloroplasts suspended in the presence of 5mM-MgCl$_2$. The values of k_α and k_β were calculated from a logarithmic plot of the growth of the area $A(t)$ over the fluorescence induction curve with time according to the equation $k = t^{-1} \ln \{[A_{max} - A(t)]/A_{max}\}$ (see Melis & Homann 1975). For other conditions see Fig. 1.

well known sigmoid behaviour characteristic for energy transfer between units. Therefore, at those temperatures a simple analysis of the fluorescence rise curve in exponential components could not be applied. For this reason we measured the change of the area above the curve as a function of time to obtain information about the kinetics of different types of system II reaction centres. As discussed by Melis & Homann (1975) this kind of analysis can be used, provided certain conditions are met, for measurements of the relative rate of charge separation when the relation between quencher concentration and fluorescence yield is non-linear. The results of such an analysis are shown in Fig. 2. As was previously demonstrated at room temperature (Melis & Homann 1976) the experiments indicated the existence of two types of reaction centres with different rates of energy conversion at all temperatures between

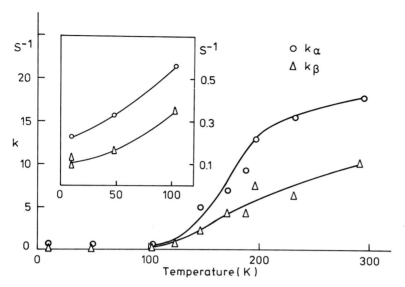

FIG. 3. Temperature-dependence of the rate constants k_α (○) and k_β (△) obtained with chloroplasts suspended in the absence of any bivalent cations. For other conditions see Fig. 1.

4.2 and 294 K. The centres with the high and low rate of conversion were called α and β centres, respectively (Melis & Homann 1976). There was little change in the rates, as expressed by the rate constants k_α and k_β, and in the relative proportions of the two types of reaction centres on cooling from 294 to 200 K, but near 190 K the two rate constants abruptly declined. This decline was accompanied by the disappearance of the S-shape in the fluorescence rise curve. Both rate constants increased with the light-intensity at all temperatures. As Fig. 1 shows, the ratio F_v/F_{max} initially showed an increase on cooling. It reached a maximum at 190 K and decreased again below that temperature.

In the absence of Mg^{2+} ions the S-shape was absent at any temperature (Fig. 1), and there was a similar but somewhat less pronounced decline of the rate constants below 200 K (Fig. 3). The effect of Mg^{2+} on the rate constants seemed to disappear below about 120 K. The relative contribution of the β centres to the maximum fluorescence increased between 200 and 120 K and remained at a level of 85%, independent of the presence of Mg^{2+} ions, on further cooling (Fig. 4).

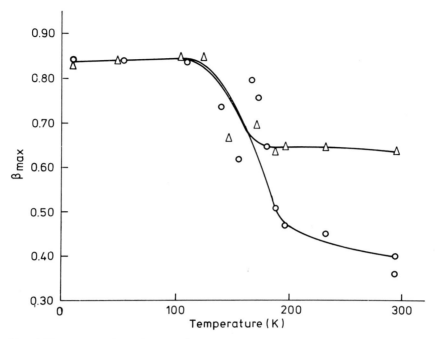

FIG. 4. Temperature-dependence of β_{max} of chloroplasts suspended in the presence ($\bigcirc$) and absence ($\triangle$) of 5mM-MgCl$_2$: β_{max} represents the amplitude of the relative contribution of the slow kinetic component (β-component) in the overall growth of the area over the fluorescence induction curve. For other conditions see Fig. 1.

Emission spectra between 200 and 4.2 K

The foregoing results indicate that two different types of system II reaction centres can be distinguished kinetically at all temperatures between 294 and 4.2 K. Above about 190 K the difference in the reaction rates of the two types of centres is probably due to a difference in size of the associated pool of antenna pigments (Melis 1975); below that temperature the rates of secondary donor reactions needed for charge stabilization seem to be an important factor as well (Rijgersberg & Amesz 1978). In addition to this it appears that changes occur in the rates of energy transfer between pigments and pigment units, as indicated by the fluorescence kinetics and by changes reported in the emission spectra on cooling (Murata et al. 1966).

Since data on the emission spectra of photosynthetic material below 77 K are scarce (Cho et al. 1966), we measured the fluorescence spectra of spinach and pea chloroplasts at various temperatures between 200 and 4.2 K. Fig. 5

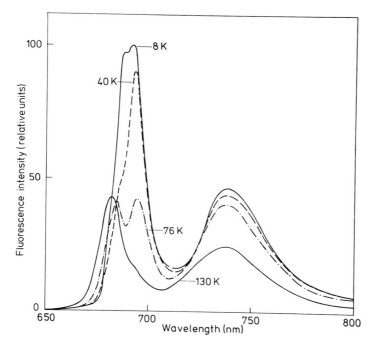

FIG. 5. Emission spectra of the maximal fluorescence yield of spinach chloroplasts (50 μg chl/ml) in the presence of 5mM-MgCl$_2$: excitation wavelength 430 nm. The absorbance of the sample was 0.04 at 680 nm at room temperature.

shows results obtained with spinach chloroplasts between 130 and 8 K. At 130 K the emission was mainly in two bands, centred at 683 and 736 nm. On further cooling, these bands intensified, the band at 683 nm shifted by a few nm towards longer wavelength, and a strong emission band at 694 nm developed in addition. Similar, but less detailed observations were reported by Cho *et al.* (1966) for *Chlorella pyrenoidosa* in the temperature region 4.2–77 K. Three emission bands at about 685, 695 and 715–735 nm have been observed in various photosynthetic organisms at liquid-nitrogen temperature (Goedheer 1972; Murata *et al.* 1966); minor emission bands have been also reported (Litvin *et al.* 1976). We shall call the chlorophyll species emitting at these three wavelengths F_{685}, F_{695} and F_{735}, respectively.

The temperature-dependence of the emission bands is shown in Fig. 6. For F_{685} and F_{695} these temperature-dependences were clearly different: the emission intensity of F_{685} was almost constant between 130 and 50 K and showed a marked increase at lower temperature, whereas the F_{695} emission

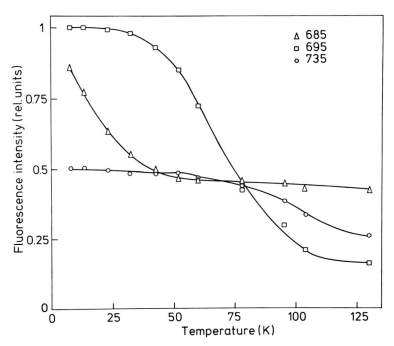

FIG. 6. Temperature-dependence of the emission intensity at 685 ($\triangle$), 695 ($\square$) and 735 nm ($\bigcirc$) of spinach chloroplasts (conditions as for Fig. 5).

showed a strong increase between 100 and 40 K and reached a constant level below that temperature. The emission by F_{735} showed the strongest increase between 180 and 130 K (not shown) and became almost constant below 80 K. Its temperature-dependence appeared to be unrelated to that of F_{695}, in contrast to what was concluded for *Chlorella* (Cho *et al.* 1966). Qualitatively similar results were obtained with pea chloroplasts (Fig. 7).

The spectra of Fig. 5 are those of F_{max}. In order to obtain information about the emission spectra of chloroplasts in the F_0 and intermediate levels and to test possible explanations for the temperature dependence of the various maxima, we measured the fluorescence kinetics on illumination at different wavelengths and temperatures. Table 1 and Fig. 8 summarize some of the results obtained. Qualitatively the results for spinach and pea chloroplasts were the same. The ratios F_v/F_{max} were similar for F_{685} and F_{695}, whereas the ratio was significantly lower at 735 nm, both at liquid helium and, as previously observed by Murata (1968) and others, at liquid-nitrogen temperature. These observations are in agreement with the hypothesis (Murata *et al.* 1966;

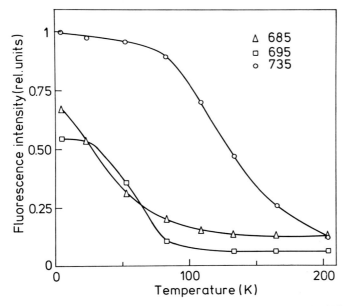

FIG. 7. Temperature-dependence of the emission intensity at 685 ($\triangle$), 695 ($\square$) and 735 nm ($\bigcirc$) of pea chloroplasts measured in the absence of $MgCl_2$. For other conditions see Fig. 5.

Gasanov & Govindjee 1974) that the emission of F_{735} comes mainly from system I and show that there is little transfer of energy from system II to this pigment, at least in the presence of Mg^{2+} ions. At each temperature, the kinetics of the variable fluorescence were the same at 685, 695 and 735 nm within the error of measurement (Fig. 8). This indicates that the bands of F_{685} and F_{695} are present in almost the same proportion in the initial (F_0) and variable (F_v) fluorescence and also that the emission spectra of the pigments

TABLE 1

Ratio of the variable over the maximal fluorescence emission (F_v/F_{max}) at various temperatures and wavelengths for spinach chloroplasts (for conditions, see Fig. 5)

Wavelength (nm)	Temperature (K)		
	102	78	5
685	0.77	0.70	0.44
695	0.76	0.74	0.48
735	0.30	0.18	0.08

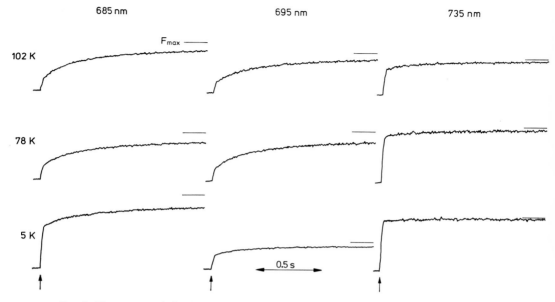

FIG. 8. Fluorescence-induction kinetics at different wavelengths and temperatures of spinach chloroplasts (conditions as for Fig. 5).

TABLE 2

Effect of artificial quenchers on the fluorescence yield: dibromothymoquinone (DBMIB) and potassium ferricyanide quenching of the relative maximal fluorescence emission F_{max} of spinach chloroplasts at different wavelengths and temperatures. For each of the conditions used, the emission intensities at 685 and 695 nm were normalized with respect to the emission at 735 nm. 0.1mM-Dibromothymoquinone gave about 50% quenching at 735 nm, both at 6 and 78 K.

Temperature	Quencher concentration (mmol/l)		Relative fluorescence yield		
(K)	DBMIB	$K_3Fe(CN)_6$	685 nm	695 nm	735 nm
78.0			91	100	100
78.3	0.1		70	69	100
78.7	0.2		76	69	100
78.2	0.4		110	100	100
79.1		0.2	65	63	100
6.0			180	220	100
6.4	0.1		200	215	100
4.8	0.2		230	220	100
5.0	0.4		200	215	100
5.6		0.2	165	185	100

associated with the two types of reaction centres discussed above resemble each other.

Addition of ferricyanide quenched the emission at 685 and 695 nm more strongly than that at 735 nm (Table 2). This finding agrees with the notion that ferricyanide lowers F_{max} by producing additional quenching at the reaction centres (Butler 1977). Dibromothymoquinone produced a strong quenching of all three emission bands, indicating that this substance also quenches system I fluorescence.

DISCUSSION OF RESULTS

Kinetics

Our results indicate that the heterogeneity observed in the fluorescence properties of chloroplasts at room temperature (Melis & Homann 1976) and that reported at low temperatures (Rijgersberg & Amesz 1978) can be observed throughout the whole temperature region. This suggests a basic property of system II that is reflected in the rate of energy conversion. As shown by Figs. 2–4 the relative contribution of the two types of reaction centres to the fluorescence and the rates of energy conversion were almost constant between room temperature and about 200 K, but below that temperature for both kinds of system II reaction centres an abrupt decline in the rate constants (k_α and k_β) occurred. In the same temperature region the S-shape in the fluorescence induction curve disappeared. Bonnet et al. (1977) reported the existence of more than two components in the fluorescence kinetics at 77 K. However, the optical properties of their samples were not well defined, giving an uncertainty about the actinic intensity throughout the sample.

Above 200 K, the difference in rates of the two types of reaction centres is probably due to a difference in the pool sizes of the antenna pigments associated with these centres (Melis 1975). The cause of the abrupt lowering of the rate constants below 190 K is not fully understood. A change in the fluorescence properties of chloroplasts below about 200 K has been observed before (e.g. Butler et al. 1973; Amesz et al. 1973) and measurements on the photooxidation of cytochrome b_{559} suggest that the effect is related to a change in the secondary donor reactions of system II and in the identity of the fluorescence quencher involved (Butler et al. 1973; Visser 1975). Our data indicate that the effect is also accompanied by a change in the efficiency of energy transfer between chlorophyll molecules and that the decrease of the rate constants can be partly explained by an increase in energy losses by other processes, such as fluorescence. However, measurements of the increase of fluorescence yield induced by saturating light flashes (Visser 1975; Amesz et al. 1973; Rijgersberg & Amesz 1978) indicate that the main effect is not due to a lower efficiency of transfer of energy to the reaction centre but to a lower efficiency of charge stabilization in the reaction centre proper. This effect is thought to be caused by a low rate of secondary electron donation as compared to the back reaction of the primary products.

It is generally assumed that the effects of Mg^{2+} ions on fluorescence are caused by an enhanced efficiency of energy transfer between chlorophyll molecules belonging to different photosynthetic units and pigment systems (Murata 1969). This may explain why the effects of Mg^{2+} on the rate constants (Figs. 2 and 3) and relative contributions of the α and β centres (Fig. 4) disappear below about 120 K, since below this temperature energy transfer between system II units appears to become negligible anyway. The change in the ratio of the two kinetic components on lowering the temperature is not well understood. It might be attributed to dissimilar changes in the efficiency of charge stabilization at the two types of system II reaction centres. Alternatively, it might indicate a differential temperature effect on the fluorescence yield or energy transfer properties of the pigment systems associated with the two kinds of reaction centres.

Spectra

The shape of the low-temperature emission spectra and the temperature-dependence of the principal emission bands indicate that drastic changes in the efficiencies of energy transfer between different chlorophyll molecules occur on lowering the temperature. First, energy transfer between photosynthetic units of photosystem II appears to cease between 190 and 150 K. In the temperature region 200 to 80 K the system I band at 736 nm (F_{735}) develops. This band may be due to a chlorophyll species absorbing at relatively long wavelength(s) and that normally transfers its excitation energy directly or indirectly to the reaction centre of system I, but is unable to do so at low temperature and acts as an energy sink emitting fluorescence instead. As discussed by Robinson (1964) 'uphill' transfer of energy will be inhibited on cooling at a temperature that depends on the difference in energy between the molecules involved, owing to lack of thermal energy.

The complicated temperature-dependences of the emission bands of the system II pigments, F_{685} and F_{695}, are more difficult to explain. Fluorescence excitation spectra obtained with *Chlorella* (Cho & Govindjee 1970) suggest that F_{685} is identical to a chlorophyll absorbing at 678 nm. This indicates that F_{695} is a chlorophyll absorbing at longer wavelengths and present in smaller amounts than the major antenna chlorophylls. Cho & Govindjee

(1970) proposed that F_{695} was a reaction-centre chlorophyll and identical to P_{680}, the primary electron donor of system II. This hypothesis is not incompatible with the strong emission by F_{695} in the F_{max} spectrum at low temperature, but it is more difficult to explain the probably equally-prominent band of F_{695} in the spectrum of the initial fluorescence (F_0) in this way. Therefore we assume, like Butler & Kitajima (1975), that F_{695} is an antenna pigment which may be closely associated with the reaction centre and that at least part of the energy transfer from F_{685} to the reaction centre occurs *via* F_{695}. The strong increase in F_{695} fluorescence between 100 and 40 K then may be explained by a similar mechanism as used to explain the increase in F_{735} fluorescence, by inhibition of 'uphill' transfer from F_{695} to F_{685} or to F_{685} and the reaction centre. This inhibition appears to be complete at about 30 K. As Table 1 indicates, the relative components of variable fluorescence are about the same in the emission of F_{695} and F_{685}, also at 5 K. This suggests that, at least at this temperature, energy transfer from F_{695} to the reaction centre does not occur, but such transfer may occur at higher temperature. The strong increase in emission of F_{685} below about 40 K (Fig. 6) or 80 K (Fig. 7) cannot be explained by an inhibition of a direct transfer of energy from F_{685} to the reaction centre (unless one assumes that there is also quenching, e.g. at the reaction centre, in F_{max}), because the increase occurs in both F_0 and F_{max}. It might be explained by inhibition of transfer to a pigment intermediary between F_{685} and the reaction centre.

The above explanations are hypothetical only and should be checked by further experiments. In particular, we do not know if the strong emissions near 687 nm at 8 K and at 684 nm at 76 K (Fig. 5) are due to the same chlorophyll species emitting at 682 nm at 130 K, or if a new emission band is created at longer wavelength at the expense of the 682 nm emission on lowering the temperature. More detailed analysis of the emission spectra of both the initial and variable fluorescence may provide an answer to these and other questions.

ACKNOWLEDGEMENTS

We thank Dr G. W. Canters and M. Noort for loan of the cryostat. The investigation was supported by the Netherlands Foundations for Biophysics and for Chemical Research (SON), financed by the Netherlands Organization for the Advancement of Pure Research (ZWO). A. Melis was supported by a long-term fellowship from the European Molecular Biology Organization (EMBO).

References

AMESZ, J. (1977) Low temperature reactions in photosynthesis, in *Research in Photobiology* (Castellani, A., ed.), pp. 121–128, Plenum Press, New York

AMESZ, J., PULLES, M. P. & VELTHUYS, B. R. (1973) Light induced changes of fluorescence and absorbance in spinach chloroplasts at —40 °C. *Biochim. Biophys. Acta 325*, 472–482

BONNET, F., VERNOTTE, C., BRIANTAIS, J. M. & ETIENNE, A. L. (1977) Kinetics of chlorophyll fluorescence at 77 K in *Chlorella* and chloroplasts. Effects of CCCP, ferricyanide and DCMU. *Biochim. Biophys. Acta 461*, 151–158

BUTLER, W. L. (1977) Chlorophyll fluorescence: a probe for electron transfer and energy transfer, in *Encyclopedia of Plant Physiology New Series*, vol. 5 (Trebst, A. & Avron, M., eds), pp. 149–167, Springer Verlag, Berlin

BUTLER, W. L. & KITAJIMA, M. (1975) Energy transfer between photosystem II and photosystem I in chloroplasts. *Biochim. Biophys. Acta 396*, 72–85

BUTLER, W. L., VISSER, J. W. M. & SIMONS, H. L. (1973) The kinetics of light induced changes of C-550, cytochrome b_{559} and fluorescence yield in chloroplasts at low temperature. *Biochim. Biophys. Acta 292*, 140–151

CHO, F. & GOVINDJEE (1970) Low temperature (4–77 °K) spectroscopy of *Anacystis*; temperature dependence of energy transfer efficiency. *Biochim. Biophys. Acta 216*, 151–161

CHO, F., SPENCER, J. & GOVINDJEE (1966) Emission spectra of *Chlorella* at very low temperature (–269 ° to –196 °). *Biochim. Biophys. Acta 126*, 174–176

GASANOV, R. A. & GOVINDJEE (1974) Chlorophyll fluorescence characteristics of photosystem I and II from grana and photosystem I from stroma lamellae. *Z. Pflanzenphysiol. 72*, 193-202

GOEDHEER, J. C. (1972) Fluorescence in relation to photosynthesis. *Annu. Rev. Plant Physiol. 23*, 87–112

JOLIOT, A. & JOLIOT, P. (1964) Etude cinétique de la réaction photochimique libérant l'oxygène au cours de la photosynthèse. *C.R. Acad. Sci. Paris 258*, 4622–4625

LITVIN, F. F., SINESHCHEKOV, V. A. & SHUBIN, V. V. (1976) Investigation of energy migration between native forms of chlorophyll at —196 °C by the method of sensitized fluorescence. *Biofizika 4*, 669–675

MELIS, A. (1975) *Properties of Photosystem II as Determined by the Kinetic Analysis of Chloroplast Fluorescence*, Ph. D. Thesis, The Florida State University, Tallahassee, Florida, USA

MELIS, A. & HOMANN, P. H. (1975) Kinetic analysis of fluorescence induction in 3-(3,4-dichlorophenyl)-1,1-dimethylurea poisoned chloroplasts. *Photochem. Photobiol. 21*, 431–437

MELIS, A. & HOMANN, P. H. (1976) Heterogeneity of photochemical centers in system II of chloroplasts. *Photochem. Photobiol. 23*, 343–350

MURATA, N. (1968) Fluorescence of chlorophyll in photosynthetic systems. IV. Induction of various emissions at low temperatures. *Biochim. Biophys. Acta 162*, 106–121

MURATA, N. (1969) Control of excitation transfer in photosynthesis. II. Magnesium ion dependent distribution of excitation energy between two pigment systems in spinach chloroplasts. *Biochim. Biophys. Acta 189*, 171–181

MURATA, N., NISHIMURA, M. & TAKAMIYA, A. (1966) Fluorescence of chlorophyll in photosynthetic systems. III. Emission and action spectra of fluorescence – three emission bands of chlorophyll *a* and the energy transfer between two pigment systems. *Biochim. Biophys. Acta 126*, 234–243

RIJGERSBERG, C. P. & AMESZ, J. (1978) Changes in light absorbance and chlorophyll fluorescence in spinach chloroplasts between 5 and 80 K. *Biochim. Biophys. Acta 502*, 152–160

ROBINSON, G. W. (1964) Quantum processes in photosynthesis. *Annu. Rev. Phys. Chem. 15*, 311–348

VISSER, J. W. M. (1975) *Photosynthetic Reactions at Low Temperatures*, Thesis, State University of Leiden, The Netherlands

VISSER, J. W. M., AMESZ, J. & VAN GELDER, B. F. (1974) EPR signals of oxidized plastocyanin in intact algae. *Biochim. Biophys. Acta 333*, 279–287

Discussion

Butler: Dr Amesz, you have saved me doing several experiments that I was planning. I am very happy with your conclusions.

Amesz: We are not completely happy ourselves, because we find it difficult to explain by a simple model both the different temperature dependences of F_{695} and F_{685} (Figs. 5 and 6) and the similar F_v/F_{max} ratios for the two pigments, even at 5 K (Table 1). The simplest explanation seems to be that energy transfer from F_{695} to the reaction centre becomes blocked between 100 and 40 K as the temperature is lowered. This may suggest a different pathway of energy transfer from F_{685} to the reaction centre, which does not involve F_{695} and is less sensitive to temperature lowering.

Butler: I assume that the 695 nm fluorescence comes from traps in the antenna chlorophyll of photosystem II (analogous to the 730 nm fluorescence of photosystem I at low temperature) and that these traps become more fluorescent at temperatures below 77 K. The increasing yield of the 695 nm fluorescence at the very low temperature should compete with the use or dissipation of energy in the rest of the photochemical apparatus. Specifically, as the temperature is decreased below 77 K, the yield of photochemistry of photosystem II should decrease (as you clearly showed in your measurements of the rate of photoreduction of C-550) and the yield of energy transfer from photosystem II to I should decrease. Your measurements of the relative extent of the variable-yield part of the 730 nm fluorescence confirm that energy transfer from photosystem II to I is considerably less at the lower temperatures.

Amesz: As judged from the rate at which the fluorescence yield increases on turning on the light (Figs. 2 and 8) the rate of system II photochemistry does not seem to drop dramatically between 100 and 5 K. But my point is that the fluorescence of F_{695} does not increase any further below 30 K. If this means that transfer to the reaction centre and perhaps also to F_{685} is completely blocked, one has to assume another pathway from F_{685} (supposedly a chlorophyll of the light-harvesting complex) to the reaction centre at low temperature.

Butler: F_{695} is a fluorescing species; it competes for all the energy available in the photosystem II unit. As that energy is used by fluorescence at 695 nm all the other yields for energy use are decreased. The yield, then, for back-transfer from photosystem II to F_{685} will decrease just as the yield for energy transfer from photosystem II and I or the yield for energy use by photosystem II is decreased for photochemistry.

Robinson: Why does the 735 nm fluorescence flatten off instead of decreasing on lowering the temperature?

Amesz: We suppose that below 75 K the fluorescence yield of F_{735} is essentially maximal because it can no longer transfer energy to other pigments. The emission intensity should remain constant on further cooling if energy transfer to F_{735} does not change significantly. But we have no independent way to check that.

Butler: The rate of oxidation of P700 as a function of temperature would be an indicator of that.

Katz: Some aspects of chlorophyll structure may be relevant to the discussion of intersystem crossing and to the fluorescence yield as a function of temperature. Lowering the temperature from 77 to 4 K affects the rotation of the methyl groups in a major way; the change is easily detected in the endor spectrum. Could some of the fluorescence changes be related to the fact that some rotations are frozen out, or that the *in vivo* chlorophyll species are immobilized to different degrees at different temperatures, i.e., that they are in a different matrix?

Porter: I should have thought not. The proportion of fluorescence and intersystem crossing is about 1/3. We have studied the temperature-dependence of triplet formation down to 193 K with no observed change in the triplet yield. The effect of rigidity on intersystem crossing is usually only encountered when the principal chromophore can change; a floppy molecule such as a cyanin dye gives an enormous change when one inserts a cross-link to make it planar. In a porphyrin, for example, the chromophore is so rigid that the effect is small. I imagine that the intersystem crossing is little changed, down to 77 K.

Katz: I was referring to experiments on the fluorescence yield of stilbene (M. Kasha, personal communication) as a function of the viscosity of the medium: an ethanolic solution of stilbene at room temperature does not fluoresce but a glycerol solution is highly fluorescent. This difference is due to restriction of *cis–trans* isomerization in the more viscous medium.

Porter: That (like the cyanins) is a special case; surely chlorophyll is one of the most rigid molecules as far as its chromophore is concerned?

Katz: Low-frequency vibrational modes can be associated with flexing of the chlorophyll macrocycle or change in the position of the central magnesium atom relative to the nitrogen atoms of the macrocycle. Probably the flexibility of the chlorophyll macrocycle is related to tetrahedral carbon atoms in ring IV; as judged from models, it is much more flexible than is the porphyrin ring. Possible changes in the coordination number of the central magnesium atom from five to six when the temperature is lowered must also be taken into consideration.

Robinson: When one cools an organic crystal from 77 to 4 K the separation of the molecules decreases tremendously. I realize that we are not dealing here with organic crystals but surely the density will increase and so tend to increase the rate of energy transfer? If this were not a uniform effect, traps could be created, as Dr Butler pointed out.

Amesz: With regard to Dr Katz's point, Avarmaa *et al.* (1977) reported that the fluorescence lifetime of chlorophyll *a* dissolved in ether depends little on temperature and increases from 7 to 9 ns between 300 and 4.2 K.

Katz: That just describes the lifetimes of the excited state; it does not tell us much about the relative amounts of radiative and non-radiative processes unless the fluorescence yields are specified.

Porter: It does; with more intersystem crossing the lifetime will shorten.

Katz: But such factors as rotation of methyl groups are non-radiative ways of dissipating energy. From Dr Amesz's comment, the increase in lifetime at 4 K might be associated with the freezing out of rotation of methyl groups.

Porter: You would see it in the lifetime. The only way the yield can change while the lifetime does not is to have more than one species.

Seely: If F_{685} fluorescence increases, but not at the expense of any other band, it must be at the expense of some non-radiative process, that is, a triplet or a photochemical reaction.

Amesz: I suppose it is photochemistry.

Porter: Why couldn't energy be transferred to a trap which has been created at the low temperature, for example—two molecules coming together to form a dimer?

Seely: The 695 and 735 nm fluorescences do not decrease. Perhaps there is an invisible trap—i.e. a non-radiative process in competition with the 685 nm fluorescence.

Porter: Is the total yield at 4 K greater than 30%?

Amesz: By comparison of the emission intensities of spinach chloroplasts at 8 K and room temperature, we obtained an estimated fluorescence yield in the presence of Mg^{2+} ions of 45–50% in the 685–695 nm bands for quanta absorbed by system II and of 55% for system I emitted in the 735 nm band(s). This estimate was based on an assumed fluorescence yield of 13% at room temperature for system II.

Porter: In other words, the probability that a photon put in results in fluorescence is 100%? I am astonished that it is more than 30%. If so, the lifetime should lengthen to about 20 ns.

Amesz: We assumed for our estimation that at 430 nm equal amounts of light are absorbed by systems I and II at 8 K.

Butler: The increase of the 695 band is tremendous. The kinetics at 730 nm shows linearity between lifetime and yield (over the liquid N_2 range).

Porter: Nobody has observed in any system *in vivo* a fluorescence lifetime of chlorophyll longer than 6.5 ns.

Seely: So there might be competition between transfer to the reaction centres of photosystems I and II.

Amesz: Below 30 K where F_{695} emission no longer increases there is no decrease in F_{735}.

Duysens: Photosystem I chlorophyll may start fluorescing at temperatures below 100 K.

Reference

AVARMAA, R., SOOVIK, T., TAMKIVI, R. & TÖNISSOO, V. (1977) Fluorescence life-times of chlorophyll *a* and some related compounds at low temperatures. *Studia Biophys.* 65, 213–218

Transfer and trapping of excitation energy in photosystem II

L. N. M. DUYSENS

Department of Biophysics, Huygens Laboratory of the State University, Leiden, The Netherlands

Abstract   The fluorescence yield of chlorophyll *a* of system II in spinach chloroplasts as a function of the fraction q^- of reaction centres in the weakly trapping state PQ^-, with reduced acceptor Q^-, and reduced primary donor chlorophyll, P, of the reaction centre, is described by the function $\phi = a/(1 - pq^-)$, *a* and *p* being constants (Van Gorkom *et al.* 1978); *p* was estimated to be 0.74.   By special treatment and additions it was ascertained that the donor complex (S-states, see below) was in the reduced state.

Three models of pigment systems have been considered: separate units; units with a boundary limiting energy transfer; and the matrix or pigment bed model, which was found to describe the experimental data. The following supplementary assumptions were made: $k_{tf} > k_t > k'_t > 0$. The rate constant k_{tf} is that for electronic excitation transfer from a chlorophyll *a* molecule (or reaction-centre chlorophyll) to the surrounding chlorophyll molecules; k_t and k'_t are rate constants for trapping at the reaction centres in the states PQ and PQ^-, respectively. From this model and additional data such as fluorescence yield *in vivo* and *in vitro*, k_t was estimated to be 4×10^{11} s^{-1} and $k'_t = 7.1 \times 10^{10}$ s^{-1}; $k_{tf} > 10^{12}$ s^{-1}.

In dark-adapted *Chlorella*, a series of curves representing changes in fluorescence yield as a function of time in a succession of six 16 μs xenon flashes spaced at 3 s crossed at one point. It is concluded from this and other observations that in the states S_2 and S_3 (with two or three oxidizing equivalents in the donor complex of system II) a certain fraction of the reaction centres occurs in a special conformational state. In this state electron transfer and, possibly, energy transfer to P^+ are appreciably decreased.

The pigment systems of photosynthesizing cells consist of a small fraction (about 1%) of so-called reaction-centre chlorophyll molecules, P, which are photochemically active, and of antenna molecules which absorb the light energy and transfer this energy in the form of electronic excitation energy to the reaction centre.

The first step after absorption of light is internal conversion into the lowest excited singlet state. If the light is absorbed by a so-called accessory pigment, it is first transferred to an antenna chlorophyll *a* molecule and subsequently transferred *via* other chlorophyll *a* molecules to P. In purple bacteria, bacteriochlorophyll plays an analogous role to that of chlorophyll *a* in oxygen-evolving organisms (see *e.g.* Duysens 1964).

During the transfer and trapping processes, losses occur amongst other things in the form of fluorescence and luminescence of chlorophyll *a*. Luminescence, sometimes called delayed fluorescence, is emission from the lowest singlet state caused by back reactions of the photochemical reactions.

In purple bacteria and in pigment system II of oxygen-evolving organisms the fluorescence or luminescence yield of chlorophyll *a* depends strongly on the state of the reaction centres. On excitation of P, electron transfer occurs to an acceptor molecule Q, probably a plastoquinone (see reactions 1):

$$Z P^*Q \rightarrow Z P^+Q^- \rightarrow Z^+ PQ^- \qquad (1)$$

P* is excited P; P is a dimer chlorophyll *a*; Q is reduced to the anion radical Q^-; Z is an electron donor which has not yet been chemically identified. Under 'normal' regimes of illumination the P^+ concentration is small because the reaction $ZP^+ \rightarrow Z^+P$ is rapid. If all reaction centres are in the state PQ^- (the reaction centre is 'closed') the chlorophyll *a* fluorescence yield is several times higher than in the state PQ. PQ is called an open trap or reaction centre, because the energy is trapped more easily by PQ than by the 'closed' reaction centre PQ^-. Thus the fluorescence yield ϕ is an increasing function of the fraction q^- of reaction centres in the state PQ^-; q^- is the ratio of the number of reaction centres in state PQ^- to the total number of reaction centres. If there are only two states, PQ (fraction q) and PQ^-, then $q + q^- = 1$. The function $\phi = \phi(q^-)$ may be determined experimentally by measuring the fluorescence yield and the concentration of PQ^-. PQ^- may be measured by means of absorption difference spectroscopy. This function on the one hand provides important information about structural aspects of the pigment system, essential for understanding details of the energy-transfer process and trapping and, on the other hand, is needed for calculating quantitatively and precisely the kinetics of PQ^- and other states of the reaction centre from measurements of the fluorescence yield kinetics. The measurement of the fluorescence kinetics is, because of its better signal-to-noise ratio, at the moment the most precise and for many applications the only method available: the fractional absorption changes due to the transition $PQ \rightarrow PQ^-$ are about 1000-times smaller than the fractional fluorescence changes.

In this paper I shall discuss various physical models. It is surprising that one mathematically simple model appears consistent with experiments on pigment system II and on pigment systems of purple bacteria, provided experimental conditions are suitably chosen so as to avoid complications, such as contribution of more than one pigment system or of more than two states of the reaction centre.

MODELS FOR PIGMENT SYSTEMS TRANSFERRING AND TRAPPING EXCITATION ENERGY

Separate units

We assume that system II consists of identical 'units' each containing several chlorophyll a molecules and one reaction centre, and that no energy transfer between units occurs. Then ϕ will be a linearly-increasing function of q^-: $\phi = \phi_0 + aq^-$ where a is a constant, and $\phi_0 = \phi(0)$ is the fluorescence yield, when all the reaction centres are in the oxidized state PQ (see Figs. 1 and 2, curve 1).

If this system were heterogeneous (e.g. consisting of two types of units, with one type having more antenna chlorophylls and a high fluorescence yield in the state PQ$^-$), then, as the following arguments show, the fluorescence yield should in general no longer be a linear function of q^-. After a period of darkness all reaction centres are in state PQ. We assume that illumination causes reduction of Q at a rate proportional to the number of antenna

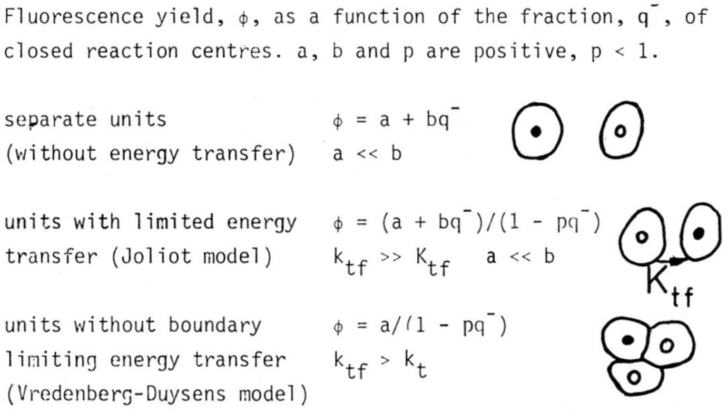

Fluorescence yield, ϕ, as a function of the fraction, q^-, of closed reaction centres. a, b and p are positive, p < 1.

separate units (without energy transfer)	$\phi = a + bq^-$ $a \ll b$	
units with limited energy transfer (Joliot model)	$\phi = (a + bq^-)/(1 - pq^-)$ $k_{tf} \gg K_{tf}$ $a \ll b$	
units without boundary limiting energy transfer (Vredenberg-Duysens model)	$\phi = a/(1 - pq^-)$ $k_{tf} > k_t$	

FIG. 1. Three models with different mutual location of photosynthetic units, each consisting of antenna chlorophyll a molecules with a trapping (open) or weakly-trapping (closed) reaction centre. The equation $\phi = \phi(q^-)$ gives the fluorescence yield as a function of the fraction, q^-, of weakly-trapping reaction centres (see text).

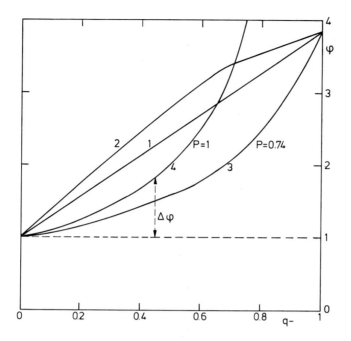

FIG. 2. Curves giving $\phi = \phi(q^-)$ with $\phi(0) = 1$ (see legend of Fig. 1) for models: (1) with identical separate units; (2) with separate units of unequal size (see text); (3) matrix of pigment bed; $k_{tt} > k_t$; value for p is taken from experimental curve of Fig. 3; (4) matrix model for three dimensions; $k_t > k_{tt}$ and $k'_t = 0$; the curve is only exact if excitation losses before trapping are small (for small q^- values).

molecules in the type of unit in which it is present. Then Q in the larger units will be reduced first. The rapid formation of the large, highly fluorescent units in the state PQ⁻ will cause an initially more-steeply-rising curve $\phi = \phi(q^-)$ (see Fig. 2, curve 2). In other words, in heterogeneous systems with units of different sizes, the $\phi(q^-)$ curve will have a negative curvature. We assume now that after the light has been shut off the rate of re-oxidation is the same for all Q. Then the $\phi = \phi(q^-)$ will be a linear function of q^- (Fig. 2, curve 1). The same linear relationship may be expected if Q is reduced chemically instead of by light.

The curves observed so far experimentally have a positive curvature like curve 3 in Fig. 2. Thus models for separate units are inconsistent with the experimental data. This is not only true for systems consisting of identical units but *a fortiori* for units of unequal size, since the effects of inhomogeneity will be qualitatively the same as for systems with separate units. If possible, homogeneity of units should be checked by causing the $Q \leftrightarrow Q^-$ transitions for the study of the $\phi = \phi(q^-)$ relation, both by light and by dark reactions.

Model with limited energy transfer between units

In system II about 200 antenna chlorophyll a molecules are present in each reaction centre. We may imagine that the reaction centre is located at the centre of a protein molecule (the 'unit') in which 200 chlorophyll a molecules are embedded more or less regularly. The rate of energy transfer, k_{tf}, from one excited chlorophyll a to its (non-excited) neighbours is assumed to be the same for all chlorophyll molecules in the unit and also for the reaction centre chlorophyll, P. The transfer rate k_{tf} decreases rapidly with the distance between the chlorophyll molecules in a unit. It is of the order of 10^{12} s^{-1} or higher (see later). Two units are at such a distance that the rate of transfer from an excited unit to another unit, K_{tf}, is not negligible. To provide some physical background for a simplified mathematical model described below we assume that $k_{tf} \gg K_{tf}$. The excitation will on the average move several times through all molecules of a unit before being transferred to another unit, so that transfer to another unit becomes more or less independent of the location of the first excited chlorophyll a molecule of a unit. If the units would approach each other so that they touch (see Fig. 1, bottom), we have a different model: the matrix, lake or pigment-bed model (see later).

As far as I know there exists no exact derivation of the function $\phi = \phi(q^-)$ for a physical model as pictured above. Joliot & Joliot (1964) postulated without proof that energy transfer in a system of units can be described with the following mathematical model. Similar postulates have been used by Butler and his co-workers (see elsewhere in this volume). With a slight generalization Joliot's postulates are as follows. After excitation of a unit in the state PQ^-, it has the probability f of emitting its excitation as fluorescence, and a probability p of transferring its energy to any other unit. Similarly for units in the state PQ, the rate constants are f' and p'; $f > f'$ and $p > p'$. The unproven hypothesis is that distances between units, or at least real physical structures, can exist to which the postulates apply.

The derivation of $\phi = \phi(q^-)$ now proceeds as follows. In equation (2), ϕ

$$\phi = q^-(f + p\phi) + q(f' + p'\phi) \tag{2}$$

is the quantum yield for fluorescence or the probability that a quantum absorbed will be emitted. The right-hand side is the sum of two terms: the first is the probability that the quantum is absorbed by a unit in state PQ^-

(probability q^-) and directly emitted as fluorescence (probability f) or is transferred with probability p to *any* other unit and is emitted as fluorescence (probability ϕ). The second term is analogous to the first. One assumption is that, after transfer, the probability of fluorescence ϕ is the same, independent of which unit the transfer occurs from. Otten (1974) has shown that this assumption is a good approximation for energy-transferring units. Putting $q = 1-q^-$, we get equation (3) by solving the preceding equation for ϕ:

$$\phi = [f' + (f - f')q^-]/[1 - p' - (p - p')q^-] \tag{3}$$

If we assume that more than 90% of the energy is trapped if an open unit is excited, then $p' < 0.1$. Joliot & Joliot assumed that the open unit was a perfect trap, and we obtain their equation by substituting $p' = f' = 0$ in equation (3). They derived this expression by means of an infinite series, a method mathematically equivalent, albeit somewhat more involved than the above one (*cf.* Otten 1974). In an analogous way expressions can be obtained for the quantum yield of trapping and other processes.

The hyperbolic equation (see Fig. 1, middle) has three constants, which can be found, for example, by three $\phi(q^-)$ values. Joliot & Joliot chose $\phi(0) = 0$. The observed fluorescence in state PQ was attributed to 'dead' fluorescence not caused by system II.

The relation $\phi = \phi(q^-)$ can also be written with rate constants like K_{tf} instead of probabilities as parameters.

Matrix model

The pigment system in the matrix or pigment-bed model consists of antenna and reaction-centre chlorophyll *a* molecules, which are regularly distributed in a two- or three-dimensional matrix. One may also consider this model as consisting of densely packed units, without boundaries for energy transfer (Fig. 1, bottom).

For this system exact solutions can be calculated by computer for various given values of rate constants for energy transfer (k_{tf}), trapping (k_t), fluorescence (k_f), and loss processes such as fluorescence *plus* internal conversion (k_l) and for special distributions of pigments (see Knox 1977 for a discussion and references). In general $\phi = \phi(q^-)$ cannot be expressed as a simple function of q^- and the parameters.

For a first discussion of the experimental data, relatively simple equations for special conditions based on results by Montroll (1969) will be written down. These equations are good approximations, if loss processes during transfer are negligible, which probably is the case if most reaction centres are in the state PQ and if the trap does not transfer back the excitation energy.

In a two-dimensional square lattice with N antenna molecules to each reaction centre and with equal probabilities of jumping to the nearest neighbouring points and no loss processes, the average number of jumps $\bar{n}$ before the reaction centre is hit is given by (4) to good approximation (Montroll 1969) and by (5) for a (three-dimensional) cubic lattice.

$$\bar{n} = (0.318N \ln N + 0.195N)N/(N-1) \tag{4}$$

$$\bar{n} = 1.516N \tag{5}$$

The number of antenna molecules per reaction centre in either state PQ$^-$ or PQ is N_0 (that is the number of antenna molecules per unit). We assume that reaction centres in state PQ$^-$ are only weakly trapping and do not affect the energy transfer. The number of reaction centres N per perfectly-trapping centre PQ is $N = N_0/q = N_0(1-q^-)$.

The excitation will be trapped in the reaction centre without back transfer, if the trapping rate k_t is much greater than the transfer rate k_{tf}. The fluorescence yield for the three-dimensional case is given by (6) for small values of q^-;

$$\phi_1 = \tau k_f \bar{n} = a/(1-q^-) \tag{6}$$

$a = 1.5\tau k_f N_0$ and the factor $\tau = 1/(k_{tf} + k_1)$ is the average time that the excitation is on each antenna molecule.

For the two-dimensional equation $\phi_2' = \phi_2'(q^-)$ obtained in an analogous way, we find that, for $q^- < 0.4$ and $N_0 = 200$, a good approximation is equation (7).

$$\phi_2 = a(N_0)/(1 - 1.153q^-) \tag{7}$$

The ratio $r = \Delta\phi_2/\Delta\phi_2'$, which is a measure for the quality of the approximation, is close to 1 (see Fig. 2 for the definition of $\Delta\phi$). The ratio $r = 1.02$ for $q^- = 0.4$ and approaches 1.00 for smaller values of q^-.

Since, also for small values of q^-, equations (6) and (7) are not consistent with experimental data on the pigment systems measured so far, we change the condition that the trap is completely trapping, and assume, as Robinson (1966) did, that $k_t < k_{tf}$. Thus excitation passes repeatedly through the reaction centre and chlorophyll a antenna molecules, even in the state PQ. Nevertheless, if $k_t \gg N_0 k_1$, efficient trapping will occur. Further we assume that the reaction centre also in the state PQ$^-$ traps energy, but with a trapping rate $k_t' < k_t$

Since the excitation visits all chlorophyll molecules, including P, repeatedly if $k_{tf} > k_t \gg N_0 k_1$, the time τ that the excitation stays on a molecule is, to a first approximation, $1/k_{tf}$.

The fluorescence yield ϕ_3 is equal to the ratio of the quanta emitted, when excitation visits each molecule of the unit once, and the energy lost in the unit and trapped in one passage of the reaction centre (see equations 8 and 9).

$$\phi_3 = N_0\tau k_f/(N_0\tau k_1 + q\tau k_t + q^-\tau k_t') \tag{8}$$

$$\phi_3 = N_0 k_f/(N_0 k_1 + k_t q + k_t' q^-) \tag{9}$$

After putting $q = 1 - q^-$ we obtain the simple hyperbolic equation (10).

$$\phi = a/(1 - pq^-) \tag{10}$$

where

$$p = (k_t - k_t')/(k_t + N_0 k_1) \tag{11}$$

and

$$a = N_0 k_f/(k_t + N_0 k_1) \tag{12}$$

In an analogous way we can find an equation for the quantum yield of trapping or of other quantities.

One can obtain a presumably somewhat better approximation than given by equations (9)–(12), by using a method analogous to that applied to units in the preceding section. In this case 'units' is replaced by 'chlorophyll a molecules' and 'reaction centres'. In the equation thus obtained one can substitute rate constants for probabilities. A hyperbolic equation like (10) is obtained with rather complicated expressions for p and a in the equations corresponding to (11) and (12). By taking the limit for large k_{tf}, equations (11) and (12) are obtained again.

I shall discuss below evidence that for system II an interpretation consistent with various experimental data is obtained by using equation (10) and the underlying assumptions.

For the purple bacterium *Rhodospirillum rubrum* an equation analogous to (10) with $p < 1$ was found to describe the experiments (Vredenberg & Duysens 1963). However, for the mutant strain R-26 of the purple bacterium *Rhodopseudomonas sphaeroides*, Clayton & Sistrom (1966) concluded from a difference in the action spectra of bacteriochlorophyll fluorescence and P oxidation that light energy directly absorbed by the reaction centre is not transferred back with high efficiency to the antenna bacteriochlorophyll. If for this species p values smaller than 1 were found, one would have to assume that trapping also occurs by traps different from the reaction centre. If the fraction of the traps assumed to trap without back transfer of energy to reaction centres is $\alpha < 1$, instead of equation (6) we should, by applying the same method as for equation (6), obtain equation (13) for a three-dimensional matrix.

$$\phi_1 = \tau k_f \bar{n} = a/(1 - pq^-) \qquad (13)$$

where
$$a = 1.5\tau k_f N_0/(\alpha + 1)$$

and
$$p = 1/(\alpha + 1).$$

Even if all photoactive traps would be closed ($q^- = 0$), a fraction α of the excitation quanta would be lost. The quantum yields for the primary reaction appear to approach 1 for cultures of rapidly growing bacteria, thus α may be expected to be close to zero.

INTERPRETATION OF EXPERIMENTAL RESULTS

In this section we shall restrict ourselves to a discussion of the $\phi = \phi(q^-)$ relation in system II.

Recently Van Gorkom *et al.* (1978) determined directly the concentration of q^- by measuring absorption changes in the ultraviolet region, caused by

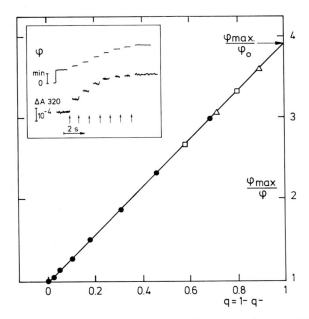

FIG. 3. Spinach chloroplasts: $\phi_0 = \phi(0) = 1$. $\phi_{max}/\phi = \phi(1)/\phi$ is plotted against $q = 1 - q^-$ for experimentally obtained points (after Van Gorkom *et al.* 1978, with a modification, see text). Insert shows absorption changes at 320 nm after light flashes, caused by reduction of the electron acceptor, Q, and corresponding changes in fluorescence yield. The ϕ_{min} was the fluorescence yield in the presence of ferricyanide.

the reduction of the reaction-centre plastoquinone to the anion radical. Also the corresponding fluorescence yields $\phi(q^-)$ were determined.

In order to ensure that only two states of the reaction centres were present, they used Tris-washed spinach chloroplasts and added DCMU (3-[3,4-dichlorophenyl]-1,1-dimethylurea) to minimize Q^- reoxidation, and tetra-phenylboron to keep the donor side of system II in the reduced state. Their results are plotted in Fig. 3; $\phi_{max} = \phi(1)$. The fact that the measured points, if ϕ_{max}/ϕ is plotted against q^-, coincide with surprisingly-good precision with a straight line shows that the simple hyperbolic relation $\phi = a/(1-pq^-)$ obtains, which is equivalent to equation (4), based on the matrix model with $k_{tf} > k_t$. For p the value $p = 0.74$ is obtained, if for ϕ_{max} the maximum fluorescence in the light is taken. The scales in the original publication were different because for ϕ_{max} the fluorescence in the presence of dithionite was taken, but the additional fluorescence increase by dithionite is presumably caused by an inactive part of the pigment (see next paragraph).

In non-treated chloroplasts and algae, there are two more or less independent subsystems of system II, which do not give the simple relationship (Melis & Homann 1975, 1976; Melis & Duysens 1978). In Van Gorkom's preparations, the fluorescence changes caused by the minor subsystem are probably suppressed, and the ϕ_0 of this subsystem is small (Van Gorkum & Melis, personal communication). This amongst other things may explain why the Joliots preferred the hyperbolic relationship $\phi = (a + bq^-)/(1-pq^-)$ and made *ad hoc* assumptions about 'dead' fluorescence in order to describe their results.

Using the equations (9) to (12), we can estimate rate constants for photosystem II. Since the *in vivo* absorption and fluorescence spectra of chlorophyll a are essentially similar to those in solution, we use the value $k_f = 6.7 \times 10^7 \, s^{-1}$, determined in solution. We use also the k_1 value for a solution: $k_1 = 2.2 \times 10^8 \, s^{-1}$. The fact that the fluorescence yield $\phi_{max} \, (q^- = 1)$ of system II is about 0.1, instead of 0.3, as in solution, may be attributed to quenching by the reaction centres in state PQ^-, thus $k_t' > 0$. Further assuming that $\phi(0) = 3\%$, we calculate k_t and k_t'.

If k_t' were zero, $\phi_{max} = \phi(1)$ would be 0.3 (the measured fluorescence yield in solution), but ϕ_0 would remain 3%. Using equation (9), we obtain (14).

$$\phi(1)/\phi(0) = 0.3/0.03 = 1/[1 - k_t/(k_t + N_0k_1)] \tag{14}$$

Solving for k_t and substituting $N_0 = 200$ and $k_1 = 2.2 \times 10^8 \, s^{-1}$, we find $k_t = 4.0 \times 10^{11} \, s^{-1}$. The values of k_t' for Van Gorkom's preparations (Fig. 3) can be calculated by means of equation (11): $p = 0.74 = (k_t - k_t')/(k_t + N_0k_1)$. Solving for k_t' we get $k_t' = 7.1 \times 10^{10} \, s^{-1}$. For system II no measurement of k_t is available, but for reaction-centre preparations of a purple bacterium

k_t was found to be larger than 2×10^{11} s^{-1}. One assumption made above was $k_{tf} > 10^{12}$ s^{-1}.

Fig. 2, curve 3, is the function $\phi = \phi(q^-)$ with $p = 0.74$, which is plotted in another way in Fig. 3. The calculated curve for complete trapping, $k_t \gg k_{tf}$ (curve 4 in Fig. 2), deviates strongly from the experimentally-found curve 3, Fig. 2 (valid for small q^-). Thus complete trapping probably does not occur, unless additional trapping in system II is assumed, lowering appreciably curve 4. In the case of complete trapping the smallest number of k_{tf} would be calculated. For $N_0 = 200$, we find $\bar{n} = 300$ (equation 5), giving as a minimum $k_{tf} = 7 \times 10^{11}$ s^{-1} for the rate constant of transfer.

In the state PQ, $\phi_0 = 0.03$; thus 3 % of the quanta is lost as fluorescence, and $(70/30) \times 3\% = 7\%$ is lost by internal conversion to the triplet state of chlorophyll. 90 % of the quanta are available for photosynthesis, but part of the energy of these quanta will be lost as heat. In the state PQ$^-$, about 12 % of the energy, absorbed by system II, is lost as fluorescence and about $(70/30) \times 12\% = 28\%$ by internal conversion to the triplet state of chlorophyll. This triplet excitation presumably is transferred to the triplet state of a carotenoid. In an indirect way (by measuring the quenching of the chlorophyll a fluorescence in system II by the carotenoid triplet), Den Haan (1977) estimated that in the state PQ$^-$ a carotenoid triplet was formed with an efficiency of about 24 %. The remaining $100 - (28 + 12) = 60\%$ of the quanta are probably trapped in the reaction centre in state PQ$^-$, and may give rise to the intermediates responsible for the 1 μs luminescence component, which are probably formed with good efficiency in the reaction centre. The evidence for this is that this luminescence component saturates at about the same intensity of the exciting laser flash at which the fluorescence increase from the state PQ is saturated (see, however, Amesz & Van Gorkom 1978), and that hydroxylamine effects on the 1 μs component and on oxygen yields in flashes are correlated (Van Best & Duysens 1977).

The hypothesis consistent with the experiments described above is that the fluorescence increase on the reduction of the acceptor Q of the major subsystem of system II is mainly determined by the rate of trapping at the reaction centre, both in state PQ and in state PQ$^-$. This subsystem may be considered as a matrix of antenna molecules in which the reaction centres are embedded.

A POSSIBLE CHANGE IN TRAPPING PROPERTIES OF THE REACTION CENTRE IN STATE PQ, CAUSED BY CHANGES IN THE S-STATE

In the experiments of Van Gorkom et al. (1978) the donor complex of the reaction centre was kept in the reduced state, so that in essence only two

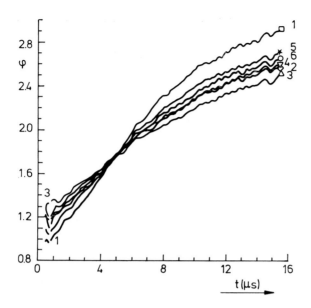

FIG. 4. Curves representing changes in fluorescence yield of chlorophyll *a* in the green alga *Chlorella*, occurring in six just-saturating 16 μs xenon flashes spaced at 3 s intervals after a dark time of 3 min. Note the isofluorescence point.

states of the reaction centre occur, S_0 PQ and S_0 PQ$^-$, in the light regime applied. Normally in the light also the states S_1, S_2 and S_3, with 1, 2 and 3 oxidizing equivalents, are present.

Quite different phenomena are observed if the conditions are such that different states of the reaction centre are present. One such experiment is shown in Fig. 4. The curves give the kinetics of the fluorescence yield in six just-saturating xenon flashes spaced at 3 s. The first flash is given after 3 min of darkness.

The most pronounced difference in kinetics is observed between the curves measured in the first and the third flash. The ϕ_0 and ϕ_{max} values in the first flash are about 1.0 and 2.9, in the third 1.3 and 2.4, respectively. This phenomenon has first been observed by Delosme (1971). A remarkable feature, which had not been observed before, is that the curves cross at one point, which may be called the isofluorescence point, in analogy with the isosbestic point for sets of absorption spectra. The ϕ value of this point is about 1.7. At a higher intensity of the flashes, the curves become steeper, but the isofluorescence point remains at about the same ϕ value.

The occurrence of an isofluorescence point indicates that the curves are in essence the sum of two curves, being caused by two 'states' of the reaction centre which have to be distinguished from the S-states and are described below. In state A, mainly present in the S-states S_0 and S_1, which are present after a few minutes of darkness, ϕ_0 is low and ϕ_{max} is high. In state B, being present in a higher proportion after one or two flashes, the ϕ_0 and ϕ_{max} have increased and decreased, respectively.

The explanation we offer for this unusual fluorescence behaviour is the following. The reaction-centre complex can have two conformations, A and B, which are different in fluorescence kinetics in a flash. We postulate that in state A more-efficient energy transfer from the antenna chlorophyll to P is possible than in state B. This explains $\phi_{0A} < \phi_{0B}$. P^+, formed in a light flash, is in state A immediately reduced by a donor Z within 1 μs; in state B the reduction of P^+ (probably by another donor D) requires 20 μs or longer. Thus in state B, P^+ remains present in the flash. For interpreting low-temperature experiments, Butler et al. (1973) assumed that P^+ is a good quencher of chlorophyll a fluorescence. This hypothesis proved also to be fruitful for interpreting correlations between the fluorescence and luminescence kinetics in a succession of xenon, respectively, laser flashes at room temperature (Duysens et al. 1975; Van Best 1977), and is used here. In state B, at the beginning of the flash, the reaction centre is in the quenching state PQ, and during the flash, owing to the slow reduction of P^+, there is a mixture of quenching states PQ and P^+Q^-. Thus the fluorescence yield of state B remains more or less at the isofluorescence value. At the beginning of the third flash not all the reaction centres are in state B, but a higher fraction than after a period of darkness. Evidence for a long-lived form of P^+ is provided by a 20 μs luminescence component with low amplitude after a first 30 ns laser flash, but of higher amplitude after the second and especially after the third flash (Duysens et al. 1975). The luminescence amplitudes after flashes were found to be inversely correlated with the ϕ_{max} after the corresponding flashes, supporting the interpretation given here.

In summary, in states S_2 and perhaps S_3 a fraction of the reaction centres is in such a state that both the energy transfer to P and the electron transfer to P^+ have become much slower. Alternatively, instead of the energy transfer to P, the electron transfer from excited P to acceptor may be slower in this fraction.

ACKNOWLEDGEMENTS

I am indebted to Mr D. J. N. Egberts for carefully doing the experiments connected with Fig. 4, and to Drs. H. J. van Gorkom and A. Melis for discussions. This investigation was supported by the Netherlands Foundation for Biophysics, financed by the Netherlands Organization for the Advancement of Pure Research (ZWO).

References

AMESZ, J. & VAN GORKOM, H. J. (1978) Delayed fluorescence in photosynthesis. *Ann. Rev. Plant Physiol. 29,* 47–66

BUTLER, W. L., VISSER, J. W. M. & SIMONS, H. L. (1973) The kinetics of light-induced changes of C-550, cytochrome b_{559} and fluorescence yield in chloroplasts at low temperature. *Biochim. Biophys. Acta 292,* 140–151

CLAYTON, R. K. & SISTROM, W. R. (1966) An absorption band near 800 mμ associated with P870 in photosynthetic bacteria. *Photochem. Photobiol. 5,* 661–668

DELOSME, R. (1971) Variations du rendement de fluorescence de la chlorophylle *in vivo* sous l'action d'éclairs de forte intensité. *C.R. Acad. Sci. Paris 272,* 2828–2831

DEN HAAN, G. A. (1977) *Chlorophyll a Fluorescence as a Monitor for Rapid Reactions in Photosynthesis,* Thesis, University of Leiden

DUYSENS, L. N. M. (1964) Photosynthesis (review). *Progr. Biophys. Mol. Biol. 14,* 1–104

DUYSENS, L. N. M., DEN HAAN, G. A. & VAN BEST, J. A. (1975) Rapid reactions of photosystem 2 as studied by the kinetics of fluorescence and luminescence of chlorophyll *a* in *Chlorella pyrenoidosa,* in *Proceedings of the Third International Congress of Photosynthesis* (Avron, M., ed.), pp. 1–12, Elsevier, Amsterdam

JOLIOT, A. & JOLIOT, P. (1964) Etude cinétique de la réaction photochimique libérant l'oxygène au cours de la photosynthèse. *C.R. Acad. Sci. Paris 258,* 4622–4625

KNOX, R. S. (1977) Photosynthetic efficiency and exciton transfer and trapping, in *Primary Processes of Photosynthesis* (Barber, J., ed.), pp. 55–97, Elsevier/North-Holland Biomedical Press, Amsterdam

MELIS, A. & DUYSENS, L. N. M. (1978) *Photochem. Photobiol.,* in press

MELIS, A. & HOMANN, P. (1975) Kinetic analysis of the fluorescence induction in 3-(3,4-dichlorophenyl)-1,1-dimethylurea poisoned chloroplasts. *Photochem. Photobiol. 21,* 431–437

MELIS, A. & HOMANN, P. (1976) Heterogeneity of the photochemical centers in system II of chloroplasts. *Photochem. Photobiol. 23,* 343–350

MONTROLL, E. W. (1969) Random walks on lattices. III. Calculation of first passage times with application to exciton trapping on photosynthetic units. *J. Math. Phys. 10,* 753–765

OTTEN, H. A. (1974) A mathematical analysis of the relation between (bacterio)chlorophyll fluorescence yield and the concentration of reaction centre traps in photosynthesis. *J. Theor. Biol. 46,* 75–100

ROBINSON, G. W. (1966) Excitation transfer and trapping in photosynthesis. *Brookhaven Symp. Biol. 19,* 16–48

VAN BEST, J. A. (1977) *Studies on Primary Reactions of System 2 of Photosynthesis by Means of Luminescence and Fluorescence,* Thesis, University of Leiden

VAN BEST, J. A. & DUYSENS, L. N. M. (1977) A one microsecond component of chlorophyll luminescence suggesting a primary acceptor of system II of photosynthesis different from Q. *Biochim. Biophys. Acta 459,* 187–206

VAN GORKOM, H. J., PULLES, M. P. J. & ETIENNE, A. L. (1978) Fluorescence and absorbance changes in Tris-washed chloroplasts, in *Proceedings of Symposium on Photosynthetic Oxygen Evolution* (at Tübingen) (Metzner, H., ed.), in press

VREDENBERG, W. J. & DUYSENS, L. N. M. (1963) Transfer of energy from bacteriochlorophyll to a reaction centre during bacterial photosynthesis. *Nature (Lond.) 197,* 355–357

Discussion

Porter: Why is the lifetime of the excitation of chloroplasts with open traps not $\tau = 1/k_t$ (where k_t is the trapping rate at the reaction centre) but a much longer time?

Paillotin: This relation is only valid if the excitation stays on one molecule. If the excitation moves through several molecules with different rates for loss of excitation, $\tau = 1/k_a$, where k_a is the 'average loss rate', which for chloroplasts in the trapping-limited case is about 200 times smaller than kt.

Clayton: Are the rate constants consistent with a 3% yield when the traps are open.

Duysens: I used the fluorescence yields to calculate the rate constants.

Joliot: For state B you assumed that there is no change in fluorescence. Is state B the sum of the concentrations of the states S_2 and S_3?

Duysens: No, in the state $S_2 + S_3$ there is a constant fraction which is in state B.

Joliot: In that case, it is analogous to the equilibrium we hypothesized between a secondary donor and P which defines the fraction of P that is oxidized. In your state B, P remains after the flash for 20 μs or more in state P^+.

Duysens: Yes.

Joliot: Then we agree. How does your model explain the increase of the F_0 level?

Duysens: We made two assumptions about state B. The first is, as you mentioned, that P^+ quenches and remains for 20 μs or more in state P^+ after the flash. But not only is it more difficult to transfer the electron to P^+, but also it is more difficult to get the excitation to the reaction centre in this state. That lowers the quenching by P or P^+ and increases F_0.

Joliot: If you assume that the quenching efficiency of state B before the flash is less than in the initial state, we agree—but it is only an assumption.

Duysens: I wanted to explain the isostilbic points in Fig. 4.

Barber: You have taken the value $p = 0.74$ (p. 332), which is the measured value, but that value, as has been reported (see Barber 1976), varies considerably.

Duysens: Yes; normally the $\phi(q^-)$ curves have been measured with the S states and the two subsystems of photosystem II present—i.e. a mixture of states. That gives a different curve.

Barber: That could be the explanation.

Joliot: Even without taking into account the S states, one cannot describe the fluorescence-induction curve by assuming only one type of quenchers or of centres; this has been established by Dr Paillotin.

Barber: Dr Duysens, one can change the induction curve by adding Mg^{2+} ions. I find it hard to imagine how a salt addition changes the state of the trap.

Duysens: Salt addition may not change the state of the trap. Addition of Mg^{2+} ions might convert the state of the pigment system (corresponding to Dr Joliot's model) into that corresponding to the matrix model, and move part of the antenna chlorophyll away from system I.

Paillotin: The connection parameter p introduced by Dr Joliot is not equal to the probability P of an exciton transfer from a closed to an open unit. But $p = P(F_v/F_{max})$ where F_{max} is the maximum yield of fluorescence and F_v the variable fluorescence. For instance, $p = 0$ if $F_v = 0$. In this condition one cannot see any induction even if p is equal to unity. If one wants to study the physical properties of the photosystem II, one has to calculate P, i.e. to correct p by the F_v/F_{max} term. Furthermore, the probability P is related to the rate of transfer T from one unit to the other by equation (1). Thus

$$P = T/[T + (k_f/F_{max})] \tag{1}$$

P depends also on F_{max}. One can imagine two experimental conditions which give different values of the connection parameter p, a small difference in P and no effect on T.

Duysens: With non-diffusion-limited transfer one may expect a strictly exponential decay of the fluorescence yield but, if transfer is diffusion-limited, the question of whether the photon is trapped far away from or close to the reaction centre is important for the decay kinetics—there is a non-exponential more-slowly-decaying transfer. The fact that the largest part of the decay in *Chlorella* (Dr Beddard's Fig. 3, p. 196) was exponential suggests that this part is not diffusion-limited.

Porter: That only starts at about 200 ps; the first part is non-exponential. Experimentally, the exponentiality is a test of what you are saying. It seems to me that you are not excluding the possibility of the excitation transfer not being infinitely fast. You can obtain the same result by having an infinitely higher probability of trapping and a finite rate of transfer to the traps. This treatment does not prevent you from increasing this trapping rate constant k_t and lowering the migration rate.

Duysens: We did the calculation for the trapping-limited case. For the diffusion-limited case we used Montroll's equations, which implied that we assumed small excitation losses during transfer.

Porter: Montroll's equations apply to a regular lattice, anyway. Do you agree that one could increase the trapping rate and decrease the diffusion rate but still reach the same conclusion?

Robinson: I do not understand why. The rates may be different depending on whether the traps are open or closed.

Porter: How can you distinguish the two experimentally except by the detailed kinetics of the exponentiality?

Duysens: The $\phi(q^-)$ relationship may not be of the Stern–Volmer type for the diffusion-limited case.

Paillotin: With different numerical values one can obtain a larger value of k_t and then introduce a smaller value of the diffusion constant, D. But the main point is that Dr Duysens' experiment shows that the rate of quenching by the reaction centres is a linear function of the number of reaction centres in the domain.

Porter: We are saying that k is a function of the diffusion rate constant (k_d), the concentration of traps (q) and the rate of trapping (k_t): $k = k_d q k_t$. Dr Duysens assumes that k_d is high and k_t is low, but the product will be constant.

Paillotin: We can describe this problem in simple terms if we consider the radius of the diffusion of the excitation. One has to compare the dimension of a domain with the radius of diffusion of an excitation. If the radius is small compared to the separation between reaction centres, there is no proportionality of the quenching rate with the number of reaction centres.

The proportionality holds only if the radius of the diffusion is not too small—i.e. comparable to or even larger than the distance between reaction centres; this gives a boundary value for the diffusion constant. If the quenching rate is proportional to the number of centres, the excitation may visit many reaction centres.

Amesz: This is a separate-units model.

Barber: The value $p = 0.74$ does not indicate a separate model.

Porter: I cannot see any distinction between this and a straight-forward diffusion-controlled reaction with a probability of reaction on encounter. The product of the two rate constants, the probability of reaction on encounter and the rate of diffusion to the reactant are the factors that matter.

Duysens: Even in the simple case of a regular two-dimensional lattice, the Montroll equation is complicated. I doubt that in the general case one would find a hyperbolic equation of the Stern–Volmer form, which would be a good approximation.

Joliot: Dr Paillotin's equation for p (p. 338) is extremely important for those people working on the effect of Mg^{2+} ions. From the fluorescence-induction curves one cannot infer any direct information about p, the parameter of excitation transfer, unless one takes into account the variable fluorescence. Photosystem I offers the best example: measuring the rate of photosystem I reaction as a function of the number of active centres, one can conclude that there is no energy transfer between photosystem I units. With the traps partially removed by mutation, Delepelaire & Bennoun (1978) measure a parameter for the energy transfer which is about the same for photosystem I as for photosystem II. In the wild type, this parameter is not accessible because there is no variable fluorescence in photosystem I; this means that P700$^+$ is as efficient a quencher as P700.

Several erroneous estimates of the value of the parameter of transfer as a function of Mg^{2+} concentration have been obtained because only the time course of the fluorescence induction was taken into account and not the change of the variable fluorescence.

References

BARBER, J. (1976) in *The Intact Chloroplast* (Barber, J., ed.) (*Topics in Photosynthesis*, vol. 1), Elsevier, Amsterdam

DELEPELAIRE, P. & BENNOUN, P. (1978) Energy transfer and site of energy trapping in photosystem I. *Biochim. Biophys. Acta 502*, 183–187

General discussion

SOME RETROSPECTIVE COMMENTS

Clayton: In some ways this symposium has been an exercise in humility for me because areas that I had thought were characterized by futility have dropped their 'f' and acquired utility!

About 20 years ago it was a scientific cliché that the primary photochemistry of photosynthesis was one of nature's great mysteries. Almost 20 years before that I was privileged to work as an unpaid technician in James Franck's laboratory for one summer and at that time the peripheral dark chemistry was subsumed under the mysterious title of the 'Blackman reaction'. We have come a long way from there, in terms of function, as we now have a fairly convincing outline of the photochemistry, at least in the case of the bacteria, and much of the peripheral electron transport is delineated—that is, until we get into the morass of phosphorylation. So it is not surprising that we have drifted from emphasis on function toward an emphasis on structure.

Structurally, we are confronted by a special problem in photosynthesis in that nobody has yet succeeded in crystallizing a hydrophobic protein. And so, with rare exceptions, we have not been able to apply X-ray diffraction to components of photosynthetic tissues. I suppose that the basis of crystallizing hydrophilic proteins is that a few hydrophobic patches offer focal points for crystallization. We could try to crystallize these hydrophobic proteins from hydrophobic solvents, but most of these solvents denature the system that we are trying to study. Consequently we have to resort to other means, especially studies with polarized light. This leads me to one of my lessons in humility because, when the picosecond (laser) spectroscopists began to look at fluorescence lifetimes and to encounter the anomalies of biphotonic interactions, I expected little of biological interest because of the high quantum

341

fluxes being used. Now I find that one can use the higher quantum fluxes to restrict excitons to a small range and in that way probe local order. This reminds me of the feeling I once had when the e.p.r. Signal 1 and Signal 2 and the bacterial signal for P$^+$ had been observed, namely, that that was the end of the use of e.p.r. in photosynthesis. So I should try to be a little more open-minded.

We have essentially been looking at structure at three levels: antenna, photochemistry, and interaction of larger functional units. With regard to models for exciton migration in the antenna, I had a sense that this was a pleasant but frivolous pastime engaged in by a few physicists. I felt that, even within the entirely disordered chlorophyll aggregate, at the density at which chlorophyll occurs *in vivo* it is easy to explain the known efficiency of energy transfer. What would be gained by sorting out the minute details? If an efficiency of 90% in a random aggregate of pigments can be improved to 98%, say, with special exciton focusing arrangements, is that especially interesting? Yes; in terms of the evolutionary time scale the difference between 90 and 98% can be overwhelming with regard to one organism replacing another. I conclude that this modelling is not a frivolous activity and I retract my earlier prejudice.

Photochemistry

The history of the study of photochemistry of chlorophyll *in vitro* has been characterized by an embarrassment of wealth. It became apparent that chlorophyll can do photochemically just about anything that one could imagine in different constructions and the problem was to identify what does happen out of the many things that could happen. Starting with the recognition of the importance of solvated dimers, Dr Katz has made a major contribution in providing a convincing interface between the *in vitro* and *in vivo* studies. We shall now be pursuing these ideas constructively.

Interaction of larger functional units

In the case of the thylakoid membrane of chloroplasts, thanks in part to electron microscopy, in part to fractionation as practised by Drs Anderson and Thornber, and in part to Dr Butler's incisive studies of energy transfer through fluorescence, we are beginning to put restrictions on the possible models. The discussions at this symposium have suggested many new experiments to all of us. This is the best criterion of the success of a conference.

The last surviving enigma is the photochemistry of oxygen evolution. If we do not learn to make practical solar cells with chlorophyll dimers, in an attempt to solve our energy problems, then perhaps when we understand the intricate chemistry of oxygen evolution we shall know how to catalyse efficiently the photolysis of water and make hydrogen in abundance.

MANGANESE AND THE EVOLUTION OF OXYGEN

Porter: With regard to your last point I might add that in our *in vitro* studies we are trying to determine how nature uses manganese in the evolution of oxygen. We are trying to mimic photosystem II; that is to eliminate oxygen from water and to make a fuel. Making hydrogen from water photochemically poses no real problem—it can be done in several ways. The real problem is to make oxygen because for a cyclic process using solar energy one has to put oxygen into the atmosphere to replace what will be removed when the fuel is burnt. Whether the fuel is hydrogen, sugar or, as in our case at present, hydroquinone (here we are taking a tip from photosystem II) does not matter because hydroquinone will react exothermically with oxygen to give quinone and water.

We have irradiated aqueous solutions of a manganese complex, chlorophyll and quinone (recognizing that we have to transfer four electrons to make one oxygen molecule) in order to generate oxygen and hydroquinone. In the 1960s Calvin did much work on manganese complexes from this point of view but not with an acceptor present. We can excite chlorophyll and transfer an electron to quinone. Some interesting problems about that arise because the triplet reaction seems to go well whereas the singlet reaction goes but then reverses. There seem to be good spin reasons for that. Our trouble lies in getting anything else to happen except the back reaction.

When we try to make excited chlorophyll react with manganese complexes in various forms we do not succeed. For example, in the absence of quinone, we cannot make Chl$^-$ and an oxidized manganese species. Chlorophyll is quenched by manganese complexes of all kinds extremely efficiently but we have not been able to transfer an electron efficiently. Plants clearly can bind these components together in the right configuration. That is what we tried to do next: to bind the chlorophyll to the manganese so that it cannot move away when it has to transfer its electron.

One way of doing that is to replace the magnesium in the middle of chlorophyll by manganese, that is, to use manganese porphyrins and manganese phthallocyanins. The use of these compounds brought us a bonus because water forms an axial ligand to the manganese complex. In that way we brought together all three elements of our system. Dr A. Harriman has done most of the work and has shown that oxygen is eliminated from water concurrently with reduction of the quinone and that the manganese porphyrin acts only as a photocatalyst.

We want to know about the relevance of this to the oxygen evolution in photosynthesis. What is the form of manganese in chloroplasts? Can anybody rule out the possibility that it is a manganese porphyrin or indeed a chlorophyll with manganese in place of magnesium?

Katz: We have ignited a substantial amount of chlorophyll extracted from spinach and subjected the ash to spectroscopic analysis. As far as we could tell, no heavy metal was present in the MgO residue. We had thought that an occasional molecule of copper chlorophyll or some other heavy metal chlorophyll might be present *in vivo*.

Porter: That method would have detected 1 molecule in 200?

Katz: Yes.

Porter: The manganese seems to be very labile. It can be inserted into and removed from the porphyrin quite easily.

Katz: The experiment I described provides no evidence for the incorporation to a significant extent or substitution by manganese or other heavy metal in spinach chlorophyll *a*.

Joliot: N.m.r. spectra indicate only Mn^{II} and Mn^{III} forms in the S states (Wydrzynski *et al.* 1976). How reliable is such n.m.r. evidence? Most people propose models which imply the presence of Mn^{IV}.

Duysens: Van Gorkom *et al.* (1978) have not found absorption changes in the visible region caused by changes in the S_1, S_2, S_3 and S_4 states—only in the ultraviolet were small changes observed.

Porter: It is hard to see how manganese could be involved in the process of electron transfer without causing absorption changes. If no absorption change due to manganese is seen, it cannot be involved in the electron-transport chain.

Duysens: There may be changes in the ultraviolet region.

Malkin: Do they have to be porphyrin complexes?

Porter: No, but the ligand tied to manganese has to accept energy in some way.

THE STATE OF CHLOROPHYLL IN PHOTOSYSTEM I, PHOTOSYSTEM II AND THE LIGHT-HARVESTING UNIT

Katz: First, let me list the absorption maxima that have been mentioned in our discussions: 660, 676, 680, 700, 705–720 and 740 nm. I shall now give a somewhat idiosyncratic view of the structural features of the chlorophyll molecule that are compatible with these maxima. First, I shall give a brief summary of chlorophyll species *in vitro*.

In monomeric chlorophyll the fifth or sixth coordination positions of the central magnesium atom are always occupied by an electron-donor group (nucleophile). In acetone, pyridine, tetrahydrofuran or any of the ordinary polar solvents (containing oxygen, nitrogen, or sulphur atoms bearing lone electron pairs) one or two solvent molecules will occupy the axial positions on magnesium to an extent that depends on the basicity of the nucleophile. In pure pyridine the magnesium atom in chlorophyll is essentially all six-coordinate, and the Mg atom will lie in the plane of the four pyrrole nitrogen atoms. When the Mg atom is pentacoordinated, X-ray evidence indicates that it sits above the plane of the four nitrogen atoms, perhaps to the extent of about 0.05 nm, which is a large distance on the molecular scale. Sometimes the chlorophyll molecule is depicted as a flat saucer with a magnesium atom protruding, or the Mg can be visualized as in the plane of the four nitrogen atoms, but with the whole of the molecule 'dished'. The 'dishing' is not as large as one might imagine because chlorophyll is a large molecule and the deviations at the extremes of the macrocycle are only about 0.03 nm. Consequently, there is access to the magnesium atom from both sides.

The spectral features of monomeric chlorophyll depend, to a certain extent, on the coordination number. Although the effect of coordination number is minimal on the red band (Q_y), it considerably affects the Q_x transition. There is reason to suppose that the Mg coordination number affects the lifetime of excited states and other photochemical properties of chlorophyll.

Any chlorophyll species that absorbs between 660 and 670 nm is considered to be monomeric chlorophyll. If chlorophyll is dissolved in a highly polarizable nucleophilic solvent such as pyridine, it absorbs maximally in the red at 670 nm; chlorophyll dissolved in octane with only a small amount of pyridine to ensure disaggregation to a monomeric form absorbs at 662 nm. The red-shifting of the absorption maximum in polarizable polar media is a general solvent effect and not the consequence of coordination interaction with the electron donor (pyridine). When chlorophyll is dissolved in a non-polar solvent (e.g. carbon tetrachloride or an aromatic hydrocarbon) and a strenuous effort is made to exclude water or other adventitious nucleophiles which may have been introduced during the extraction and purification of the chlorophyll, the chlorophyll will be present as a dimer [Chl a]$_2$. The equilibrium constant for dimerization (2Chl $a \rightleftarrows$ [Chl a]$_2$) is very large, probably considerably greater than 10^6. Consequently very little monomer is present in such a solution. Infrared, nuclear magnetic resonance and other spectroscopic techniques indicate that this dimer is formed by a coordination interaction between the magnesium of one chlorophyll molecule and the keto carbonyl group of another. The two chlorophyll molecules are then orthogonal. N.m.r. and other data suggest that in solution the two chlorophyll molecules in the dimers exchange, i.e., on the n.m.r. time-scale the chlorophylls are equivalent. They are not equivalent on the electronic transition time-scale. The line-width of electronic transitions of [Chl a]$_2$ will be determined by the distribution of various conformers present in the system.

In a hostile organic solvent such as octane (which tends to exclude the chlorophyll macrocycle), it can be shown by direct molecular-weight measurements that chlorophyll oligomers are formed by additional keto C=O------Mg interactions between dimers. These may have molecular weights greater than 20 000 in aliphatic hydrocarbon systems. The absorption maximum in the red of dimeric or oligomeric chlorophyll a depends on concentration and in concentrations of about 10^{-2} mol/l (comparable to lower estimates of chlorophyll concentration *in vivo*) chlorophyll a solutions in n-octane is about 680 nm. The visible absorption maxima in the red of the oligomers are only slightly shifted relative to the dimer and are quite insensitive to the size of the oligomer. This suggests that the red shift in chlorophyll–chlorophyll aggregates reflects the relative amounts of donor and acceptor chlorophyll in the aggregate (for a detailed discussion, see Shipman *et al.* 1976). The red shift in chlorophyll self-aggregates is not a measure of the size of the aggregates and is not considered to arise from π–π interactions.

Whenever a chlorophyll a species with an absorption maximum below 670 nm is observed, we consider it to be monomeric chlorophyll; chlorophyll a species absorbing at about 675–680 nm we consider to be aggregates formed by chlorophyll–chlorophyll interactions. Any chlorophyll a species absorbing at a wavelength longer than 680 nm is most likely to be an aggregate of chlorophyll, but aggregated by a bifunctional ligand (H$_2$O, RXH, where R = H or alkyl and X = O, N, S), which cross-links the chlorophyll. In such aggregates the orientation of the chlorophylls is most often parallel rather than orthogonal as in chlorophyll–chlorophyll aggregates. Water is a good cross-linking ligand (it can form hydrogen-bonds to the oxygen functions of another chlorophyll molecule and its oxygen atom can coordinate to the magnesium atom of one chlorophyll), and large aggregates can be formed by repetition of the cross-linking interaction. Thus, when wet nitrogen is bubbled through solutions of oligomeric chlorophyll in aliphatic or cyclo-aliphatic hydrocarbon solvents large changes in the visible absorption maximum can be observed, indicative of the formation of a new chlorophyll species. The λ_{max} of the chlorophyll *a* oligomer shifts from 680 to 740 nm on hydration. On the basis of the Strouse X-ray crystal structure for ethyl chlorophyllide a·2H$_2$O, the red shift in chlorophyll–water aggregates can be related to the formation of–Chl·H$_2$O·Chl·H$_2$O·Chl– stacks further cross-linked to form two-dimensional sheets and three-dimensional crystals. For chlorophyll *a* a monolayer sheet (this has been observed experimentally) absorbs at 735 nm. A stack of sheets absorbs near 743 nm. In these species excitonic interactions are important because the π-systems of the chlorophyll molecules in the stacks are in close contact, being at their van der Waals' radii. The chlorophyll transition dipoles are linked up in parallel and so the conditions for strong interaction obtain. Calculations indicate that two chlorophyll molecules linked in this way will absorb at 695 nm, a linear stack of three will absorb at 705 nm and an infinite linear stack at 720 nm (Shipman & Katz 1977). Species of chlorophyll that have absorption maxima between 695 and about 720 nm may therefore be linear Chl·H$_2$O·Chl·H$_2$O··· stacks of different lengths. Interaction of chlorophyll *a*, pheophytin *a*, and water in an aliphatic hydrocarbon solvent is another way to form species absorbing between 705 and 720 nm. This particular system replicates the red-shifted chlorophyll absorption spectra that have been observed in old cultures of *Euglena* that have a high content of pheophytin. Our synthetic photo-active 'dimer' has a fundamentally different structure from the 'true' dimer formed by a keto C=O----Mg interaction between two chlorophyll molecules. Our covalently linked dimer in its folded configuration absorbs at 695 nm and fluoresces at 730 nm. The red-shift in the covalently linked folded dimer to 695 nm is consistent with a parallel arrangement of the macrocycles; the true [Chl a]$_2$ absorbs at wavelengths shorter than 680 nm and has an orthogonal arrangement.

When chlorophyll *a* is hydrated to form the chlorophyll–water adduct absorbing at 740 nm, the observed red shift is the largest seen for any chlorophyll system. Hydration of bacteriochlorophyll *a* in n-octane forms bacteriochlorophyll *a*–water aggregates that have all the wavelength maxima charcteristic of intact photosynthetic bacteria. The chlorophyll *b*–water aggregate absorbs in about the same region of the spectrum as does oligomeric chlorophyll *a* (i.e., about 680 nm). It is thus difficult to differentiate by visible spectroscopy alone between hydrated chlorophyll *b* and anhydrous oligomeric chlorophyll *a*.

Chlorophyll *b* has two donor groups: a formyl group and a keto carbonyl group. From molecular weight determinations as a function of concentration, it appears that chlorophyll *b* exists as a trimer in conditions where chlorophyll *a* is dimeric. One can imagine a 'natural' sandwich in which one chlorophyll *a* molecule is coordinated to the formyl group and another to the keto carbonyl group of the chlorophyll b to form a unit a–b–a. The ratio of two chlorophyll a to one b chlorophyll is near the physiological ratio.

I must emphasize that both *in vitro* and *in vivo*, temperature-sensitive and concentration-sensitive equilibria are involved. Consider the behaviour of water:

$$Chl_2 + 2H_2O \rightleftharpoons 2Chl.H_2O$$

(An additional equilibrium involves the formation of stacked structures from Chl $a \cdot H_2O$.) At room temperature in non-polar solvents and at low concentrations of water, nearly all the chlorophyll may be present as Chl $a \cdot H_2O$. As the temperature of the system is lowered, water is forced from the organic solvent, and the least hospitable site for it to go is the magnesium atom of a chlorophyll molecule. On cooling a dry solution of chlorophyll a in toluene, the solution almost invariably develops long-wavelength absorption, which to us is indicative of the formation of hydrated, aggregated species. Without having more information about the environment of the chlorophyll in the chloroplasts than is at present available, we should not exclude the possibility that new, hydrated species are generated at low temperature that may not be present to a significant extent at room temperature.

Porter: What are the fluorescence quantum yields of the orthogonal dimer and the parallel dimer?

Katz: The true dimer is not fluorescent at all at room temperature. On cooling a solution of dimer in toluene, a red shift in the absorption spectrum and long-wavelength emission usually are observed. Presumably new chlorophyll species are formed from traces of water (or other nucleophile) on cooling. The covalently linked folded dimer—our synthetic P700—fluoresces at 730 nm with about the same fluorescence yield as does monomeric chlorophyll a.

Porter: You leave us with the problem of what is the light-harvesting chlorophyll. It fluoresces with a high yield but at too long a wavelength for it to be a monomer.

Katz: We should consider the possibility that P680 and F685 and F695 arise from hydrated chlorophyll b species.

Knox: That it is too long a wavelength for the monomer I have to reject if chlorophyll may interact with itself and absorb at a few nm longer wavelength —why not if it interacts with something else?

Katz: I know of no experimental situation in which monomeric chlorophyll a has an absorption maximum greater than 670 nm.

Knox: It may be improbable but not impossible.

Katz: True. On the basis of known results no obvious way for such a red shift to occur in monomer chlorophyll *a* has been advanced. It is difficult to invoke hydrogen-bonding, because dissolution of monomeric chlorophyll *a* in trifluoroethanol (one of the strongest hydrogen-bonding compounds that does not remove the magnesium from the chlorophyll) causes no observable red shift.

Barber: Throughout this symposium I have been convinced that chlorophyll is intimately related with proteins, as indicated by the extraction of chlorophyll–protein complexes. Why do you ignore this in your various considerations of *in vivo* chlorophyll organization?

Katz: Protein side-chains contain nucleophiles such as the OH group of serine, the NH_2 group of lysine, the guanidinium group, the imidazole groups, etc. We have used, for instance, butylamine, ethanol, ethanethiol, imidazole, and other analogues of protein side-chain nucleophiles that can coordinate to the magnesium in chlorophyll, and in no case have we observed a red shift in the chlorophyll absorption. Coordination to a protein *per se* as a cause of chlorophyll red-shifts, therefore, does not appear to me to be a likely possibility. I do not, however, exclude such a possibility completely. However, I find it difficult to avoid the conclusion that red-shifts in chlorophyll–protein complexes arise from chlorophyll–chlorophyll rather than chlorophyll–protein interactions.

Cogdell: Binding by the protein could distort the porphyrin ring so that the structure is no longer the same as you associate with the porphyrin in red or blue.

Duysens: What effects do ionic charges in the solvent have?

Katz: Chlorophyll is not soluble in water, so it is difficult to check this.

Duysens: Charges on amino acids in a protein may change the absorption spectrum of the chlorophyll bound to this protein.

Katz: Addition of a quaternary ammonium salt to chlorophyll solutions in cyclohexane has no obvious effect on the spectrum. As the salt forms tight ion pairs with its *gegen* ion, it may not interact strongly with the chlorophyll.

Paillotin: Can you select a protein complex with one chlorophyll molecule to see the *in vivo* absorption spectrum of a monomer?

Katz: No. All the chlorophyll–protein complexes I know of, both natural or synthetic, have more than one chlorophyll molecule per protein.

Junge: There is still a long way to go before we know the structure of the photooxidizable dimer *in vivo*. Dr Katz and his colleagues have provided sound data on related structures *in vitro*. However, the difficulties in recognizing them *in vivo* are perhaps best illustrated by an example from his own group.

Dr Shipman found it difficult to explain the observed absorption spectrum of the crystallized bacteriochlorophyll–protein for which the spatial distribution of the porphyrin rings is known with high precision (Fenna & Matthews 1975).

An important question seems to me why nobody has yet succeeded in stripping off more than some 20 chlorophylls from photosystem I reaction centres without changing the functional characteristics of the system? Is there any functional reason why so many chlorophylls are required in chloroplasts (over and above the antenna function)? Moreover, we do not know to what extent the dimer is a supermolecule, on the time scale of e.p.r. experiments or on the time-scale of optical experiments. We have no convincing reason why a chlorophyll-dimer is required. Is it because the formation of a radical-pair is advantageous?

Porter: Could it be that because it has lower energy it is a trap?

Junge: The reaction centre in photosystem II operates at 680 nm.

Clayton: The reaction centre behaves as an anti-trap in *Rhodopseudomonas viridis.*

Katz: So many problems still remain in the interpretation of the optical properties of systems consisting of only chlorophyll and solvent that I am not surprised that the behaviour of a chlorophyll–protein complex is difficult to explain. The notion that any observation can be blamed on an otherwise unspecified protein interaction should be resisted. As to the protein described by Fenna & Matthews, I should point out that it is not photoactive, it is a bacterial not a green plant constituent, and, as far as I know, it has optical properties that nobody can explain in specific terms.

Clayton: Why don't you add a protein?

Katz: Because protein is not soluble in hydrocarbons and chlorophyll is not soluble in water.

Clayton: You could try detergent micelles for hydrophobic proteins.

Katz: Detergents are nucleophiles and would compete for coordination to the chlorophyll.

Porter: The biggest problem with the light-harvesting chlorophyll is your inability to explain the high fluorescence yield in photosystem II.

Barber: The yield is high for chlorophylls serving photosystem II but not for the light-harvesting chlorophyll of photosystem I.

Porter: Those of photosystem II absorb at too high a wavelength for Dr Katz to allow us to say that it is a monomer.

Katz: I should be inclined to assign a 677 nm absorption to a chlorophyll *a* dimer or short oligomer, or to a hydrated chlorophyll *b* species. Neither of these is monomeric.

Porter: If you were to allow us to believe that photosystem II chlorophyll is a monomer, then we should be happy.

Katz: Why does a chlorophyll species have to be fluorescent for energy transfer? All that is required is that the energy transfer be faster than non-radiative decay.

Porter: But it is.

Robinson: In principle resonance Raman spectroscopy should disclose something about the coupling.

Katz: There has been much talk about the random walk and the lattice as if energy transfer from one chlorophyll to another can be a perfectly random process. Because chlorophyll is anisotropic, the orientation of the transition dipoles has to be proper for rapid energy transfer.

Porter: But if chlorophyll sits in a rigid matrix in a random orientation, energy is transferred efficiently, from chlorophyll *b* to chlorophyll *a*, for example, as long as they are close enough.

Katz: Nevertheless, efficient energy transfer between two orthogonal chlorophyll molecules is difficult to attain.

GROSSER STRUCTURE OF CHLOROPHYLL–PROTEIN COMPLEXES

Thornber: I have been delighted at this symposium to notice how the organic and physical chemists have listened to, learnt from and usefully commented on what the cell biologists and the biochemists are doing, and *vice versa*. This will result in new and useful collaborative interactions.

There now seems to be a general acceptance of the notion that chlorophyll–protein complexes are necessary for the organization of chlorophyll *in vivo*.

Such a view has not always been popular; some of us have fought for several years to win general acceptance for the notion. The chlorophyll–protein story started 70 years ago but began to advance rapidly in the middle 1960s with work of (*a*) Olson (cf. Olson 1978) on the green bacterial chlorophyll *a*–protein, who provided the first convincing evidence for the existence of such protein complexes, (*b*) Dr Clayton and the late Dan Reed who isolated the first reaction-centre complex which was later shown to be a chlorophyll–protein complex, and (*c*) N. K. Boardman, J. S. C. Wessels, L. P. Vernon and myself who have been developing fractionation techniques for plant and bacterial systems to obtain pigmented fractions of the photosynthetic apparatus. There is now general agreement about the nature of the isolated complexes which I shall describe below.

The discussions at this symposium have confirmed that eukaryotic organisms contain complexes which are more difficult to obtain in solution and which are less stable in detergent solution than those from the prokaryotes. We still do not know why this should be so. Much less free pigment is found in photosynthetic bacteria after dissolution of their photosynthetic membranes than with membranes from higher plants (cf. Cogdell & Thornber, this meeting). An exciting advance that has just come to light at this meeting and one that will, I hope, greatly aid studies on eukaryotic chlorophyll-proteins is the discovery by Drs Wessels, Anderson and myself of better gel systems for separating the chlorophyll-protein complexes from higher plants. I shall summarize these new gel systems and the information that is being derived from them.

The old gel systems gave three green bands which, for over 10 years, we have been trying to characterize (cf. Thornber *et al.* 1977). The basic components are CP I, CP II and free pigment. With the new system we have discovered that the amount of chlorophyll associated with CP I is now about 30% (compared to 10% in the older electrophoretic systems) of the total chlorophyll in the plant (see discussion after Dr Anderson's paper). The complex contains chlorophyll *a* and probably no chlorophyll *b*. Whether we are now isolating more chlorophyll–protein like CP I in the old system, or whether the same amount of protein is obtained but more chlorophyll molecules are associated with it, remains to be determined.

On these new gel systems CP II represents 50–55% of the total chlorophyll and contains chlorophyll *a* and *b*, the proportion of which varies between investigating groups: from a chlorophyll *a/b* ratio of 1.0 (Dr Knox, Dr. Van Metter and my group) to 1.3 (Dr Wessels and Dr Anderson). Drs Knox (this symposium) and Van Metter (1977) have proposed a model for this complex.

Dr Knox suggests a trimer of chlorophyll *b* molecules located in the centre of the folded polypeptide with three chlorophyll *a* molecules arranged around the trimer but within the protein. If it turns out that the correct chlorophyll *a/b* ratio of the complex is 1.3, then the ratio of 1.0 may represent some denaturation of the complex—i.e. one chlorophyll *a* molecule may have been removed from each molecule of the complex. That is one point of disagreement amidst our general agreement.

The free pigment zone now accounts for only about 10% of the total chlorophyll. With the decrease in total chlorophyll in this zone, the new gel systems are revealing pigmented bands due to other chlorophyll–proteins (e.g. Hayden & Hopkins 1977) and one of these, I imagine, is chlorophyll a_2 (frequently referred to as photosystem II at this meeting). Some additional bands are aggregates or oligomeric forms of components containing chlorophyll *a* and *b*, perhaps reflecting how the 30 000 unit of chlorophyll *a/b*–protein occurs *in vivo*. A point about the chlorophyll *a/b*–protein that should be mentioned is that the threonyl residues of the CP II chlorophyll–protein are phosphorylated in the light and dephosphorylated in the dark (Bennett 1977). Why this occurs is not known.

Let us now consider how the chlorophyll–proteins observed on gel electrophoresis might be related to the so-called digitonin fractions. Dr Wessels' complexes, prepared with digitonin, are designated F_I, F_{II} and F_{III}. Perhaps we tend more than is justified to suggest that F_I corresponds exactly to CP I, F_{III} to CP II etc.; the size of F_I, F_{II}, F_{III} is probably much greater than that of the chlorophyll–proteins. The correspondence of F_I with CP I is best— we all agree on that. F_{III} may represent an aggregated form of the chlorophyll *a/b* bands that we see (Fig. 1 in discussion after Dr Anderson's paper), but other minor pigmented components may be present in F_{III} (i.e. it may not be made up entirely of an aggregated form of those 30 000 units). F_{II} probably has some relationship to the photosystem II reaction-centre band that has just been detected on gel electrophoresis.

Some other thoughts connected with what we have heard that come to mind are: (*a*) can the reaction centre of P700 be isolated with about the same enrichment of reaction centre as has been done for the photosynthetic bacterial component? We automatically tend to think so, but this may not be correct. At present we are stuck with preparations having about 20 chlorophyll molecules per P700, and it is conceivable that we cannot lower the ratio and purify at the same time a complex without destroying the precise organization of chlorophyll *in vivo* that gives the P700 entity. (*b*) Some polypeptides of antenna pigment–proteins of plants and photosynthetic bacteria are soluble in organic solvents. Our discussions have indicated a need for a polypeptide that can be added to chlorophyll *in vitro* if we are to study pigment–protein interaction spectrophotometrically. The protein part of CP II is soluble in chloroform–methanol (Henriques & Park 1976) as is the B890-conjugating polypeptide we described before (Cogdell & Thornber, pp. 61–73).

Turning to Dr Staehelin's structures, I want to try to correlate information on the pigment–protein complexes observed in higher plants with electron micrographs of freeze-etched membranes and models of the membrane derived therefrom. In the Staehelin–Arntzen model the reaction centre of photosystem II and those pigments we refer to as chlorophyll a_2 coexist as a 8.0 nm particle. The light-harvesting complex (presumably equivalent to F_{III} and largely made up of the chlorophyll *a/b*–protein) interacts with the 8.0 nm particle in units of 1, 2 and 4 (see Dr Staehelin's paper, pp. 147–169). According to my calculations (below) there is probably only one photochemically-active molecule in the reaction centre (8.0 nm particle); Dr Joliot has pointed out that there may be more. Thus, if there is one photochemically-active molecule present and if the light-harvesting complex is largely composed of the chlorophyll *a*/*b*–protein, we (Thornber *et al.* 1977) calculate that the four entities in the light-harvesting complex would contain some 220 chlorophyll molecules, of which 110 would be chlorophyll *a* and 110 would be chlorophyll *b*. These 220 pigments will be divided into units of six or seven, i.e., the chlorophyll *a*/*b*–protein monomeric unit. About 40 such chlorophyll–protein units will be contained in the four light-harvesting entities that surround the 8.0 nm core. Each of the four contains some oligomeric form of about 10 chlorophyll *a*/*b*–protein molecules giving a molecular weight for each entity of about 300 000 and containing about 60 chlorophyll molecules. It appears that we can have only one P680 in this photosystem II particle observed by electron microscopy since it would be difficult to pack in more than 220 chlorophyll molecules in that sized particle; one would need about 200 chlorophylls for every trap molecule present.

Porter: Where is photosystem I?

Thornber: The 7.6 nm particle (see Fig. 18, p. 166) probably represents the same aggregated form of the CP I that we observe on gels.

Staehelin: If the plane of the membrane is at right angles to that of the figure with the lobes predominantly on the left and right of the photosystem II core, photosystem I will be at the back and front of the core.

Porter: Is this big enough to account for the particles seen by electron microscopy?

Staehelin: Yes.

Porter: According to such pictures, they are often separated from each other by distances that are too great to allow energy transfer.

Staehelin: Yes.

Porter: So this photosynthetic unit is a separate package, unless under special circumstances they coagulate in a regular form.

Joliot: I have nothing against the idea of only one reaction centre per particle, but as Dr Paillotin shows that at least five photocentres are connected, it is difficult to reconcile this conclusion with the structure observed. Can you rule out the possibility that chlorophyll in the lipid phase establishes a communication between several particles? We have to assume the existence of such a link, as the distance shown in Dr Staehelin's picture (20 nm) is too great to allow the jump of the excitation.

Porter: Could 10–15% of the chlorophyll be in the lipid phase?

Thornber: I cannot object to that notion.

Clayton: Is that the free chlorophyll?

Thornber: Yes.

Barber: That cannot be true for the bacterial system where no free chlorophyll is found after extraction of the light-harvesting pigment–protein complexes with detergent.

Clayton: But then we did not have this morphological problem.

Thornber: In the bacteria, whole conglomerates of entities containing the 890 complex with their reaction centre, etc., form islands in a sea of proteins that absorb at shorter wavelength, so one has many interacting photosynthetic units over which excitation energy can be distributed. In plants one would expect similar interactions between four to 10 of the photosystem II units; this is not observed in electron micrographs. Dr Staehelin, is it possible that the actual preparation for freeze-etch samples pushes the photosystem II particles further apart than they really occur *in vivo*?

Staehelin: I am convinced now, after using the best freezing methods available for structural preservation, that the photosystem II particles are as far apart as we see them. However, as our techniques have improved, the ratio of big particles to the smaller particles that fracture with the stroma leaflet (among which we must have the photosystem I units and probably also the cytochrome units) has decreased to 1/5–6 in stacked regions—that is, each big unit is surrounded by five or six smaller particles.

Barber: I find it difficult to visualize what goes on in the membrane at room temperature. Our fluorescence measurements encourage me to consider a dynamic system in which changes in energy transfer occur as a result of changes in membrane organization induced by proton pumping and ionic movements.

Porter: But not during the photosynthetic process.

Barber: I agree; I am talking about a slow reorganization in response to electron transport and I am trying to emphasize that the membrane is probably not a 'rigid' system.

Clayton: When one considers that volume is the third power of the linear dimension and remembers the vicissitudes of electron microscopy (with regard to both diameter and shape), how much uncertainty do we have in estimating the number of reaction centres in one presumed system II particle?

Butler: Dr Staehelin's freeze-etch micrographs appear to show photosystem I particles separated from photosystem II particles.

Staehelin: We see them physically separated owing to the technique used for their visualization. When the membrane is split, the photosystem II particles stay with the luminal leaflet, while the photosystem I particles and other components remain with the other leaflet. To obtain a picture of the real system we have to integrate the two leaflet images.

Butler: It would be very satisfying if we could obtain fractions enriched in the two halves of the membrane and thereby separate photosystems I and II. Even small amounts would suffice for fluorescence measurements at low temperature.

Staehelin: In the long run one will probably be able to do so but at present the technical problems are severe.

Duysens: Is it possible to measure freeze-fractured membranes at —100 °C?

Staehelin: It should become possible soon. Dr O. H. Griffith in Oregon is developing a photoelectron microscope and has achieved a resolution of 40 nm but is aiming for 10 nm next year. With that microscope we hope to see where the chlorophyll is located with respect to the particle.

Porter: Dr Breton, do you see any difficulty about how your orientation curves of the chlorophyll molecules can happen in a thing like this?

Breton: No; there is no difficulty in explaining our polarized spectroscopy data with a model in which the pigments are bound to proteins, as long as these proteins are oriented with respect to the membrane plane.

Amesz: Did you calculate the amount of chlorophyll that should be in single particles on the basis of the chlorophyll content of the thylakoid membrane? This could be done by dividing the estimated amount of chlorophyll present per unit area of membrane by the number of particles as determined from electron micrographs.

Staehelin: We have not done that.

Thornber: It is difficult to know the particle density over all the membrane in a chloroplast.

Anderson: I have tried to do these sorts of calculations by two methods. (1) The molecular weights of the freeze-fracture particles were calculated from their molecular volumes divided by an assumed partial specific volume. The mass of chlorophyll and its associated intrinsic protein of a photosynthetic unit was calculated from the composition of membranes which were stripped of most of the protein except for the pigment complexes. The photosynthetic unit could be housed in one large particle containing the reaction centre and chlorophylls of photosystem II and one or two small particles containing the P700 and chlorophylls of photosystem I. (2) The surface area of the membrane covered by lipid and freeze-fracture particles was calculated and the number of lipid molecules per large particle was calculated for the stacked membrane region. The number of lipid molecules deduced per large particle corresponds to the number of lipid molecules available for each photosynthetic unit. Although the calculations involve different assumptions, both indicate that a photosynthetic unit in terms of its chlorophyll–proteins could be accommodated in one large particle and one to two small particles.

Joliot: The proportion of photosystem II particles in the unstacked region is low. In your calculations, did you take into account the total area of the membrane or only the stacked region?

Anderson: The first calculation was independent of particle location. The second calculation was done with stacked membranes.

Junge: What is the average area?

Staehelin: The density of the particles in the stacked regions on electron micrographs is about 1200 μm^{-2}. About 60% of the membranes are stacked and about 80% of the photosystem II particles are in the stacked regions.

Junge: Assuming an area of 2 nm^2 per chlorophyll molecule (Thomas *et al.* 1956) I calculate one big particle per 450 chlorophylls.

COMPARISON OF THE PAILLOTIN–SWENBERG KINETIC THEORY AND THE SEARLE–TREDWELL EMPIRICAL TREATMENT

Knox: In comparing the essential elements of the two treatments that have been put forward I shall try to be scrupulously even-handed but I must admit that Dr Swenberg is an old colleague of mine, in fact a thesis student, and so I shall probably overcompensate. Table 1 sets out the basis of both the Paillotin–Swenberg 'collision' theory and the Searle–Tredwell '$t^{1/2}$' treatment. First, let us be clear that two kinds of data are to be described: (1) the fluorescence yield as a function of intensity, which decreases sigmoidally on a log(intensity) plot (see, e.g., Fig. 1 of Paillotin & Swenberg, p. 205) and (2) the intensity of fluorescence as a function of time, which increases after the pulse and then decays, frequently in an apparent $\exp(-At^{1/2})$ curve. Integration of the time-dependent results and the use of different total pulse strengths would give the first-mentioned curve.

Basis

The basis of each theory is as follows. According to Drs Paillotin & Swenberg collisions between excitons reduce the fluorescence yield at high intensity. That means that the kinetics are affected at short times, because after a short time the high density (of excitons) has subsided and ordinary kinetics ensue. On the empirical $t^{1/2}$ treatment, the transfer kinetics in any system to which Drs Searle & Tredwell want to apply the theory are strongly affected by the disorder of system; that is, they are talking about disordered systems. Without disorder they would use a different treatment.

TABLE 1

Comparison of various features of the Paillotin–Swenberg theory and the Searle–Tredwell treatment of fluorescence yield

Property	Treatment	
	Paillotin–Swenberg ('collision' theory)	Searle–Tredwell ('$t^{1/2}$')
Principal basis	Exciton collisions occur; lower fluorescence yield at high intensities, affecting kinetics at small t	Transfer kinetics strongly affected by disorder.
Structure	Details ignored, but requires uniform sampling of domains by excitation *Question:* would randomness prevent this sampling?	Assumed random *Question:* would order really change the predictions?
Mechanism of energy transfer	None particular, but must be rapid	Mechanism implies specific time dependence in fluorescence
Sensitivity to intensity (I) of initial pulse	Theory deals with I entirely; at low I, always exponential or linear combination of exponentials	$t^{1/2}$ not dependent on I, but A, in exp $(-At^{1/2})$, is empirically; $t^{1/2}$ still predicted at low I
Rate equation	$\dfrac{d[S^*]}{dt} = -K[S^*] - \gamma[S^*]^2$ Good pedigree	$\dfrac{d[S^*]}{dt} = -\frac{1}{2}At^{-1/2}[S^*]$ Differentiate solution; by analogy with the Schmolukowski solution for diffusional quenching in solution
Limits of validity	$(6Dt)^{1/2} \approx 15$ nm $t \geqslant 10$ ps	$[S^*] \leqslant A/\gamma t^{1/2}$ $\approx 10^{18}$ cm^{-3} at 10 ps

Structure

In the collision theory the detailed structure—the location of the chlorophyll molecules with respect to each other—is ignored but uniform sampling of the domain by the excitation is required. On the other treatment the structure is assumed to be random. At this stage we may ask two questions: on the collision theory, does randomness invalidate the results because of the assumption made about uniform sampling? On the $t^{1/2}$ theory, does order or partial order invalidate the results because of the assumption of random structure?

Mechanism of energy transfer

For Drs. Paillotin & Swenberg the transfer mechanism does not matter as long as transfer is fast enough but Drs. Searle & Tredwell tie the mechanism closely to the power, n, of t. If the rate depends on the inverse of r^6, then the intensity is a function of $t^{1/2}$. For other interactions, n would have different values. (Here r is the intermolecular separation of two particular participants in energy transfer.)

Sensitivity to the intensity of the initial pulse

The collision theory effectively deals only with the intensity and at low intensity the decay is always exponential or a combination of exponentials. (Incidentally, I am painting a picture in broad strokes so that people can argue with it later.) The $t^{1/2}$ aspects *per se* do not depend on the intensity but the rate constant A which appears in the Searle–Tredwell kinetic equation as $\exp(-At^{1/2})$ depends on the intensity, empirically.

Rate equation

Dr Paillotin did not quote a rate equation but implicit in his treatment in a limiting case is the 'Swenberg' equation (see Swenberg *et al.* 1976) for the rate of change in concentration of the excited singlet, $d[S^*]/dt$, which has a first-order term and a bimolecular term (see Table 1) and has a good pedigree. The equation put forward by Drs. Searle & Tredwell is $d[S^*]/dt = -\frac{1}{2}At^{-1/2}[S^*]$. The Swenberg equation derives from the Schrödinger equation, the Liouville equation, the generalized Master equation, the Pauli Master equation and so to the kinetics. The $t^{1/2}$ rate equation is obtained by differentiating the solution, but it is correct for that particular solution.

Limits of validity

I am sure that there will be some argument about my following comments. In order that the whole domain be sampled, the root-mean-square displacement during the time t is, roughly, $(6Dt)^{1/2}$, which should be about 15 nm; D is the exciton diffusion constant. If we assume that $D = Fa^2$ (i.e. a two-dimensional Förster expression for the exciton diffusion constant, where F is the nearest-neighbour transfer rate on a lattice with lattice constant a), the time has to be greater than or equal to 10 ps; in other words one would not want to apply the Paillotin–Swenberg theory for times shorter than about 10 ps (this time may be off by a couple of factors of 2) at which time the exciton could not have sampled the whole domain. On the $t^{1/2}$ treatment the lack of any $[S^*]^2$ term in the rate equation means that the $[S^*]^2$ term is assumed to be smaller than the linear term at all times, which leads to the requirement $[S^*] \leqslant A/\gamma t^{1/2}$, where γ is the bimolecular coefficient that one *would* put in the Searle–Tredwell equation if one did put in a collision term. At $t = 10$ ps, the requirement for safely omitting such a term is therefore $[S^*] \approx 10^{18}$ cm^{-3}, which I compute with the assumed values of $\gamma = 10^{-8}$ cm^3 s^{-1} and $A = 0.047$ (ps)$^{-1/2}$. This might seem pretty safe, but large densities of excitons are produced in many of these experiments and the consistency of the solution deserves checking in each case.

Beddard: The first statement you give as the basis for the Paillotin–Swenberg theory is the same as that for the other treatment. The difference is that, according to the latter, the term with the rate constant k (or γ) is supplemented by another term that depends on t—by analogy with what happens on collisional quenching in solution. First, consider what happens at low intensity if one ignores the $\gamma[S^*]^2$ term in the collision theory's rate equation. To explain the data which show the time-dependence at short times, we have used equations (1)–(3) by analogy with what happens in solution:

$$k = k' + k''t^{-\frac{1}{2}} \tag{1}$$

$$k' = 1/\tau_0 + 4\pi RDN[Q] \tag{2}$$

$$k'' = 4R^2(\pi D)^{\frac{1}{2}}N[Q] \tag{3}$$

where R is an encounter radius, D a diffusion coefficient and $[Q]$ the quencher concentration. Because we could not solve these equations for the random system, we assumed a relation $k = a + bt^{-1/2}$ to see how well it fitted the data. We then discovered that k' was very small and hence we obtained an equation of the form $k = bt^{-1/2}[S^*]$. Secondly, the Paillotin–Swenberg equations describe how the fluorescence yield varies with excitation light intensity but do not accurately describe how the fluorescence intensity varies with time. We supposed that the rate constants k and γ both depend on time. We found that both k and γ' were small compared to the corresponding terms $k''t^{-1/2}$ at short times. So we had an equation which had two time-dependent terms (Beddard & Porter 1977). This was a semi-empirical way of fitting the data in order to try to estimate the rate constants for the annihilation of singlets with singlets and of singlets with a trap.

Porter: We all agree that there is a high-intensity effect in picosecond work which is due to exciton–exciton annihilation and causes deviations from exponentiality. At low intensities, however, when that has been eliminated and at short times, the decay is still not exponential but follows the $t^{1/2}$ term That did not surprise us because we expected the $t^{1/2}$ term either with an ordinary diffusion process, because of the time-dependent diffusion term in the diffusion equation, or for a single-step Förster-type transfer where this is the equation. Quite apart from its genealogy, the equation has been tested extremely well experimentally in ordinary Förster-type transfer.

Beddard: I should like to dissociate the single-step transfer from what we are talking about because we are here considering how the excitation migrates toward a trap in the same way that a molecule in fluid solution migrates towards a quencher.

Porter: Mechanistically, if Dr Paillotin is right and if the excitation homogenizes throughout the whole light-harvesting bed before it is trapped, then decay would be exponential and the $t^{1/2}$ term would not apply.

Paillotin: We should not be trying to decide which theory is better but whether we can derive any biological information from the picosecond experiments. From that point of view two problems must be distinguished.

First what happens when only one exciton is created? According to the Royal Institution Group, the plot of $\ln F$ against the time shows, at short times, a $t^{1/2}$ dependence. I prefer a description of that dependence in terms of exponential modes: at short times there are several exponential modes of deactivation which correspond to a phase of relaxation. When this relaxation is achieved one observes a pure exponential mode. Such a process was first studied theoretically by Bay & Pearlstein (1963). It would be interesting to study experimentally this relaxation phase which is related to a relaxation within the chlorophyll–protein complexes.

A second point refers to the purpose of the theory we presented. We were trying to elucidate the fusion process. It seems to be clear that it must depend on the ratio of normal deactivation rate of the excitons to the rate of fusion, i.e. K/γ (which we call r). It is important to determine this ratio, so as to obtain some information about the fusion rate, which depends on the diffusion constant. On the other hand exciton fusion is not a linear process: it depends on the dimension of the system. Our theory tries to allow for this.

Porter: The rate of transfer implicit in your model means that the excitation has to cover every chlorophyll molecule in this array of 200 in 10 ps (that is the lower limit). Does that worry you?

Paillotin: The values given to these parameters cause much discussion. The photosynthetic apparatus contains many complex elements. I am not convinced, for example, that two states of the reaction centres are not involved in your experiments. If there are, one may have to analyse two exponential decays for the two populations of the reaction centres. Also if one particle is excited on the outside of the unit, one ought to see some relaxation process. Maybe, you are exciting one particular part of the pigment array.

Beddard: This is what we have proposed as the source of the time dependence.

Paillotin: Our knowledge of the structure of photosystem II suggests that it is difficult to excite one particular place in photosystem II. That is the main problem. The different chlorophyll–protein complexes seem to be built from rather similar chlorophyll molecules, according to the absorption spectra.

Tredwell: The actual mechanism appears to be halfway between the two cases: our results suggest that the exciton population randomizes over the whole antenna system on a picosecond time scale. Consequently, Dr Paillotin's theory might only become valid 100 ps after excitation. If we assume that the migration kinetics in the phycobiliproteins are the same as those in the chlorophyll antennae, excitons require 120 ps to cross the pigment bed (by analogy with the results from *Porphyridium cruentum*). Even assuming a more efficient transfer process in the chlorophyll antennae, we should still expect to see a fast transient at the beginning of the fluorescence decay curve.

Paillotin: I agree with the results for *Porphyridium* but I do not agree with the concept of time-dependent transfer rate. Since one excites only one part of the system one may see the diffusion of excitation across the system. If we were able to excite one chlorophyll molecule in particular in the unit, an accurate analysis of the relaxation would be most interesting.

Tredwell: How would your theory be affected if one could not assume an extremely rapid randomization of the exciton population over all the antenna systems?

Paillotin: Basically it would not have to be modified much.

Tredwell: If the process of exciton migration is relatively slow, would it not be possible to observe the rapid relaxation of the exciton population immediately after excitation?

Paillotin: You detect the process of relaxation of the primary excitation. That is an interesting process; but it depends on the initial state, on the transfer rate, on many parameters which are still unknown.

Knox: An addendum to the Table, which I omitted, should read that the authors of each theory seem to regard the basis of the others' theory as a nuisance! The supporters of the $t^{1/2}$ treatment would seem to prefer to do without the collisions and the protagonists of collisions do not want to let randomness affect their results.

Porter: I shall not attempt to summarize; we have just had four excellent summaries of the state of the subject. At least we have been able to agree what the important areas of disagreement are and this I hope will be a useful guide for future work.

References

ANDERSON, J. M. & BARRETT, J. (1979) The preparation and characterization of different types of light-harvesting pigment–protein complexes from some purple bacteria, in *This Volume*, pp. 81–96

BAY, Z. & PEARLSTEIN, R. M. (1963) A theory of energy transfer in the photosynthetic unit. *Proc. Natl. Acad. Sci. U.S.A. 50*, 1071–1078

BEDDARD, G. S. & PORTER, G. (1977) Excited state annihilation in the photosynthetic unit. *Biochim. Biophys. Acta 462*, 63–72

BENNETT, J. (1977) Phosphorylation of chloroplast membrane polypeptides. *Nature (Lond.) 269*, 344–346

BUTLER, W. L. (1979) Tripartite and bipartite models of the photochemical apparatus of photosynthesis, in *This Volume*, pp. 237–253

COGDELL, R. J. & THORNBER, J. P. (1979) The preparation and characterization of different types of light-harvesting pigment–protein complexes from one purple bacteria, in *This Volume*, pp. 61–73

FENNA, R. E. & MATTHEWS, B. W. (1975) Chlorophyll arrangement in a bacteriochlorophyll protein from *Chlorobium limicola. Nature (Lond.) 258*, 573–577

HAYDEN, D. B. & HOPKINS, W. G. (1977) A second distinct chlorophyll *a*–protein complex in maize mesophyll chloroplasts. *Can. J. Bot. 55*, 2525–2529

HENRIQUES, F. & PARK, R. B. (1976) Compositional characteristics of a chloroform-methanol soluble-protein fraction from spinach chloroplast membranes. *Biochim. Biophys. Acta 430*, 312–320

KNOX, R. S. & VAN METTER, R. L. (1979) Fluorescence of light-harvesting chlorophyll *a/b*–protein complexes: implications for the photosynthetic unit, in *This Volume*, pp. 177–186

OLSON, J. M. (1978) in *The Photosynthetic Bacteria* (Clayton, R. K. & Sistrom, W. R., eds.), Plenum, New York, in press

PAILLOTIN, G. & SWENBERG, C. (1979) Dynamics of excitons created on a single picosecond pulse, in *This Volume* pp. 201–209

SEARLE, G. F. W. & TREDWELL, C. J. (1979) Picosecond fluorescence from photosynthetic systems *in vivo*, in *This Volume*, pp. 257–277

SHIPMAN, L. L. & KATZ, J. J. (1977) Calculation of electronic spectra of chlorophyll *a*–water and bacteriochlorophyll *a*–water adducts. *J. Phys. Chem. 81*, 577–581

SHIPMAN, L. L., COTTON, T. M., NORRIS, J. R. & KATZ, J. J. (1976) An analysis of the visible absorption spectrum of chlorophyll *a* monomer, dimer, and oligomers in solution. *J. Am. Chem. Soc. 98*, 8222–8230

STAEHELIN, L. A. & ARNTZEN, C. J. (1979) Effects of ions and gravity forces on the supramolecular organization of chloroplast membranes, in *This Volume*, pp. 147–169

SWENBERG, C. E., GEACINTOV, N. E. & POPE, M. (1976) Bimolecular quenching of excitons and fluorescence in the photosynthetic unit. *Biophys. J. 16*, 1447–1452

THOMAS, J. B., MINNAERT, K. & ELBERS, P. D. (1956) *Acta Bot. Neerl. 5*, 314–321

THORNBER, J. P., ALBERTE, R. S., HUNTER, F. A., SHIOZAWA, J. A. & KAN, K.-S. (1977) The organization of chlorophyll in the plant photosynthetic unit. *Brookhaven Symp. Biol. 28*, 132–148

VAN GORKOM, H. J., PULLES, M. P. J. & ETIENNE, A. L. (1978) Fluorescence and absorbance changes in Tris-washed chloroplasts, in *Proceedings of an International Symposium on Photosynthetic Oxygen Evolution* (Tübingen 1977) (Metzner, H., eds.), in press

VAN METTER, R. L. (1977) Excitation energy transfer in the light-harvesting chlorophyll *a/b*–protein. *Biochim. Biophys. Acta 462*, 642–658

WYDRZYNSKI, T., ZYMBULYADIS, N., SCHMIDT, P. G., GUTOWSKY, H. S. & GOVINDJEE (1976) Proton relaxation and charge accumulation during oxygen evolution in photosynthesis. *Proc. Natl. Acad. Sci. U.S.A. 73*, 1196–1198

Index of contributors

*Entries in **bold** type indicate papers; other entries refer to discussion comments*

Indexes compiled by William Hill

Subject index